ISRAELIS AND PALESTINIANS IN THE SHADOWS OF THE WALL

Shedding light on the recent mutations of the Israeli separation policy, whose institutional and spatial configurations are increasingly complex, this book argues that this policy has actually reinforced the interconnectedness of Israelis and Palestinian lives and their spaces. Instead of focusing on the over-mediatized separation wall, this book deals with what it hides: its shadows. Based on fieldwork studies carried out by French, Italians, Israelis, Palestinian and Swiss researchers on the many sides of the Israeli–Palestinian divide, it highlights a new geography of occupation, specific forms of interconnectedness and power relations between Israeli and Palestinian spaces. It offers a better understanding of the transformation of people's interactions, their experiences and the ongoing economy of exchanges created by the separation regime. This heterogeneous regime increasingly involves the participation of Palestinian and international actors. Grounded in refined decryptions of territorial realities and of experiences of social actors' daily lives this book goes beyond usual political, media and security representations and discourses on conflict to understand its contemporary stakes on the ground.

AF833664

Border Regions Series

Series Editor: Doris Wastl-Walter, University of Bern, Switzerland

In recent years, borders have taken on an immense significance. Throughout the world they have shifted, been constructed and dismantled, and become physical barriers between socio-political ideologies. They may separate societies with very different cultures, histories, national identities or economic power, or divide people of the same ethnic or cultural identity.

As manifestations of some of the world's key political, economic, societal and cultural issues, borders and border regions have received much academic attention over the past decade. This valuable series publishes high quality research monographs and edited comparative volumes that deal with all aspects of border regions, both empirically and theoretically. It will appeal to scholars interested in border regions and geopolitical issues across the whole range of social sciences.

Israelis and Palestinians in the Shadows of the Wall

Spaces of Separation and Occupation

Edited by

STÉPHANIE LATTE ABDALLAH
French Institute of the Near East, CNRS, Palestinian Territories

CÉDRIC PARIZOT,
IREMAM, CNRS, Aix Marseille Université, France

Routledge
Taylor & Francis Group
LONDON AND NEW YORK

First published 2015 by Ashgate Publishing

2 Park Square, Milton Park, Abingdon, Oxfordshire OX14 4RN
711 Third Avenue, New York, NY 10017

Routledge is an imprint of the Taylor & Francis Group, an informa business

First issued in paperback 2018

Copyright © 2015 Stéphanie Latte Abdallah and Cédric Parizot

Stéphanie Latte Abdallah and Cédric Parizot have asserted their rights under the Copyright, Designs and Patents Act, 1988, to be identified as the editors of this work.

All rights reserved. No part of this book may be reprinted or reproduced or utilised in any form or by any electronic, mechanical, or other means, now known or hereafter invented, including photocopying and recording, or in any information storage or retrieval system, without permission in writing from the publishers.

Notice:
Product or corporate names may be trademarks or registered trademarks, and are used only for identification and explanation without intent to infringe.

British Library Cataloguing in Publication Data
A catalogue record for this book is available from the British Library

The Library of Congress has cataloged the printed edition as follows:
À l'ombre du mur. English
Israelis and Palestinians in the shadows of the wall : spaces of separation and occupation / [edited] by Stéphanie Latte Abdallah and Cédric Parizot.
pages cm. – (Border regions series)
Includes bibliographical references and index.
ISBN 978-1-4724-4888-0 (hardback) 1. Arab-Israeli conflict. 2. Israel–Boundaries. 3. Palestine–Boundaries. I. Latte Abdallah, Stéphanie, editor. II. Parizot, Cédric, editor. III. Title.
DS119.76.A444913 2015
956.9405'4–dc23

2014031720

ISBN 978-1-4724-4888-0 (hbk)
ISBN 978-1-138-54730-8 (pbk)

Contents

List of Figures and Maps

Figures

Maps

Notes on Contributors

Emilio Dabed is a lawyer and PhD in political sciences specialized in constitutional matters. His previous research focused on the constitutional process in Palestine. Currently, he is a director of the Human Rights Program at Al-Quds University/Bard College, Jerusalem. His latest research looks at the relations between, on the one hand, legal processes and discursive practices, and, on the other hand, political and social changes, subjectivity, and identity formation. It aims to shed light on the role that juridical phenomena plays in sociological and anthropological questions. He has published numerous articles on a variety of issues regarding politics, law, and constitutionalism in Palestine.

Yaakov Garb is a senior lecturer at Ben-Gurion University. He brings his three main teaching and research areas (environmental studies, social studies, and science and technology studies) to the study of border processes in the Israeli-Palestinian space. He has authored a series of empirical studies on the flows of freight and people across boundaries in this region, and is currently working on fundamental examinations of cross-boundary water and waste value chains. His other publications include studies on environmental discourse in the US, India, and Israel, on planning and transport, and on technological closure, translation, and change.

Lev Luis Grinberg is an associate professor in the Department of Sociology and Anthropology at Ben Gurion University, Israel. He is the founding chairman of the department (2006–2009) and former director of the Humphrey Institute for Social Research (1998–2003). He is a political economist and political sociologist and has specialized in the history of the Zionist Labor Movement, Israel's political economy, the sociology of the Israeli–Palestinian Conflict, and resistance movements. Among his books are: *The Histadrut Above all* (1993, Nevo), *Introduction to Political Economy* (1997, The Broadcast University), *Politics and Violence in Israel/Palestine* (2010, Routledge) and *Mo(ve)ments of Resistance* (2013, Academic Studies Press).

Ariel Handel is a postdoctoral fellow at the French Research Center in Jerusalem (CRFJ) and a research fellow at the Minerva Humanities Center, Tel Aviv University, Israel. His research interests are mobilities in the West Bank and the Gaza Strip, mapping and spatial representations, and the political philosophy of geography. His publications include *Protest: A Political Lexicon* (ed., 2012, Hakibbutz Hameuhad Publishing), *Geographies of Occupation* (forthcoming), and several other journal papers and book chapters.

Shira Havkin is a PhD candidate in political sociology at the CERI/Sciences-Po Paris and an assistant professor at Paris Ouest Nanterre University. Her doctoral dissertation examines recent reforms in the management of Israeli border control apparatuses: the outsourcing of the military checkpoints in the Occupied Territories and the "statization" of the asylum procedure. Her research lies at the intersection of border and migration, state redeployment, and economic sociology. Within this general area, she studies the articulation of nationalism and neoliberalism in contemporary Israel.

Marc Hecker is a research fellow at the French Institute for International Relations (IFRI) in Paris. He is the deputy editor of the quarterly *Politique* étrangère. He holds a PhD in political science from University Paris 1 Panthéon-Sorbonne. He has published several books including *War 2.0: Irregular Warfare in the Information Age* (co-authored with Thomas Rid, 2009, Praeger) and *Intifada française?* (2012, Ellipses).

Karine Lamarche submitted her PhD dissertation in 2011 at EHESS (Paris) about Israelis actively involved against the occupation during and after the Second Intifada. In 2014, she was a postdoctoral researcher in the "Labex Structuration des Mondes Sociaux" in Toulouse and a research fellow at the Laboratoire des Sciences Sociales du Politique (LaSSP). Her current research deals with the political dimensions of both Israeli migration to Europe and Israeli-European dual citizenships. She recently published *Militer contre son camp? Des Israéliens engagés aux côtés des Palestiniens* (2012, PUF).

Stéphanie Latte Abdallah, PhD, is an historian and political scientist. She is a research fellow at the CNRS (IFPO, Palestinian territories) and has published numerous works on Palestinian refugee social history, broader issues of gender, civil society mobilizations, and feminisms in the Middle East. She coordinated (together with Cédric Parizot) two international research programs on borders and circulations in the Israeli-Palestinian spaces (2007–2011) and is currently researching on confinement and imprisonment in Palestine. She also worked on the connection between image, history, and politics and developed film projects. Among her publications are *Femmes réfugiées palestiniennes* (2006, PUF), *Islamic Feminism Today* (ed., 2013, *Critique Internationale*), *Gender Transformations in the Arabian Peninsula* (edited with Blandine Destremau and Marina de Regt, 2013, *Arabian Humanities*), and *Parentalités Enfermées*(edited with Coline Cardi, 2014, *Champ Pénal*).

Dganit Manor is an anthropologist and external lecturer at Achva College and Sapir College in Israel. She obtained her PhD in sociology and anthropology from Ben-Gurion University. It is titled *Guarding Home, Volunteering for the Civil Guard: A View towards the Recruited Civil Discourse in Israel*. Her research focuses on security, military, marginality, bureaucracy, and nation-state issues.

Elizabeth Marteu has a PhD from the Paris Institute of Political Studies (Sciences Po) in cooperation with Ben-Gurion University, Israel. She specializes in Palestinian civil organizations, women's activism, and transnational mobilizations in the Middle East. Among her publications are *Civil Organizations and Protest Movements in Israel: Mobilization around the Israeli-Palestinian Conflict* (ed., 2009, Palgrave Macmillan).

Esmail Nashif is an anthropologist, writer, and art critic. His main research interests focus on language, ideology, and aesthetics. Currently, he works as a lecturer at Ben-Gurion University of the Negev, Israel. Some of his recent books include: *Palestinian Political Prisoners: Identity and community* (2008, Routledge), *The Architectonics of Loss: The question of contemporary Palestinian culture* (2012, Al Farabi), *Thagharat: Short Stories* (2012, Raya), and *Gradus for Opening the Episteme* (2014, Al Farabi).

Basel Natsheh is an economist, assistant professor, and Head of the Department of Business Administration and Islamic Banking in Al Khawarizmi International College, in Al-Ain (UAE). In 2010, he worked as macroeconomic policy advisor for the Palestinian minister of economy in Ramallah. Between 2008 and 2010, he worked for the French Ministry of Finances for the establishment of the Public Finance Institute in the Palestinian territories. Then, he was dean of the Finance and Management Faculty at Hebron University till 2014. In 2012, he received an EU award as a young Mediterranean leader. His research focuses on development, labor, regional economics, and youth entrepreneurship.

Cédric Parizot, PhD, is an anthropologist and a research fellow at the Institute of Research and Studies of the Arab and Muslim World (CNRS/Aix-Marseille University) in Aix-en-Provence, France. His research focuses on borders and mobility in the Israeli-Palestinian spaces. From 2007 till 2011, he coordinated, together with Stephanie Latte Abdallah, two international research programs "Mobility and Borders in the Israeli-Palestinian Spaces" (MOFIP) and "Appraising the Israeli Palestinian Conflict through Cross Border Mobility." He has recently published *Borders, Mobilities and Migrations: Perspectives from the Mediterranean, 19–21st Century* (edited with Lisa Anteby-Yemini, Virginie Baby-Collin, Sylvie Mazzella, Stéphane Mourlane, Céline Regnard, and Pierre Sintès, 2013, Peter Lang).

Nicolas Pelham is a journalist and writer on the Middle East affairs for *The Economist* and the *New York Review of Books*, and has spent 20 years studying and working across the Middle East. From Rabat to Baghdad, he has worked as a correspondent for the BBC, the *Financial Times* and *The Economist*. He has worked as a senior analyst based in Jerusalem for International Crisis Group, and has reported on the political economy of Gaza for the United Nations. He is the author of *A New Muslim Order* (2008, I.B. Tauris), which maps Shia resurgence in the Middle East, and co-author of *A History of the Middle East* (2010, Penguin).

Valérie Pouzol is an assistant professor of history at Paris 8 University (Centre de recherches historiques: Histoire des Pouvoirs, Savoirs et Sociétés). She has published on Israeli and Palestinian women's peace activism, with a special focus on gendered identities in both nationalisms. More recently, as member of the International Research Program MOFIP, she focused on Palestinian and Israeli LGBTQ activism – questioning and underlining the political dimension of this form of activism. Since 2011, she has conducted fieldwork and researches on Jewish orthodox feminism and protest activism against gender segregation in Israeli public space. Her publications include *Clandestines de la paix: Israéliennes et Palestiniennes contre la guerre* (2008, Complexe; HTP/CNRS).

Acknowledgments

We first wish to thank our Palestinian, Israeli, Swiss, Italian, and French colleagues who have been involved, since 2007, in the research programs[1] from which this book is the outcome: Lisa Anteby, Gilbert Benhayoun, William Berthomière, Riccardo Bocco, Véronique Bontemps, Philippe Bourmaud, Sylvaine Bulle, Emilio Dabed, Jackie Feldman, Yaakov Garb, Honaida Ghanim, Lev Grinberg, Ariel Handel, Shira Havkin, Marc Hecker, Wasfi Kailani, Adriana Kemp, Karine Lamarche, Dganit Manor, Elisabeth Marteu, Daniel Meier, Eleonore Merza, Emmanuelle Moustier, Esmail Nashif, Basel Natsheh, Nicolas Pelham, Valérie Pouzol, Pierre Renno, Caroline Rozenholc, Irène Salenson, and Adoram Schneidleder. Their contribution to the reflections presented in this book was crucial.

Our warm acknowledgments also go to Eberhard Kienle, Ghislaine Alleaume, Thierry Fabre, Olivier Tourny, Robert Ilbert, Bernard Morel, Christiane Laye, Brigitte Marin, Dionigi Albera, and the late Sophie Mesguisch-Kessler, for their moral and institutional support. We are also very grateful to Myriam Laakili who took charge of the administrative coordination and the management of these programs. We would like to thank Marjolaine Barazani, Tareq Natsheh, and Mathieu Coulon for their help in the production of the maps and figures edited in this volume.

Finally, our greatest obligation is to our contacts and friends who keep on working, circulating, imagining, and inventing despite the difficult conditions created by the Israeli occupation.

1 *Appraising the Israeli conflict through cross border mobility* (2008–2010) implemented in the frame of the Network of Research Centres in Human Sciences on the Mediterranean (FP6) and *Mobility, Borders and Conflicts in the Israeli Palestinian Spaces* (2007–2011) funded by the Agence Nationale de la Recherche, the CNRS, and the Regional Council Provence Alpes Côte d'Azur.

This volume is published with the support of the Directorate General for Research of the European Commission, in the framework of the RAMSES[2] Network of Excellence, funded by the 6th Framework Programme (contract CIT3-CT-2005-513366); the support of the Agence Nationale de la Recherche (contract ANR-07-JCJC-0083—CSD 9); and the support of the Regional Council of the Region Provence Alpes Côte d'Azur (APO 2007-2008-2009).

This volume is solely the responsibility of the authors. The European Commission, the Agence Nationale de la Recherche, and the Regional Council Provence Alpes Côte d'Azur cannot be held responsible for its content or for any use which may be made of it.

Niveau 2:

Preface

This book is an updated and augmented version of the French edition *A l'ombre du Mur: Israéliens et Palestiniens entre séparation et occupation*, that was published by Actes Sud in September 2011. It is the outcome of two international research programs initiated by Cédric Parizot in 2007 and coordinated with Stephanie Latte Abdallah at the Institute for Research and Studies of the Arab and Muslim World (IREMAM/CNRS, Aix-en-Provence): "Mobility and Borders in the Israeli Palestinian Spaces" (2007–2011)[1] and "Appraising the Israeli Palestinian conflict through Cross Border Mobility" (2008–2010).[2]

It was important for us to translate into English, update, and expand the original manuscript in order to share our approach and analysis of the dynamics that have been structuring the conflict over the last 20 years with a wider public. For the conclusions we reached in 2011 are even more relevant today. While the Israeli separation policy has created the illusion of initiating a bordering process between Israelis and Palestinians, it has actually reinforced the interconnectedness of their lives and their spaces. This dynamic has not merely inscribed the Israeli occupation in the long run and compromised the creation of a viable Palestinian state in the near future; it has also generated new forms of governmentality and territoriality that challenge the imaginary of the modern nation state. Israel's increasingly sophisticated management systems of populations' mobility and relations to space and time have generated complex forms of sovereignty. The role of the Palestinian Authority, of international agencies, NGOs, as well as of informal Israeli and Palestinian actors in the daily functioning of the occupation regime has radically transformed its nature. Chains of political and juridical responsibility have been profoundly readjusted and a new architecture of violence has emerged.

In order to demonstrate further our main thesis and to broaden our perspective, we have decided to enrich our analysis on the geographies of occupations (Part I) not merely by adding a chapter on their juridical ramifications (Emilio Dabed) but also by closing this part with a chapter deconstructing the narratives through which observers, political actors, NGOs, and some researchers have criticized the occupation regime (Ariel Handel). We have decided to expand Part II, on the economy of separation, by supplementing it with a chapter by Nicolas Pelham on the tunnel economy between the Gaza Strip and Egypt. Part III, dealing with people's experiences of separation at the margins, was enlarged with a contribution by Valérie Pouzol on the changing relations between Israeli and Palestinian

---

1 This program unfolded from 2007 to 2011, and was funded by the Agence Nationale de la Recherche, the Centre National de la Recherche Scientifique (CNRS) and the Regional Council Provence Alpes Côtes d'Azur.

2 This program was implemented between 2008 and 2010 in the frame of the Network of Research Centers in Human Sciences on the Mediterranean (FP6). Both programs benefited from the partnership of the Mediterranean House for Humanities (CNRS, Aix Marseille University), the Institute for Comparative European and Mediterranean Ethnology (IDEMEC), the French Institute of International Relations (IFRI) in Paris, the Graduate Institute of International and Development studies (Geneva), the French Research Center in Jerusalem (CRFJ) and the French Institute of the Near East-Palestinian Territories (IFPO-TP).

LGBTQ movements over the last 20 years. Finally, we added to Part IV, dedicated to political crossings, a chapter by Karine Lamarche studying the new forms of Israeli activism across the Green Line and a chapter by Esmail Nashif on Palestinian suicide bombings.[3]

STÉPHANIE LATTE ABDALLAH and CEDRIC PARIZOT
Jerusalem and Aix-en-Provence
May 2014

3 Called "martyrdom operations" by the actors.

Lists of Abbreviations

ABL	Amended Basic Law
ABSI	Association pour le Bien-Etre du Soldat Israélien
AFPS	Association France Palestine Solidarité
AIC	Alternative Information Centre
AJC	American Jewish Committee
AJPF	Association pour la Promotion des Jumelages entre Camps de Réfugiés Palestiniens et Villes Françaises
BDS	Boycott, Divestment and Sanctions (Campaign)
BIG	Brand Israel Group
BL	Palestinian Basic Law
CAPJPO	Coordination des Appels pour une Paix Juste au Proche-Orient
CBSP	Comité de Bienfaisance et de Secours aux Palestiniens
CCIPPP	Campagne Civile Internationale pour la Protection du Peuple Palestinien
COM	Council of Ministers
CPA	Crossing Points Authority
CRIF	Conseil Représentatif des Organisations Juives de France
DFLP	Democratic Front for the Liberation of Palestine
ECF	Economic Cooperation Foundation
GDP	Gross Domestic Product
GS	Gaza Strip
GUPS	General Union of Palestine Students
ICP	Israeli Communist Party
ICRC	International Committee of the Red Cross
IDF	Israel Defense Forces
JOH	Jerusalem Open House
KLAF	Lesbian Feminist Community
LGBTQ	Lesbian, Gay, Bisexual, Transgendered or Queer
NIF	New Israel Fund
OCHA	(see UNOCHA)
OG	Official Gazette
OPT (or OT)	Occupied (Palestinian) Territories
PA (or PNA)	Palestinian (National) Authority
PALTRADE	Palestinian National Non-profit Trade Promotion Center and Business Membership Organization (established in 1998)
PFLP	Popular Front for the Liberation of Palestine
PLC	Palestinian Legislative Council
PLO	Palestine Liberation Organization

PLO-EC	Executive Committee of the Palestinian Liberation Organization
PNC	Palestinian National Council
PQBDS	Palestinian Queers for BDS
RCDP	Réseau de Coopération Décentralisée pour la Palestine
SHABAK or SHIN BETH	Israel Security Agency
SHABAS	Israel Prison Service
SPPR	Society for the Protection of Personal Rights
TIP	The Israel Project
TQM	Total Quality Management
UEJF	Union des Êtudiants Juifs de France
UNOCHA	United Nations Office for the Coordination of Humanitarian Affairs
UNRWA	United Nations Relief and Works Agency for Palestine Refugees in the Near East
UPJF	Union des Patrons et Professionnels Juifs de France
USAID	United States Agency for International Development
WB	West Bank
WB&G	West Bank and Gaza Strip
WCLAC	Women center for legal aid and counseling

Introduction
The Shadows of the Wall: Reappraising the Israeli Occupation Regime

Stéphanie Latte Abdallah and Cédric Parizot

Erected by Israel in 2002, the West Bank Wall is the most imposing, visible, and costly control edifice built since the occupation of the West Bank and the Gaza Strip in June 1967. Built to embody the Israeli policy of unilateral separation (*hafrada*), it has become both the venue and target for local and international disputes. Now that it has received intensive media exposure, it has become the emblem of the Israeli–Palestinian conflict: concrete walls or sections of barriers put up by Israel in the West Bank now appear on the covers of publications targeting both the general public and the scientific community. Changes in the situation and the issues in the conflict are often summarized only through discussions revolving around the Wall.

But this edifice can also be seen as a trap. Mesmerized by the Wall, many local and international observers have lost touch with the processes and changes in Israeli occupation policies in the West Bank and the Gaza Strip. Considering the separation as complete, some observers even consider that the terms of the debate and conflict have changed; the Wall therefore hides as much as it reveals.

By bringing together anthropologists, sociologists, historians, political scientists, and economists, this book attempts to shift the focus from the Wall itself to the shadows it casts. It attempts to analyze the reconfigurations of Israeli occupation policies and therefore understand the nature of the separation implemented in the West Bank and Gaza over the past 20 years. Our perspective highlights the role of the local and international actors and institutions that have contributed to redeploying these systems of control, whether by participating in their administration or by circumventing or appropriating them for their own ends.

We will see how, whilst playing on the image of the border, the implementation of the Israeli separation policy causes a profound reorganization of the economic, social, and political relations of domination between Israeli and Palestinian populations. By perpetuating and increasing their relations of interdependence, the occupation regime is compromising the creation of a viable Palestinian State in the near future.

A World-Famous yet Unfinished Structure

Promoted by its partisans and detractors alike, the excessive media coverage of the Wall has contributed to making it the chief focus for local and international confrontations. But all this media attention makes the world forget that the principal role this barrier was designed to perform has not been achieved. In the context of local and international pressures that are difficult to reconcile, the planners have had to revise the route of the wall several times.

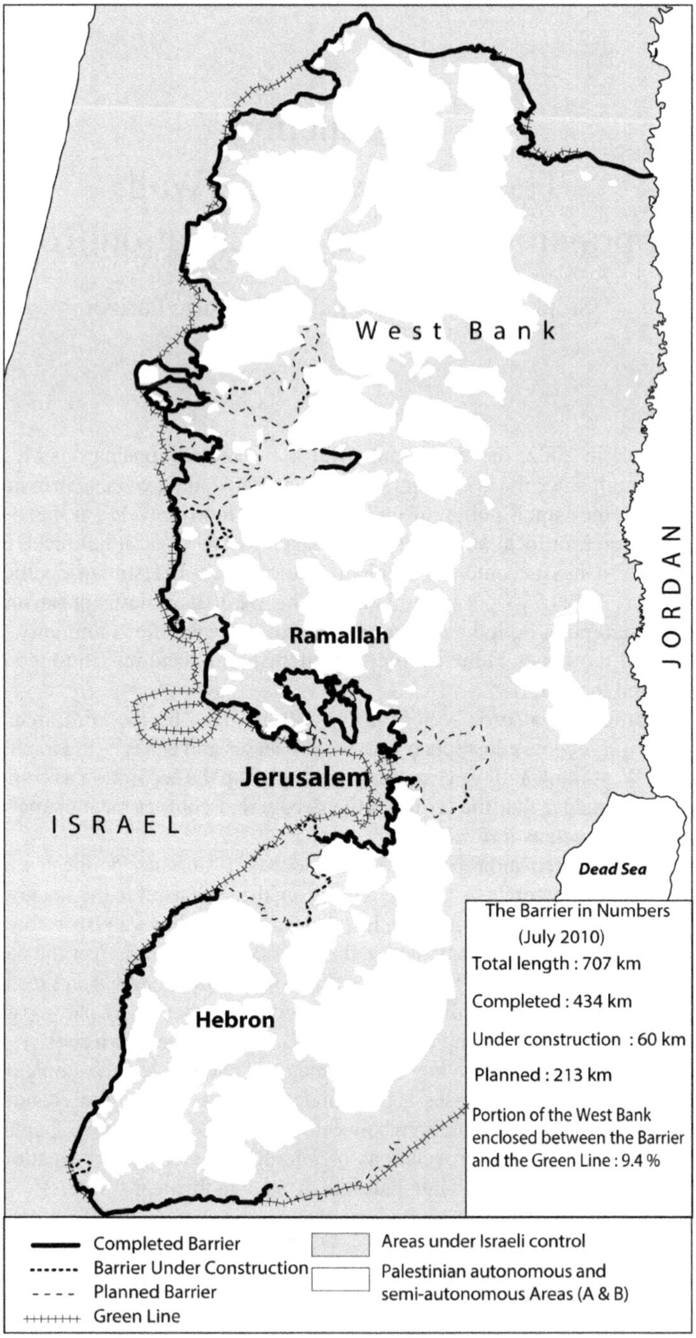

Map I.1 Separation Wall, July 2010

Map realized by M.Barazani (CRFJ) and M. Coulon (LAMES).

Source : OCHA Information Management Unit; Map produced June 2009; Data Base and Statistics: OCHA, PA, MoP.

Completion of the project has been postponed so many times that in 2014 it remained largely unfinished, and at that time, at least, created no territorial separation between Israelis and Palestinians.

The Materialization of a Security Policy

In Israel the building of the Wall was launched in response to increasing popular pressure as a result of an escalation of Palestinian suicide bombings that peaked in spring 2002 (Kershner 2005). The Israeli population demanded concrete, tangible measures from the Sharon government (Arieli and Sfard 2008). Apart from creating a climate of terror, these bombings challenged the ability of the state to defend its citizens (Dieckhoff 2003). The Wall was thus conceived as a way of ending the bombings and restoring the sovereignty of the state over its territory.

The left wing parties which included some of the chief promoters of the project (Rabinowitz 2003) depicted the construction of the Wall as a way of avoiding the reversal of the demographic balance in favor of the Arabs. Certain observers even thought that the Wall would complete the building of the nation by giving Israel borders worthy of a modern state (Halper 2003; Arieli and Sfard 2008; Rabinowitz 2003). And following 9/11, the building of the Barrier was seen as erecting a border that many people considered, to use a neologism, "civilizational," a rampart between the "free world" and "obscurantism" (Rabinowitz 2003).

Coming from the Israeli population and political class, this pressure continued and persuaded Ariel Sharon, who had hitherto been very skeptical about the Wall, to implement its construction. Sharon only agreed to the project on the condition that its path would incorporate the most significant groups of settlements and a large amount of land into Israeli territory (Arieli and Sfard 2008, p. 49). He saw this as a way of shifting the border of Israel past the Green Line[1] and making legitimate the land acquired for Israeli settlement (Snegaroff and Blum 2005). At the elections in 2006, in his "convergence" plan for the withdrawal of Israeli settlements from the east of the Wall, Ehud Olmert, head of the center-right Kadima party, presented the Wall as the de facto future border between the two states. Eight years later, on January 2014, a prominent Israeli think tank, the Institute for National Security Studies, also suggested relying on the Barrier route to fix the limit of Israel's withdrawal from the West Bank if talks sponsored by the Americans failed. The Eastern limit of that "disengagement" would be the Jordan Valley (Cohen 2014).

In Israel, the Wall project had thus gradually brought together the political agendas of the left and the right. By combining elements of security, demographics, annexation and, to a certain extent, border strategy (Parizot 2009a), it had attracted consensus from all but extreme left activists fighting the occupation (Lamarche 2009, 2013).

Symbol of a Policy of Predation and Confinement

For the Palestinians the Wall is just one more way of stealing from them; its construction has resulted in numerous spoliations and destructions that have had disastrous economic consequences. These are regularly recorded by Palestinian NGOs as well as Israeli[2] and

1 The Green Line is the 1949 armistice line between Israel and Jordan. It runs through the heart of Jerusalem and divides the East (Palestinian sector) from the West, and the West Bank from Israel inside its 1949 borders.

2 On the Palestinian side, see, for example, PENGON (2003); on the Israeli side, see the reports on the B'Tselem site (http://www.btselem.org/english/accessed January 7, 2015) and Ir Amim (http://

international NGOs and agencies.[3] Subject to repeated border closures and long periods of curfew since the beginning of the second Intifada (Bocco et al. 2002), the Palestinians have experienced the building of the Wall as a new way of imprisoning them. In fact, the Wall embodies the limits on movement progressively imposed on Palestinians in the West Bank and the Gaza Strip since the 1990s (Abu Zahra 2007). The Wall imprisons the Palestinians in a ghetto whose size has been gradually reduced to almost nothing. Lastly, it also fragments the Palestinian zones on the West Bank and isolates communities that have remained to the west of its path from those to the east. In 2013, for example, 11,000 Palestinians living in 32 communities found themselves trapped between the path of the Green Line and that of the Wall (UNOCHA 2013); if we add the 248,400[4] Palestinians in East Jerusalem, we get a total of 259,400 people.

The Wall has not only taken farmers away from their land, it has also profoundly disrupted the economic and social relations between neighboring populations, between centers and their peripheries, just as it has reduced levels of access to health and education for certain communities. By doing so, the Wall has created more obstacles to the construction of a viable Palestinian economy and state.

A Theatre of Local and International Conflicts

In order to seek international aid, Palestinian NGOs have launched a number of media campaigns. An example is the Stop Wall Campaign supported by PENGON, a federation of several local NGOs. Palestinians have been backed up by Israeli NGOs such as B'Tselem and HaMoked,[5] as well as international NGOs. The information published on the internet by these organizations offers a counter narrative to the Israeli Ministry of Foreign Affairs.[6]

Local populations directly affected by progress in the building work also mobilized to launch non-violent types of opposition; on numerous occasions these campaigns have seen such groups taking their cases to the Israel Supreme Court. These cases have enabled some plaintiffs to have building work suspended for a time or, in a few cases, to redraw the path of the Wall (Kershner 2005). Palestinian populations have also demonstrated every week, for instance at Bil'in and Na'alin, or Nabi Saleh focusing mainly on settlers' land grabs in their villages and surrounding areas. These villages have attracted considerable media attention as the focus of clashes between the Israeli army on one side and Palestinian, Israeli (such as Anarchists against the Wall), and international demonstrators[7] on the other (Lamarche 2011, 2013).

---

www.ir-amim.org.il/eng/accessed April 8, 2014).

3 Regarding international agencies and teams, see, for example, the work of the Office for the Coordination of Humanitarian Affairs, occupied Palestinian territory (http://www.ochaopt.org/), and also the series of reports drawn up by HPEG (2003) and Bocco et al. (2003).

4 See "Special Statistical Bulletin on the 65th Anniversary of the Palestinian Nakba," May 14, 2013. Available at: http://www.pcbs.gov.ps/site/512/default.aspx?tabID=512&lang=en&ItemID=788&mid=3171&wversion=Staging, accessed February 4, 2014.

5 Website: http://www.hamoked.org, accessed January 7, 2015.

6 Website of the Israeli Ministry of Defense: http://www.seamzone.mod.gov.il/pages/eng/purpose.htm, accessed January 28, 2014.

7 Because these populations have again enjoyed the support of Israeli NGOs such as Ta'ayush, Anarchists Against the Wall, Gush Shalom, etc., and international associations such as the Internal Solidarity Movement or Les Missions civiles.

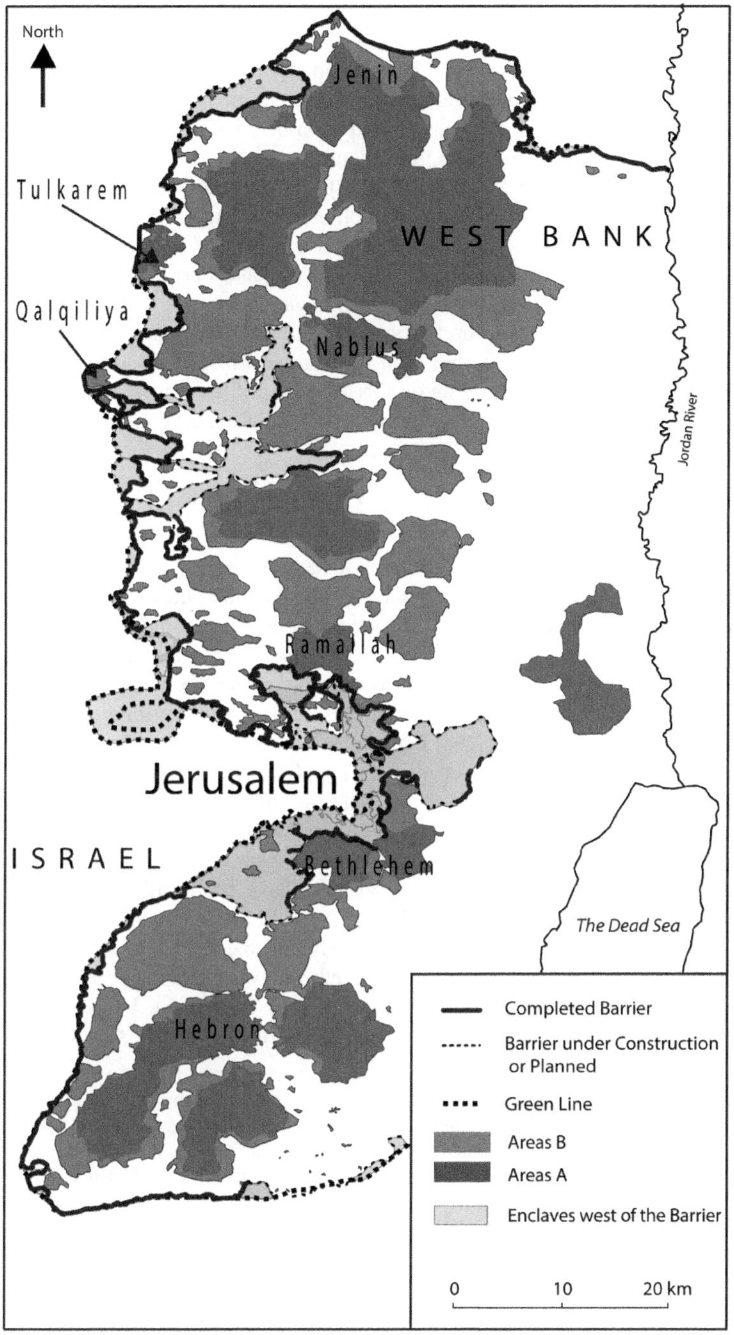

Map I.2 A, B and C areas in the West Bank

Map realized by M.Barazani (CRFJ) and M. Coulon (LAMES).

Source: OCHA Information Management Unit; Map produced June 2009; Data Base and Statistics: OCHA, PA, MoP.

The Palestinian Authority (PA) has ended up using diplomatic channels. The international community has been moved to act several times. While the project to build a "security barrier" to prevent Palestinian suicide bombers has not been challenged as such, the main international actors involved in the conflict (the United States, Europe, the United Nations, etc.) do not support Israel's attempts to annex additional territory. In 2003 the United States intervened to halt the construction of certain portions which, in their opinion, directly threatened the process of building a Palestinian State;[8] in July 2004 in an advisory opinion sought by the United Nations, the International Court of Justice at The Hague declared the route of the Wall in the West Bank illegal.[9]

In addition to pressure from the Palestinians and international community, action has also come from Israeli settlers: rather than halting the project, they have made efforts to ensure that their settlements would be on the right side of the Barrier. Their lobbying and appeals to the Israel Supreme Court have been successful on several occasions (Blum and Snegaroff 2005; Backman 2006, p. 238–59).

An Unfinished Project

The irreconcilable nature of local and international pressure has led the planners to revise the route of the Wall several times. They have gradually had to move it nearer and nearer to the Green Line and its path has become very winding and discontinuous.

These inconsistencies have created a rather absurd situation: from a strictly territorial viewpoint the barrier does not create any separation between Israeli and Palestinian territory, nor does it distinguish inside from outside. It also often separates some parts of Israel from others. Crossing the Wall does not necessarily mean a change of jurisdiction (Parizot 2009c): for example, Israeli drivers travelling from Jerusalem to the Dead Sea have to cross the Wall, but remain on a road that runs through an area controlled exclusively by Israel. Furthermore, the pursuit of settlement building behind the Wall has maintained Israeli enclaves on the Palestinian side. In order to protect these settlements as well as certain roads leading to them, "in-depth barriers" have been built, thereby maintaining "extraterritorial Israeli zones" and breaking up the Palestinian territories even further. The more the route of the Wall has approached the Green Line,[10] the more "in-depth barriers" have been built and the more the Wall has created enclaves (Weizman 2007, p. 176).

The Wall also divides Palestinian areas from the rest of the Palestinian territories. It firstly created a number of Palestinian enclaves on the Israeli side; then it defined

---

8 This was why Ariel Sharon had to cancel the construction of the first portions of the Barrier between the Jordan valley and the region of Jenin. The original plan was to build the Wall not just to the west, but also to the east of the main Palestinian-occupied zones of the West Bank. The eastern wall would have made it possible to keep the Jordan valley and its settlements inside Israel (Ariel and Sfard 2008, p. 43).

9 The Court ruled by fourteen votes to one that the construction of a barrier in the Palestinian-occupied West Bank and around east Jerusalem was in breach of international law. It asked Israel to halt building work, demolish those sections located in the West Bank and make reparation for the damage caused. By thirteen votes to one, the court asked states not to recognize the de facto situation or assist Israel in maintaining or pursuing the construction (Finkelstein 2005, p. 204–5).

10 Shaul Arieli and Michael Sfard (2008, p. 42) stressed that initial forecasts caused Ariel Sharon to envisage the possibility of unilaterally annexing 45 percent of the West Bank. Successive re-estimates made under local and international pressure have reduced this area to less than 10 percent, i.e. to an area almost equivalent to what the negotiators envisaged at Taba (2001) and during the Geneva initiative (2003).

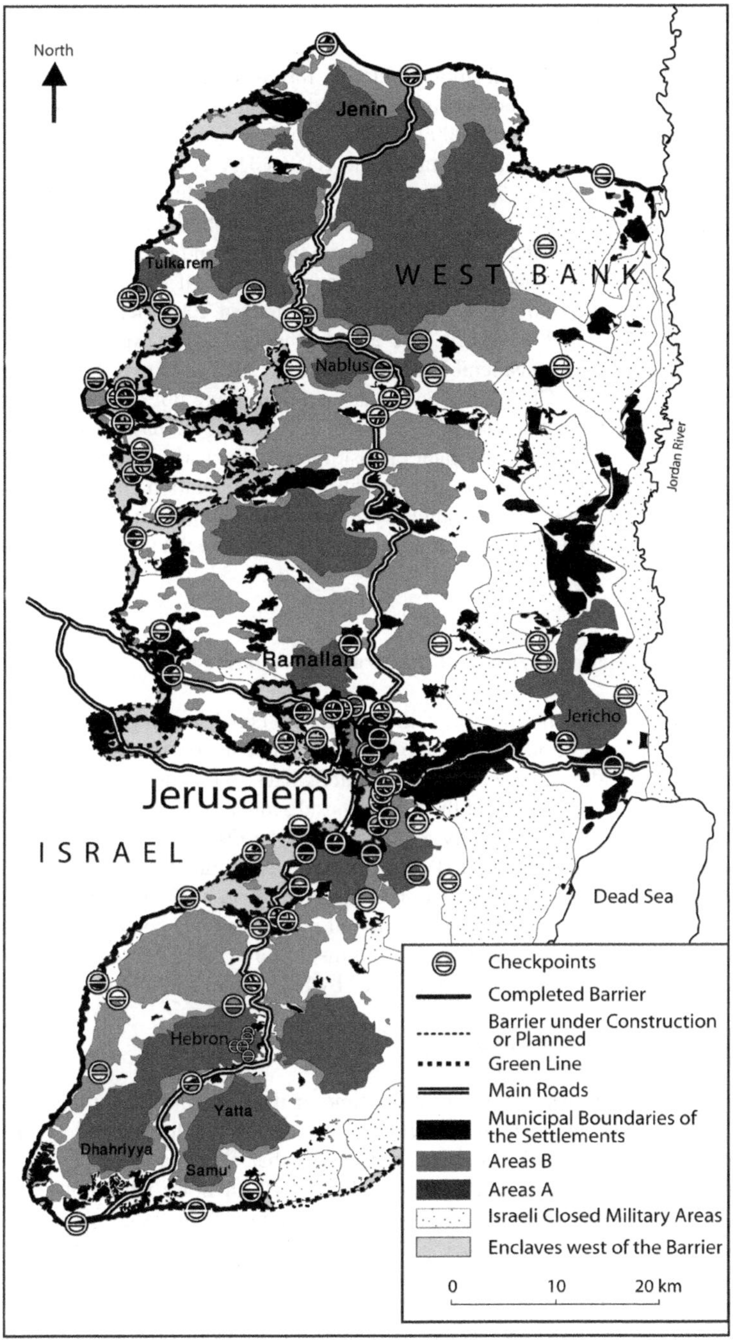

Map I.3 Palestinian and Israeli areas in the West Bank, 2009

Map realized by M.Barazani (CRFJ) and M. Coulon (LAMES).

Source: OCHA Information Management Unit; Map produced June 2009; Data Base and Statistics: OCHA, PA, MoP.

Palestinian enclaves on the Palestinian side. Its tortuous path, attempting to include the maximum number of settlements on the Israeli side, created pockets encircling Palestinian communities on several sides. In 2009, to the east of the Wall, the United Nations Office for the Coordination of Humanitarian Affairs (UNOCHA 2009) estimated that 125,000 Palestinians were surrounded by the barrier on three sides and that 26,000 were completely surrounded and only able to leave the enclaves by specially built bridges or tunnels.

Lastly, the inability of the Israelis to reconcile the local and international pressures has considerably slowed the construction of the barrier. While half the planned structure was completed between the summer of 2002 and the summer of 2006, between 2006 and 2012 only a further 12 percent was built. The initial schedule, which set the end of building work for 2008, has now considerably overrun. In July 2012 only two-thirds of the barrier had been finished (UNOCHA 2013) and many sections were not operational. Located in the heart of the West Bank, the sections that still have to be built have provoked—and will certainly provoke in the future—greater opposition from the various parties. Since 2008 the Wall appears to have lost its status as a priority for the population and the government of Israel. In this context in which the suicide bombings have stopped and/or the separation is considered to have been effected and acknowledged, the Israelis no longer appear to care about it (Parizot 2009a).

Separation: The Reorganization of the Israeli Occupation

If we are to understand the nature of the separation Israel has imposed on Palestinians, its territorial and institutional implications and its influence on the directions the conflict has taken and the stakes involved, we have to look back to the moment the separation was introduced in the early 1990s and then trace its subsequent readjustments. The separation policy was implemented differently at the time of the Oslo Accords (1993–2000) from the subsequent period (2000–2014). This policy has gradually reorganized the Israeli modes of civilian and military occupation to the extent that, by the beginning of the second decade of the twenty-first century this regime of occupation had come to seem permanent, taking complex territorial, administrative and institutional form. Moreover, the cost of the occupation appears to be increasingly covered by Palestinian and international actors.

1993–2000: The Oslo Negotiations

The separation policy was launched at the time of the First Intifada (1987–1993). Since December 1987 the confrontations between the occupying forces and the Palestinian population have revived the idea of borders in that they have given the landscape a line separating the areas the Palestinians lived in from those where the Israelis lived (Grinberg 2010). The confrontations highlighted the failure of the system of occupation deployed since June 1967 by Israel. The separation policy had been promoted by Itzhak Rabin, the minister of Defense, then prime minister of the State of Israel from 1992 to 1995 (Arieli and Sfard 2008). The policy was based on the introduction of restrictions on movement including travel permits (Handel 2009a; Hanieh 2006; Hass 2002) that gradually put in place a system for filtering the Palestinians willing to enter into Israel (Parizot 2010). The boundaries imposed were no longer those of 1967; instead they confirmed the annexing of East Jerusalem and the surrounding areas. These regions had been forbidden to Palestinians from other regions of the West Bank at the beginning of 1993 (Abu Zahra 2007).

The Oslo negotiations followed by the signing of the Declaration of Principles in Washington in September 1993 reinforced this process of separation while giving it an administrative, negotiated dimension. Israel was able to delegate the administration of the occupied population to the PA created in 1994. Between 1993 and 2000, as successive agreements were signed, the Israeli army withdrew from the zones in the West Bank and the Gaza Strip that had passed into Palestinian autonomy. In the euphoria of the first years, some saw these withdrawals as the prelude to a full disengagement of Israel from the West Bank and Gaza.

Nevertheless, at the end of the 1990s, due to the failure of the Oslo process, the occupation remained in place. Yet, its character had changed since the Palestinians and their international financial backers found themselves de facto involved in managing or supporting its costs. The PA quickly acquired the appearance and symbols of the quasi-state it had been supposed to become at the end of the interim period (1998). This process had been encouraged by the intervention of international organizations and institutions (European Union, United Nations, World Bank, cooperation with various countries and many NGOs, etc.) who got involved very early on to support the negotiation process and the construction of the economy and State of Palestine.[11] But this direct international aid to the budget of the PA was due to end in 1996, by which time it was thought that the Palestinian economy would have been relaunched and political and territorial sovereignty would be on the way to realization (Brynen 2000). But the rapid deterioration of the political situation has prevented the development of an independent Palestinian economy that was sufficiently robust to meet these costs. Sustained international aid has in fact become a way of ensuring the functioning of a PA and economy that could not survive independently—an authority which nevertheless took over in 1994 some portfolios and costs previously paid for by the Israeli authorities: health, education, police, taxation, etc. Therefore, international actors and the PA found themselves constrained to sub-contract part of the Israeli occupation (Bocco and Mansouri 2008; Latte Abdallah, 2011; Ophir, Givoni, and Hanafi 2009).

In territorial terms, the Oslo Accords led to the division of the West Bank and the Gaza Strip into three types of zone, known as A, B, and C. In the A zones, Israel delegated security and civil control to the PA; in the B zones, the PA was responsible for public order and the internal security of the Palestinians while Israel reserved the right to act on any questions of external security. Lastly, the C zones remained under Israeli control.

The gradual deterioration of relations between the parties and the successive failures of negotiations have caused the withdrawal of the Israelis to be postponed on several occasions. The Israelis have also increased the number of *fait accompli* on the ground so that they will be in a position of strength when negotiations on the final status take place. Whilst agreeing to abandon some of the territories occupied in 1967, they have reorganized their civil and military occupation to increase their presence in the C zones.

On the eve of the Second Intifada (2000), these redeployments had left a patchwork of Palestinian enclaves that were isolated from one another. In the West Bank, the A zones at the time only accounted for 17 percent of the West Bank, the B zones 23 percent, and the C zones 60 percent. In the Gaza Strip, the independent Palestinian zones covered 65

11 At this time, 43 countries committed the sum of four billion dollars to support the building of institutions by the Palestinian Authority, to develop the economy, infrastructure, and civil society up to the end of the interim period. Part of this sum was intended directly to fund the Palestinian Authority and contributed largely to setting up its administration, its ministers and services (education, health, etc.), its security forces and police (Brynen 2000; Lia 2007).

percent of the territory, the Israelis maintaining control of the remainder of zones in which there were settlements.

The isolation of the enclaves was reinforced particularly as a response to Palestinian suicide bombings in Israeli towns, and in order to pursue the separation policy the army increased the number of closures and drastically increased controls on Palestinian workers employed in Israel (Farsakh 2002, 2005; Kelly 2006; Parizot 2008). This period was therefore seen by the Palestinians as the affirmation of a policy of confinement and hardening of the occupation mechanisms. It was also in this context and that of the failure of the Oslo negotiations that the Second Intifada broke out in September 2000.

2000–2014: Separation and Interconnectedness

The readjustments made by the Israeli occupation regime at the beginning of the twenty-first century were radically different in character from those made during the previous period. And for good reason: the Israelis no longer saw the future in the same way as they did during the Oslo Accords period. Since the start of the Second Intifada most members of the Israeli ruling class along with its ordinary citizens were convinced that a negotiated solution to the conflict was now impossible (Cypel 2005). The Israeli redeployments during this period therefore tried both to regain long-term control over security in the Palestinian enclaves and move unilateral separation forward.

During the Second Intifada (2000–2004) the Israeli army entered regularly the autonomous Palestinian areas to attack the armed groups; in 2002, in an operation codenamed "Defensive Shield" (*homat magen*), the Israeli army massively invaded these zones. It directly targeted the PA's forces and infrastructure, accusing it of being mainly responsible for the uprising and Palestinian suicide bombings in Israeli towns. The Palestinian security forces were besieged in their barracks along with the President of the PA, Yasser Arafat who, until his death in 2004, was confined in his compound in Ramallah. Furthermore, the Palestinian populations in the Gaza Strip and the West Bank were subjected to unprecedented closures and curfews (Bocco et al. 2002a, 2002b).

At the same time, in the face of Israeli popular pressure demanding that the state take tangible measures to end the suicide bombings and impose a unilateral solution to the conflict, the political leaders have opted to pursue and implement the policy of separation from the Palestinians. It was also during summer 2002, a few months after organizing the renewed invasion of the Palestinian enclaves, that the Sharon government agreed to launch the building of the Wall (Arieli and Sfard 2008).

The combination of these two approaches led to the implementation of new Israeli control mechanisms. But the way these mechanisms operate is very different in the West Bank from the Gaza Strip. In the West Bank the army reinforced its long-term presence, maintaining a solid encirclement around the Palestinian enclaves by setting up a large number of outposts around the zones and increasing the number of checkpoints and obstacles on the roads linking them (trenches, road blocks, earth mounds, concrete blocks, watchtowers, etc.).[12] In this way it sought to reduce interaction with the Palestinians while maintaining tight control over their movements (Ben Ari et al. 2004)[13] and reserving itself the right to

12 Since 2002 the number of obstacles controlling the movement of Palestinians has constantly risen. In June 2009, the United Nations listed 698, including 76 permanent and 23 partial checkpoints (OCHA 2009).

13 Between 1994 and 1999 the Israeli army imposed 443 closure days, an average of two and a half months each year. These measures had serious repercussions on Palestinian employment

intervene regularly in the heart of their living space (Amidror 2007). These operations included targeted assassinations, arrests, and intelligence operations (Cohen 2009; Latte Abdallah, Natsheh and Parizot, in this volume; Razoux 2006; Weizman 2007).

Such controls have become more oppressive as the number of settlements kept on increasing. In 2011, the number of settlers in the West Bank (including East Jerusalem) reached 520,000 people (UNOCHA 2012). The rhythm of increase was equivalent on both sides of the Wall.[14] If communal areas and those under the jurisdiction of regional councils are included, the 122 Israeli settlements alone control 41.9 percent of the West Bank or nearly 80 percent of the C zones under Israeli jurisdiction. This area works out even larger if the bypass roads are included. Although they ease travel for the settlers and the army by making it unnecessary for them to pass through Palestinian settlements, they have fragmented the Palestinian territories and limited both urban and rural development (B'Tselem 2004, p. 6–7; Handel 2009a, p. 204–7).

The security cooperation between Israelis and Palestinians gradually resumed after the death of Arafat in 2004. On the West Bank this only became a reality in 2007 when, under the leadership of the Americans, and particularly General Dayton, the Israeli government agreed to the redeployment of Palestinian forces in the major towns of the West Bank (Legrain 2010). The authorization of this redeployment was also due to the Israeli desire to counter Hamas, which took power in the Gaza Strip in 2007. Besides, once redeployed, the forces of Fatah played a determining role in disbanding the armed groups, and Hamas in the West Bank.

The new systems of control introduced in the West Bank were envisaged as long-term, the Israeli authorities making every effort to reduce their political and financial cost (Havkin, and Latte Abdallah, in this volume). The building of "crossing points" (Hebrew: *ma'avarim*) along the Separation Wall, which were presented as "border terminals," and the use of private companies to manage the crossing points were all part of this strategy. The architecture, location, facilities, and operation as well as the terminology used to describe these new checkpoints confer a less obviously violent appearance to control. The use of private security companies to ensure the operation of the crossings and perform security checks depoliticized the control (Havkin 2008, in this volume). Some military experts responsible for planning in the Palestinian zones have even used the concept of "invisible occupation" (Weizman 2009).

The adoption of the Gaza disengagement project (*hitnatkut*) by Ariel Sharon fulfilled the same strategy (Signoles 2005). Moreover, Sharon saw in it a way to escape from the framework laid down by the Quartet[15] and the Road Map: drawn up in 2003, the Road Map set out the plan that the conflict should end in the creation of a Palestinian State by 2005 (Grinberg 2010). Ariel Sharon's advisor Dov Weiglass explained that the aim of

---

and economy and had a decisive effect on the flow of labor into Israel. Between 1992 and 1996 the number of Palestinian workers crossing the Line fell by 51 percent. It started to rise in 1997 and reached 145,000 people in August 2000 (Parizot 2008).

14 See Foundation for Middle East Peace (2007), "Settler Population Growth East and West of the Barrier, 2000–2009," and Nadav Shragaï (2007), "Most Settlements Lie East of Fence, Most Settlers West". Available at: http://www.fmep.org/settlement_info/settlement-info-and-tables/stats-data/population-growth-east-and-west-of-the-barrier, accessed December 30, 2010.

15 The Quartet is an international diplomatic body founded in 2002 to act as a mediator in the Israeli-Palestinian peace process. It is made up of the United States, Russia, the European Union, and the United Nations.

the operation was to divert the attention of the international community and the Palestinian population while the West Bank was being settled (Signoles 2005, p. 120).

This withdrawal kept the Palestinians under a different type of occupation. While it certainly resulted in the departure of 8,000 settlers and the military bases protecting them, Israel introduced new systems of remote control. The army maintained its control over air and sea space and forbade the movement of residents in a kilometer wide corridor along the demarcation line. Lastly, the Israeli authorities kept their control over the crossing points for people and goods, thereby controlling the flow of imports and exports as well as the movement of Palestinians trying to enter or leave the Gaza Strip. By tightening its grip on the coastal strip, the army set up a veritable siege around Gaza; using very few resources it was able to control or halt supplies of goods, electricity and fuel oil. The blockade imposed since 2005 has kept the population on the brink of a humanitarian disaster (Ophir, Givoni, and Hanafi 2009, p. 19). This siege, together with Israeli's increasingly bloody offensives on Gaza in June to July 2006, December 2008 to January 2009,[16] November 2012 and July to August 2014[17] have nevertheless damaged the image of Israel diplomatically.

Reappraising the Conflict's Trajectories and the Occupation Regime

By strengthening the interconnectedness of the Israeli and Palestinian zones, ensuring the long-term character of the occupation whilst offloading some of its cost onto the Palestinians and the international community, these changes suggest that the trajectories of the conflict and the functioning of the separation regime should be seen in a new light.

The Trajectories of the Conflict

Observers and researchers working on the region often appear to be blind to certain aspects of the present situation (Ophir, Givoni, and Hanafi 2009, p. 16). The period we are now living through is often seen as a period of transition between a time of confrontation and a political solution in the form of two states. Every event and process tends to be analyzed in the light of the hypothetical future envisaged by some people in the 1990s after the signing of the Declaration of Principles (1993): that of the establishment of a Palestinian State at Israel's side. The failure of the Oslo Accords, the start of the Second Intifada and the continuing deterioration of the situation did not really affect this transitory view of the conflict. Only a few social scientists have lately started to adopt a more critical analysis towards the negotiations (Turner 2014). Some contest the definition of the situation as a conflict preferring the concept of settler colonialism to describe the reality on the ground (Collins 2011).

This transitory reading and the illusion of an end to the crisis which underpins it have been encouraged by the lack of precision of the Declaration of Principles signed on 13 September 1993 by the Israeli foreign minister Shimon Peres and Mahmoud Abbas under the supervision of chairman of the PLO Yasser Arafat and the Israeli prime minister Itzhak Rabin. The postponement of negotiations on the final status of the refugees, the borders,

16 Immediately after Israel unilaterally declared a ceasefire on January 17, 2009, casualty figures on the Palestinian side were more than 1,300 dead and over 5,000 injured as against 13 dead on the Israeli side.

17 During the 2014 war, more than 2,100 Gazans were killed and around 11,000 were injured as against 72 dead on the Israeli side and approximately 700 wounded.

Jerusalem, and the settlements allowed everybody to imagine their own version of peace without having to take account of other people's versions (Grinberg 2007b).

The Oslo Accords actually ushered in a new period of misunderstanding and confrontation. On the one side, the Palestinians expected a process of decolonization: the withdrawal of troops and the repatriation of Israeli settlers to the other side of the Green Line ought to bring independence and the creation of their own state with East Jerusalem as its capital, this vision being shared by a large number of international actors. It was reinforced by the actions of the United Nations and the European Union as well as by financial investment by countries which supported the process. On the other side, the Israelis had no intention of discussing the status of all the territories occupied in 1967. Jerusalem should remain the unified capital of the State of Israel, the limits of which should include a large part of the settlements (Ben Ami 2006, p. 246–7). No government, including those formed by the Labour Party, has wanted to dismantle the settlements. The Rabin government even encouraged settlers who wanted to return to Israel (the "Returning Home" movement) to stay put in order to constitute a bargaining chip in the negotiations with the Palestinians (Grinberg 2010). Rabin and Peres hoped that the accords would lead to the creation of a confederation with Jordan rather than the founding of a Palestinian State (Smith 2007, p. 454). It was not until May 1997 that the Labour Party officially adopted the idea of a Palestinian State with a certain number of conditions (Ben Ami 2006, p. 246–7). Although in a speech at Bar Ilan University in 2009 the Israeli prime minister Benyamin Netanyahu formally accepted the principle of the creation of a Palestinian State[18], he has never stipulated clearly the conditions under which he sees this taking place.

Since the period of the Oslo Accords, the political goal of a Palestinian State has constantly been reactivated by political and media discourse. But the positions of the various players on the details of how such a state can be brought into existence have constantly changed under the influence of the deteriorating relations between Israelis and Palestinians and failed initiatives to re-launch negotiations.[19]

The absence of progress in these negotiations set the background to Mahmoud Abbas making a unilateral application to the United Nations for Palestine to be recognized as a state. On November 29, 2012, the UN General Assembly upgraded Palestine to a non-member observer state: 138 states voted in favor, 41 abstained, and 9 voted against. In the West Bank, the news lead to scenes of jubilation; even in the Gaza Strip a mass turnout on the streets greeted the news with expressions of joy.

While the UN General Assembly vote provided a political victory for the Palestinian president, this vote did not change the reality on the ground: Palestine is today a UN member state deprived of any territorial continuity, and devoid of economic and political control. Finally, on the diplomatic level it did not change the balance of power (Parizot 2012). The recent decision made by Mahmoud Abbas and his government to adhere to most UN

18 View the speech on YouTube at: http://www.youtube.com/watch?v=NY6fGMC0VtQ, accessed May 2, 2014.

19 The Taba negotiations in 2001; the Arab Peace Initiative in 2002 promoted by king Abdullah of Saudi Arabia proposing the recognition of the State of Israel by all the Arab countries in exchange for the creation of a Palestinian State inside the 1967 borders; the Geneva Initiative in 2003; the Road Map in 2003, the Annapolis initiative in 2007; and the recent failed attempts by the US state secretary John Kerry to restart the negotiations. While certain plans, such as those of Taba and Geneva, have tried to give clear proposals on the final status of refugees, settlements, borders, and Jerusalem, none of them have been agreed by both sides. All the other plans have postponed discussions on some of these issues.

agencies and institutions, and to join the International Criminal Court[20], coupled with the expansion of the boycott, divestment, and sanctions campaign (BDS) and joint International, Israeli, and Palestinian civil society mobilizations mark a clear shift towards a complete internationalization of the conflict that might have some effect on Israeli policies in the long run (Latte Abdallah 2014b, 2014c).

The transitory approach to the conflict encourages mistaken readings of the political reality and its challenges. There are three reasons for this; firstly by positioning the two parties on an equal footing, this approach gives a distorted perception of the power relations between them. While the conflict was seen during the First Intifada (1987–1993) as a confrontation between an army of occupation and a population trying to resist with derisory weapons such as boycotts and rocks, from the middle of the 1990s it was seen as two opposing parties on an equal footing: a state versus a quasi-state. Secondly, by focusing on the prospect of the creation of a Palestinian State it stops us thinking about the present and therefore about what needs to be done to bring it about. While the recognition of a Palestinian State is of political, legal—particularly in terms of the ultimate recourse possible under international law—and symbolic importance, the profound changes of position on the ground over the last 20 years raise serious doubts about its viability and sovereignty over the West Bank and Gaza. And thirdly, focusing on the notion that we are in a short transitional period that is preparatory to the construction of a Palestinian State will not stand the test of time: it is now 21 years since the Oslo Accords were signed. This "transitional period" has now lasted longer than that with which it is often compared, namely the period between the occupation of the West Bank and the Gaza Strip by Israel in 1967 and the start of the First Intifada in 1987.

It is by taking into account the territorial and institutional changes and the difficulty of separating two territorial, political, and economic entities that Palestinian and foreign scholars have cast doubt on the possibility of ever witnessing the political prospect of two states (Khalidi 2006; Hilal 2007; Clot 2010). For example, in the early years of the twenty-first century there was a revival of other projects formulated well before the Oslo Accords period: Palestinians suggesting the establishment of a single two-nation state (Abunimah 2006) while certain Israelis proposed a confederation with Jordan (Morris 2009). Others considered much more complex solutions capable of meeting the political, administrative, and territorial obstacles currently present on the ground (Grinberg 2010).

Rather than seeing this period as transitional, the present work suggests we analyze the functioning and changes in the occupation regime over the last 21 years. It is by taking account of these readjustments that we might come to a better understanding of the types of territoriality and government to which they have given rise; it is also on this basis that we may come to a better understanding of their consequences for the future of the conflict, the new challenges it hides and the ways it might be resolved.

A Contemporary, Post-Modern Occupation Regime

A number of works have tried to model how the Israeli occupation regime operated between the 1990s and 2000s. They provide rich documentation on the legal measures defining the status and rights of the populations and their unequal access to resources and mobility (Zureik 2001; Kelly 2006; Gordon 2008, 2009a; Azulay and Ophir 2008; Ophir, Givoni, and Hanafi 2009; Grinberg 2010). They also question how the lack of rights and limited

20 Which should be effective in March 2015.

types of sovereignty imposed on the Palestinians affect the operation and nature of the Israeli political regime itself (Yiftachel 2009; Azulay and Ophir 2008; Gordon 2008; Ophir, Givoni, and Hanafi 2009; Grinberg 2010). Other research from a more strictly geographical and architectural perspective has explored the complex ways in which the area between the Mediterranean and the River Jordan has been restructured (Efrat 2006; Weizman 2007; Petti 2008; Handel 2009a).

These researchers rightly stress that the lack of clearly defined territorial borders has had the corollary, not to say objective, of effacing a whole set of other distinctions, particularly between occupation and non-occupation, annexation and non-annexation, temporary and permanent, as well as the exception and the rule (Ben Naftali, Michaeli, and Gross 2009), which has meant that the zone behind the Green Line is indeterminate in terms of both time and legality. This indeterminate character deprives the Palestinians of the protection granted by international law to occupied populations and substituted a system of government using a series of regulations, decrees and procedures.

It was in the perspective of modeling and conceptualizing the situation that in the 1990s some researchers started comparing the Israeli political regime with the apartheid regime in South Africa, seeing the imposition of restrictions on movement imposed on Palestinian labor and the creation of autonomous enclaves administered by the Palestinian Authority as reproducing the system of bantustans (Farsakh 2002, 2006; Legrain 1996, 1997; Abu Zahra 2007; Hanieh 2006). Comparisons with apartheid increased from 2000 onwards, stimulated particularly by the construction of the Wall embodying discrimination and separation (Bishara 2002; Peteet 2009; Yiftachel 2009; Olmsted 2009; Dayan 2009; Bôle-Richard 2013; Lebrun and Salingue 2013). These researchers' objective was not simply academic but also political since they were denouncing and mobilizing against an unjust regime (Toensing 2009).

While such comparisons are helping to understand the Israeli occupation regime, they also tend to oversimplify the situation: comparing the Israeli regime in the West Bank and the Gaza Strip with frontier situations, or even ghettos or apartheid tended to make the regime look like an anachronistic colonial system. But research into the systems of territorial control used by Israel over the last 20 years stresses the very contemporary character of the occupation mechanisms. Indeed, the separation policy simultaneously mobilizes a territorial imagination appropriate to the modern state and to systems of control that are characteristic of the neoliberal globalization period.

The promotion by the Israeli authorities of metaphors based on territorial separation—of walls and borders—is aimed at meeting the expectations of the Israeli public and international stakeholders who conceive territorial control in the framework of the modern nation state, i.e. a homogenous and clearly delimited territory over which prevails the state *exclusive* sovereignty. But as we have already stressed, in practice, Israeli systems of control challenge any clear delineation of territory. The successive reorganizations of the occupation regime have been in total contradiction with border logic (Shamir 2009). Implemented unilaterally by Israel, the separation policy implies no principle of symmetry between two states. Nor is it envisaged as a way of separating the Israeli population from another population that is perceived as statutorily equivalent: its objective is to contain the Palestinian "other" who is seen as highly dangerous.

The Israeli policy of separation operates more as a mechanism for managing risk in a context where the two populations live in close proximity with one another and where their living spaces increasingly interpenetrate one another (Shamir 2005). It operates as a system for excluding a Palestinian population located inside an area that has remained under

Israeli control (Ophir, Givoni, and Hanafi 2009). In order to manage the close proximity and interpenetration of the living spaces of the Israeli and Palestinian populations as well as to ensure the security of the Israelis, the control techniques attempt to dissociate the trajectories of the two populations. They keep the Palestinians at arm's length in order to facilitate the smooth flow of the Israelis in a fluid, uninterrupted space. The Palestinians are confined in fragmented areas that are riddled with obstacles and in which movement is hampered and/or in which routes cannot be planned due to the frequent changes made to the obstacles (Weizman 2007). In this configuration the Israeli settlements and Israel itself constitute an "archipelago" of perfectly linked islands while the Palestinian "enclaves" are isolated from one another (Petti 2008).

In this context the relations between the two populations and their experience of time and space have become increasingly asymmetric (Collins 2008; Peteet 2008; Handel 2009; Petti 2008). The Israeli–Palestinian conflict should not therefore be seen solely as a territorial conflict but also as a conflict about the use of space (Handel 2009a). This inequality in the experience of space has major political implications since it gives rise to perceptions of the conflict that are increasingly disparate between the various actors (Parizot 2009c, 2010). The Israelis have, since the end of the Second Intifada (2004), experienced a normalization of their movements and everyday life; some even imagine that the conflict has been moved "to the other side of the wall." Forbidden entry by the Israeli Army to Palestinian enclaves (A zones), they are unaware of the degree to which the Palestinians spaces are fragmented and the current impossibility of separating two territories without one remaining fragmented and without territorial cohesion. In contrast, the Palestinians are constantly confined and controlled and experience the continual reinforcement of the occupation and its violence.

The Actors of the Occupation Regime

Any study of the Israeli occupation regime has to be dynamic. Neve Gordon (2008) suggests that the transformations of the occupation regime should be studied as the product of the interactions, the excesses and the contradictions created by the various modes of control deployed by Israel. Using an approach derived from Foucault, he considers modes of control not only as the infrastructures, techniques and policies of coercion deployed by Israel, but takes into account all the institutions, legal measures, bureaucratic apparatus, social practices and material infrastructures that act both on the individuals and the population in order to produce new behaviors, new habits, interests, tastes, and aspirations. Working along similar lines, Adi Ophir, Michal Givoni, and Sari Hanafi (2009) have published a collection of articles in which the authors attempt to analyze the occupation regime as an unstable assemblage of state and non-state apparatus and institutions, of ways of thinking and of a series of political technologies (Ophir et al. 2009, p. 15–17).

But while these researchers stress the role of the many actors and elements involved in the functioning of the regime, they have limited their analysis to the operation of the Israeli systems of control and the way it transforms the behavior of actors who are subjected to them. They do not envisage how such actors, whether they be Palestinian, Israeli, or international, can react to, subvert, or take over these systems of constraints and thus contribute to their readjustments.

French research, with only a few exceptions (Legrain 1997; Débié and Fouet 2001) has not focused on the Israeli occupation regime since the period of the Oslo Accords. Scholars studying Palestinian society have concentrated more on the social, economic and political changes it has undergone, sometimes in relation to their diaspora. In this way they have offered a series of very rich studies (Botiveau and Signoles 2004; Al-Husseini and Signoles 2011; Picaudou 2006; Picaudou and Rivoal 2006). Some have also insisted on the need to distance themselves from the conflict and its overt effect in order to offer a more nuanced, in-depth approach to Palestinian society (Botiveau, Conte and Signoles 2005). Others have also upheld this argument in their approach to the changes in Israeli society over the past 20 years (Dieckhoff 2009).

Starting from the point of view of social actors and setting it alongside the institutional perspective, this book offers an alternative view of how the occupation regime operates. We will examine both the power deployed by these control mechanisms and the (re)actions of certain groups or individuals. We approach power beyond its solely conflictual dimension in order to focus also on its productive capacity. Hence, we will analyze how not only the contestations but also the adaptations and reappropriations made by these mundane actors when faced with the Israeli modes of control contribute in turn to changing the way the mechanisms operate. In a word, we will consider these actors as historical subjects.

Mobility and Interactions in the Israeli-Palestinian Space

In order to highlight the role of these actors in the transformations of the occupation regime we have decided to focus on changes in mobility and interactions between Israelis, Palestinians, and international actors over the past 20 years. These various actors (individuals, groups, and institutions) experience this fragmented territory and its regulations daily and also contribute to constructing and changing them. Observing the transformations of their interactions enables us, beyond the hypothetical political goal defined during the period of the Oslo Accords, to understand the current territorial and social reality of the conflict and its concrete challenges. We focus our analysis not solely on the Palestinian or Israeli side, but on the two at once, and particularly on their interfaces. This approach therefore decompartmentalizes research on Palestine and Israel.

Changing Israeli-Palestinian Interactions

Analyzing mobility confronts researchers with the limits encountered by mundane actors and with the more or less coherence of their functioning. It highlights the social, economic, and political adaptations these people develop in their daily lives to adapt to the new systems of constraints imposed by the separation and the degradation of the situation.

By reorganizing their everyday life, they rework their spaces of social, economic, and political interactions at their own level. Before the First Intifada (1987), Israelis and Palestinians met almost every day as they moved within the same territories. During the 1990s, the deterioration of the situation and the enforcement of the first movement restrictions considerably reduced such interactions. But while Israelis stopped visiting Palestinian areas, Palestinians remained very present in the Israeli landscape. Palestinian-registered taxis and private cars continued to use Israeli highways and Palestinian workers could still be seen. The situation changed again at the end of 2000 onwards when the Second Intifada broke out. The two populations no longer met apart from in limited and specific places: working sites (in Israel and the settlements), checkpoints, West Bank bypass roads, demonstrations, and new commercial places built in C Areas.

Joint Palestinian and Israeli political activism was also affected by the restrictions of movement and the radicalization of the two sides (Pouzol, in this volume). In order to pursue their cooperation some activists have developed virtual forms of political actions and networking as well as renewed uses of law (data sharing, concerted legal action) (Latte Abdallah 2009, 2010a, 2011). Newly created groups focusing on land grab issues and organizing joint events associating Israelis, Palestinians, and Internationals activists (Anarchists Against the Wall, Ta'ayush, Fighters for Peace), have invented new practices and habitus that sharply contrast with those of their predecessors in the 1980s and 1990s (Lamarche, in this volume).

Finally, some actors have grasped the opportunities generated by this system of constraints. The restrictions on movement introduced since the 1990s have forced Palestinian workers employed in Israel to turn towards networks of smugglers to help them cross into Israel. These networks became increasingly organized and were able to develop very lucrative economic activities (Parizot 2014). The Israeli police and intelligence services have allowed this to develop in order to infiltrate these groups of traffickers and expand their intelligence networks. By being de facto integrated into the system of mobility control, these smugglers have directly contributed to its functioning and its readjustments.

The Locations of Power

The chapters of this volume play on different scales. While some adopt a macrosocial approach to examine the influence of economic interests in political decision-making and changes in the peace process in Israel (Grinberg), others develop microsocial perspectives by studying civil volunteers in the police (Manor) or post-2000 activists' trajectories (Lamarche). Others take an intermediary stance by tracing both changes in the political relations between Palestinians of Israel and Palestinians of the West Bank (Marteu) or LGBTQ movements on both sides of the Green Line (Pouzol). While some writers concentrate on the actors, others are more interested in describing new types of governmentality. Latte Abdallah and Havkin focus on the influence of new institutional and economic practices at precise key points in the systems of control, respectively on managing prisons and on the outsourcing of checkpoints.

Alternating between these different scales, contributors reconsider the many locations of power inside and beyond the Israeli-Palestinian spaces. They highlight the roles of a large number of actors in tandem with the state in the working of and the changes to Israeli systems of control: formal institutions such as international agencies like USAID (Garb), private companies (Havkin) or civil guards (Manor). The actors may also be informal such as the smugglers trafficking consumer goods between Israel and the West Bank (Natsheh and Parizot) or between Egypt and the Gaza Strip (Pelham). Studying how people work around Israeli mechanisms of control or use them for their own ends highlights the fact that even marginal groups contribute to the working of and readjustments in such systems. Finally, we scrutinize the construction of discourses and representations on the separation and the conflict through the practices and experience of NGOs and institutions (Handel), mundane actors, Israelis (Manor), Palestinians (Marteu and Nashif), as well as internationals activists (Hecker).

Book Structure

This book is organized around four parts. Part I considers the transformations of the geography of the occupation. Chapters 1 and 2 focus on the practices and devices by which Israel controls mobility and confines Palestinians: the checkpoints infrastructure (Havkin) and the prison system (Latte Abdallah). They show how these transformations are strongly shaped by neoliberal thought and to what extent they normalize or make invisible the occupation. For, they contribute to redraw the limits between spaces and time, contradicting the declared objectives of the separation policy. They blur the limits between the military and the civil, the inside and the outside, between past, present and future. They also readjust hierarchies and status between Israelis and Palestinians, as well as between Palestinians themselves. Chapter 3 deals with the juridical dimensions of the geography of occupation. Emilio Dabed shows that in the context of the absence of Palestinian territorial and political sovereignty, the drafting of the Palestinian constitution was strongly influenced by the asymmetrical power relations between the PA and Israel as well as between international actors and experts and the PA. Chapter 4 concludes this part by providing a counter intuitive approach in which Ariel Handel deconstructs the narratives by which occupation is usually analyzed and criticized. He demonstrates how the built-in utilitarian biases of these languages actually create misunderstanding of the space Palestinians use and the specific relations and emotional links they develop towards it.

Part II scrutinizes the economic and commercial exchanges between Israeli and Palestinian territories during the post-Second Intifada. Chapter 5 studies the crossings handling the formal transit of goods between the south of the West Bank and Israel (Garb); Chapter 6 analyzes the smuggling from Israel to the West Bank (Natsheh and Parizot) and Chapter 7 provides an analysis of the tunnel economy between the Gaza strip and Egypt (Pelham). The authors highlight the complex configurations of power emerging along Israeli-Palestinian "borders." The new mechanisms of regulations and models of territoriality they highlight challenge the imaginary of the modern nation state. In order to better situate these forms of economic and territorial control in an historical perspective, Chapter 8 analyzes the changes in Israeli economic policies towards the Palestinian Territories since 1967. Lev Grinberg shows how patterns of the military-economic domination regime were shaped by the interests, power relations and compromises between the military, the dominant economic groups and the ruling party.

Part III decenters the gaze to the margins of Palestinian and Israeli society by considering how the separation has been experienced among different groups: the volunteers of a peripheral town in the Israeli police (Chapter 9); Palestinians of Israel (Chapter 10); and among Israeli and Palestinian LGBTQ activists (Chapter 11). Israeli Palestinians cross the separation lines more than other Israeli Jewish citizens and more than Palestinians of the OPT, thus carving a specific place in both national arenas. Similarly, police volunteers, *mizrahim* ("oriental") Israelis from a development town, play a special role in building the separation by reconstructing the stereotyped image of the Arab enemy, i.e., of the "terrorist." Lastly, Israeli LGBTQ mobilizations show how sexual minorities and sexual identities are embedded in national considerations, and in "homonationalism." In this context, Palestinian LGBTQ organizations (mostly formed by Palestinians from Israel or Jerusalem) have defended at the sexual rights and Palestinian political rights at the same time. Being part of the most influential Palestinian popular resistance movement they are drawing new political boundaries where marginal sexualities are no longer associated with political deviance.

Part IV continues this reflection on the experiences and effects of crossings taking place within specific political actions: the travels of organized tours of French pro-Palestinian and pro-Israeli activists (Hecker); the clandestine crossings of Anarchists against the Wall within Palestinians enclaves; and those of Palestinian suicide bombers (Nashif).[21] While

21 Called "martyrdom operations" by the actors.

these crossings and actions are radically different from one another, they all contribute, in their ways and at their different scales, to adjust and construct the boundaries of the Israeli-Palestinian spaces. These practices do not really challenge the separation regime and are rather shaped by the very frames imposed by Israel. Yet, they do contribute to the definition and the reproduction of these groups' collective identities.

The Politics of Research

Like political and media discourses, researchers' narratives are also significant for the parties in conflict. In this highly polarized verbal minefield, researchers have to be cautious and show greater courage than in other research fields. The role of research is not to produce arguments backing one party or both, but to create explanatory models capable of making sense of the reality of a conflict that has changed greatly over the last 20 years. But it is not an easy task as both the definition the research objects and scientific collaborations developed in this context have political dimensions.

Working on the interaction and interconnectedness between Israeli-Palestinian spaces necessarily highlights the current obstacles to the creation of a viable Palestinian State in the near future. It also means questioning the current representations/definition of the State of Israel. Deprived of borders with Palestinians, Israel cannot be conceived as a democracy just like other democracies and neither as a state whose majority is Jewish. Though, it breaks away from the political horizon defined by the Oslo process. We are conscious that this scientific position clearly comes into conflict with national perceptions that have been forged and perpetuated by ideologies and collective imaginings, but we need to analyze the social, political and territorial reality prevailing today on the ground. As researchers, our perspective is not of course to take a stand in the discussion about one or two states—this is clearly for the Palestinians and the Israelis to decide—but rather to consider the concrete impact of the redeployments of the occupation over the last 20 years on such political projects

Working on this conflict also raises issues of scientific partnership. Our work has been undertaken within the framework of two research programs "Appraising the Israeli Palestinian Conflict through Cross Border Mobility" and "Mobility and Borders in the Israeli Palestinian Spaces." Initially conceived by Cédric Parizot as an extension of his research in the mid-1990s on mobility in the south of the Israeli-Palestinian spaces (West Bank, Israel, Gaza), these projects where coordinated jointly with Stéphanie Latte Abdallah between 2007 and 2011. They brought together French, Palestinian, Swiss, Italian, and Israeli researchers working on both sides of the Israeli and Palestinian divide, or between the two, in the interspace. We stress the fact that we are talking about researchers working on both sides and not about researchers coming from the two sides. Our aim was not to bring Israelis into discussion with Palestinians but rather to acquire the means for a better understanding of the mechanisms of the Israeli occupation since the Oslo Accords.

Each researcher took part in the program individually. We decided to avoid any institutional cooperation with Israeli or Palestinian universities or research centers. This was firstly to avoid any political obstacles or orientations that our scientific approach might have aroused; secondly, because we refuse any attempt to promote dialogue or the normalization of relations between Israelis and Palestinians which cannot be among the objectives of a scientific program. In institutional terms, the projects were financed out of

European funds, as part of the Ramses² European Excellence Network, and French funds. They benefited from the partnership of French laboratories[22] and a Swiss research institute.[23]

We should point out that despite our very clear position on institutional scientific cooperation, it remains very difficult to work in this interspace due to scientific compartmentalization, political obstacles, and the unequal capacity of mobility of the project members.

Firstly, the strict compartmentalization of French research into Palestinian studies and Israeli studies contributes to these difficulties. On the one hand, French students and researchers work on one side or the other and are only rarely in the same institutions. On the other, the historical processes in which Israelis and Palestinians have been involved have created different research agendas within each society. This trend has been accentuated because academics have approached their histories as exceptional trajectories, inviting few comparisons with other contexts (Tamari 1997, p. 20). Moreover, the rapid deterioration of the situation after the signing of the Oslo Accords first legitimized separate scientific approaches before placing additional political and material obstacles to dialogue between researchers and institutions working on Israel or Palestine. We are not denying the autonomy of the two research fields, but simply stressing the need to leave room for an approach to the interspace, the only one that is capable of making sense of the redeployments of the occupation since the Oslo Accords.

These political obstacles are all the more sensitive in a context of the radicalization of positions since the Second Intifada and the ongoing violence that has marked the post Intifada period. We are referring here to the summer 2006 Israel–Lebanon war and the military offensives in Gaza (summer 2006, December to January 2008–2009, November 2012 and July to August 2014) which, with their declared dissuasive aim, involved the use of ever greater violence. The continuation of the occupation and these particularly destructive Israeli military offensives have reinforced not only in Palestine but elsewhere, the efforts to boycott Israeli institutions: the BDS (boycott, divestment, and sanctions) movement has gradually mobilized groups and activists all over the world.

On the Israeli side, the situation has hardened considerably and many within the Israeli population now reject the idea of a new withdrawal from the Occupied Territories. As for those who criticize the occupation and the military operations, they have often prompted virulent reactions that have made activists groups who reject the national consensus forged from 2000 onward more and more marginalized (Marteu 2009a). Furthermore, a certain number of Israeli academics have faced considerable hostility and been subjected to pressure from within their institutions and some have even left the country.

Finally local and foreign researchers on the ground have encountered obstacles to their movements. Apart from the deterioration of the conflict and tensions between the two parties, restrictions on movement and security measures have naturally affected research in the field. While it is difficult, indeed dangerous, for Israeli researchers to carry out

22 Institute for Research and Studies of the Arab and Muslim Worlds (Institut de Recherches sur les Mondes Arabes et Musulmans—IREMAM), Institute for Mediterranean European and Comparative Studies (l'Institut d'Etudes Méditerranéennes Européennes et Comparatives—IDEMEC), Mediterranean Institute for Humanities (Maison Méditerranéenne des Sciences de l'Homme) in Aix-en-Provence (USR 3125), French Research Centre, Jerusalem (Centre de Recherche Français à Jérusalem—CRFJ), and the French Institute for International Research (Institut Français de Recherches Internationales—IFRI).

23 Graduate Institute of International and Development Studies (Institut des Hautes Etudes Internationales et du Développement—IHEID) in Geneva.

research on the Palestinian side, it is virtually impossible for Palestinian researchers from the Occupied Territories, unless they are natives of Jerusalem. Our team was also unable to carry out research in the Gaza Strip because of the blockade and ban on entry that the Israeli authorities impose on Israelis, West Bank Palestinians and foreigners, the only exceptions being Nicolas Pelham. Thanks to his press card, he could enter the Strip and conduct a fine analysis of the Gaza Tunnels.

Confronted with the complex reality of the occupation each observer has to take responsibility for his or her own position. Depending on our contacts, our political environment and our scientific career, we can experience considerable political, personal or material difficulties in undertaking this type of fieldwork investigation. Incidentally, we could have laid more emphasis on these problems and thought jointly about our biographies, mobility, approaches and scientific tools that they have led us to adopt.

To the difficult nature of the research practices in this context we must add the equally complex question of the concepts and terms used by the different contributors. Defining and harmonizing the concepts used is a challenge the authors of any collective work have to deal with. In this case the problem is heightened. More than elsewhere, people and groups in the Israeli-Palestinian territories are identified by the words they use. The extreme polarization and the tidal wave of political and media arguments that this conflict has prompted identify them immediately, sometimes even in spite of their authors. It is now clear that moving from one space to another or taking up a position in the interspace makes it particularly difficult to choose the words and concepts needed to describe a reality scientifically without immediately being classified as a stranger or an enemy.

We agreed on the more frequent use of the term "Wall" in preference to "Barrier." This choice was clearly not neutral as the term "Wall" evokes more clearly the massive, violent nature of what is being built and its territorial impact (confiscation, expulsion, and annexation of Palestinian lands) as well as the project of separation and its multiple demographic and symbolic dimensions. The terms "Barrier" or "Fence," on the other hand, seemed to us euphemisms for the structure. Despite certain editorial choices, we have to accept that the use of certain sometimes problematic terms has not been harmonized. In the last analysis each author is free and responsible for his or her text and the words and concepts used therein. The different terminologies relate to distinct frames of reference and existing areas of discussion, and sometimes to very different personal and political stances.

Lastly, we should stress that the terms may vary, not always because of a consciously adopted stance or policy, but most often according to what the terms signify for the different actors: this is the case, for example, with the terms "occupied territories" or "Palestinian territories" and that of "Palestinian enclaves." If one is referring to the shape of these territories the term "enclaves" appears more appropriate but when one is describing the perception of a political and symbolic experience or the internationally recognized legal reality, the term "occupied Palestinian territories" is more relevant. While the absence of terminological uniformity may disturb, it is inherent to the very subject and approach of this book which tackles the Israeli–Palestinian conflict concretely from the points of view of its many actors and the representations of the conflict they communicate.

PART I
Geographies of Occupation

Chapter 1
Outsourcing the Checkpoints: When Military Occupation Encounters Neoliberalism[1]

Shira Havkin

Introduction

The process of outsourcing the control over checkpoints along the boundaries of the West Bank and the Gaza Strip, which began during the 2000s, creates a complex entanglement of military and neoliberal logic. The study of this process aims to understand the part privatization plays in reorganizing modes of governing and redeploying forms of power and violence in Israel and the Occupied Palestinian Territories (OPT).

In 2003 the Israeli government decided to outsource the management of checkpoints, which until then had been managed by the police or the army, by appointing private security firms. The decision applied solely to checkpoints located in the "Seam Zone" (*merhav hatefer*), along the Separation Wall in the West Bank and at the entrance to the Gaza Strip. Whereas the checkpoints at the entrance to the Gaza Strip were built directly on the Green Line, in the West Bank most of them were built in the territories occupied in 1967. These therefore cannot be considered "border" checkpoints as no agreement has yet defined the boundaries between Israeli and Palestinian territories in international law.[2] These checkpoints reinforce the unilateral institutionalization of the lines separating Israelis and Palestinians.

The declared objective of the promoters of the privatization policy is as follows: "To reduce the friction that currently exists in the checkpoints and improve the quality of service without decreasing the level of security screening. These crossings will be considered as official border crossings and will resemble the terminals found in every country in the world" (Tal 2006). The outsourcing, which is defined as transmission of control to the civil sector, or the "civilianization" of the "crossing points" (*izruah hamaavarim*),[3] is considered to be a form of demilitarization. The first checkpoints were privatized in January 2006.

---

1 I would like to thank Eilat Maoz who began this research with me (Maoz, 2009), and also Béatrice Hibou, Cédric Parizot, Inna Michaeli, Ariel Handel, and Dganit Manor for their careful rereading, comments, and valuable observations.

2 However, in order to highlight the process of spatial reorganization they are implementing and to distinguish them from the checkpoints on the other side of the dividing wall, in this chapter I will refer to these points of control as "border" checkpoints and not by the official name used in Israel—"Seam Zone checkpoints" (*mahsomei kav hatefer*), "Seam Zone crossing points" (*ma'avarei kav hatefer*), or simply "crossing points" (*ma'avarim*). Nevertheless, I will put the word "border" in brackets as a reminder that these borders are, in most cases, not official ones.

3 The Hebrew word *izruah* is a new word the root of which means "civil" or "citizen." It appears in the dictionary with another meaning: attributing citizenship, or naturalization. In this chapter I use the terms "civilianization," "to civilianize," and "civilianized" when I refer to official discourses. In other cases I refer to the process in question as "privatization" or "outsourcing."

Since then, the management of 13 of the 33 "border" checkpoints have been fully outsourced and put under the official responsibility of the Crossing Points Authority. Their operation was delegated to private security firms. The remaining 20 checkpoints, located in the Jerusalem area, have a more ambiguous status in terms of privatization, managed by a heterogeneous "assemblage" of military and police officers and employees of private firms. Five firms were selected[4] on the basis of a call for tender in 2006. Besides security and surveillance tasks, some of these temporary employment agencies also provide casual cheap labor for cleaning and maintenance tasks for private and public bodies. The Crossing Points Directorate, which later became the Crossing Points Authority (CPA), was specially created for this purpose by the Ministry of Defense, and is the "customer" of these firms.

Firstly, studying the privatization of checkpoints in Israel enables us to understand how the role of the Israeli state has changed in setting up modes of occupation. Unlike the approaches that tackle privatization in terms of state withdrawal or abandonment (Strange 1996; Swann 1988), I understand this process rather as a redeployment of its means of intervention (Hibou 1999; Bayart 2004). In actual fact, the related administrative procedures are such that the state of Israel maintains its power and control over these locations while off-loading certain forms of responsibility.

Secondly, the study examines the effective operation of the new power apparatus set up in the "terminals" and the way in which it restructures the practices of domination. Once privatized, the checkpoints are comparable to what Michel Foucault referred to as the power apparatus proper to neoliberal governmentality (Foucault 2004a, 2004b). Their "rational" and "modern" character, as well as their stated aim of "reducing friction," tend to inscribe the conflicting power relations that prevail between occupiers and occupied in strategies, practices and discourses. These institutionalize them and make them seem "natural" and "normal," in other words, acceptable and consensual. It is in this sense that the privatization of checkpoints may be understood as an attempt to depoliticize, a depoliticization which—according to the definition given by Jacques Rancière—is not the dissolution of politics but its very exercise: "Depoliticising is the oldest task of politics, the one which achieves its fulfilment at the brink of its end, its perfection on the brink of its abyss" (Rancière 2004 [1990], p. 19). This chapter therefore analyzes this polishing-like process which seeks to conceal the oppressive power relations at work at the checkpoints and the new forms of domination and arbitrariness to which they give rise.

This analysis also examines the new spatial and symbolic relation between the *inside* and *outside* produced by the creation of the "border terminals." These places, which are intended to "resemble border terminals like in every country in the world" according to the official text quoted above, simulate borders where there are actually none, as for the state of Israel the OPTs are neither an inside nor an outside. The construction of the "terminals" is obviously part of an Israeli decision to separate and externalize these territories. Nevertheless, parallel processes of extension of the settlements, maintaining the presence of the Israeli army and the omnipresent possibility of military incursions inside the OPTs reveal an opposite policy: that of appropriating the territories and maintaining them under Israeli control. Moreover, the Israeli decision makers refuse to consider the path of the Separation Wall as a border between Israel and a future Palestinian state. In this context, the apparatus of the outsourced checkpoints draws on a logic of sedentarization of a "border"

4 In a second tender launched in November 2008, only two security firms were selected: Modi'in Ezrahi for the north and center of the West Bank, and Sheleg Lavan for the south of the West Bank and the Gaza Strip.

which isn't one, as a further attempt to frame and structure the perplexing situation of territoriality without borders (Rabinowitz 2003, Parizot 2009a).

This research is based on fieldwork carried out in the West Bank between 2003 and 2010, mostly conducted as participant observation. From 2003 to 2005 frequent trips were made to the checkpoints as part of the Machsom ("checkpoint") Watch movement, an Israeli women's organization that conducts daily observations around several dozen checkpoints in the West Bank in order to watch the behavior of soldiers, document violations of Human Rights and intervene where possible.[5] This research is also based on texts dealing with the outsourcing: Israeli press articles and official documents (parliamentary commission protocols, annual reports by the state comptroller's office, calls for tender, etc.). A close reading of these documents reveals valuable information about the elaboration of the privatization process and the decision-making it involved. A critical analysis of these official discourses enables an examination of the common senses and specific rationalities of agents and institutions. The intermingling of written sources with field observations allows an analysis of the process both through its discourse and its practice, official and unofficial.

This analysis does not deal with the effects of this transformation on the daily lives of Palestinians nor with the resistance it arouses. Focused on the political strategies, the discourses and the effective strategies of Israeli control, the aim of this research is rather to shed light on the process through the prism of power, its mechanisms, structures and rationales. The chapter begins by analyzing the genesis of the outsourcing of the checkpoints, through two parallel processes that emerged in the 1990s: on the one hand, the restructuring of forms of governmentality in Israel, and, on the other, the transformation of the occupation regime through the setting up of the separation policy. The second section of this chapter analyses the impact of the outsourcing of the checkpoints on the implementation of Israeli control apparatus in the OPTs.

Political History of Outsourcing

The decision to outsource the checkpoints was taken in a historical and political context where two processes converged: the first was the spread of neoliberal doctrines in Israeli society leading to the adoption of strategies of privatization, outsourcing and "good governance" in the public sector and in the army; the second the change in the occupation regime and the institutionalization of a separation policy, which, in the context of the occupation always comes along with mechanisms of control and appropriation.

Privatization and "Good Governance"

The outsourcing of the checkpoints took place together with a massive privatization of institutions and state-run services in Israel, beginning in the second half of the 1980s and becoming more widespread during the 1990s (Bichler and Nitzan 2001; Filc and Ram 2004; Swirski 2005; Hason 2006; Ram and Berkovitch 2007; Maman and Rozenheck 2009). The restructuring of relations between the state and the private sector does not affect Israel alone. It is part of the rise of neoliberalism characterized by heightened mistrust of state-run structures and civil servants, which Michel Foucault referred to as "State phobia" (2004b). It is in the name of this mistrust that the reform of the public sector is justified. Inspired by

5 Machsom Watch website: http://www.machsomwatch.org, accessed April 22, 2014.

the practices of private management, it imposes efficiency as an assessment criterion and competition as the main instrument of management. The neoliberal reform of the public sector was achieved by introducing competition between the public and private sectors and the adoption of new strategies of public management guided by the principles known as good governance. Competition, downsizing, outsourcing, regulation by specialist agents, staff flexibility and the creation of performance indicators all constitute instruments that administrators and political decision makers will import and distribute throughout the public sector in the name of "good management" (Dardot and Laval 2009).

As many researchers have pointed out, despite their stated objective, privatization and outsourcing are not necessarily rational in strictly economic terms of efficiency and productivity. Whether there is an economic justification for the outsourcing of the checkpoints remains an open question. It is difficult to assess the costs incurred in building the new "terminals," setting up the sophisticated technological infrastructure and creating new administrative branches such as the CPA attached to the Ministry of Defense, particularly since the budget intended for this reform comes from separate sources: national (Ministry of Defense, a special budget for the "Seam Zone," a special budget for the "civilianisation of the crossing points"), private sector capital, and international players (mainly the generous support of the United States government). The protocols of the Knesset Commission on Internal Affairs and the Environment stress the difficulty in calculating the costs and mention the disagreements between the representatives of the various ministries on the subject (protocol n. 17, 20 June 2006). In a report by the State Comproller at the time, the drawbacks and dangers of outsourcing military tasks and industries are openly discussed (State Comptroller 2004; Maoz 2009).

Rather than tackling the efficiency of privatization in strictly economic terms, I prefer to approach it in terms of its political rationale. In general, privatization transforms political investment into economic power and redistributes power and profits between actors (Hibou 1999). It thus opens up new horizons for private investment in the security sector, a substantial product in the Israeli economy and a growing field of export (Nevo and Shur-Shmueli 2004; Gordon 2009a; Hever 2010; Levi 2010). The process strengthens the penetration of the neoliberal entrepreneurial rationale into the public sector, particularly the army. In order to create the conditions for competition between Tzahal and private industries, the army, for whom until recently labor and goods have not come with a price tag, is now forced to adopt the same calculation methods as the private sector.

The Israeli army began this "managerial revolution" in the early 1990s. In 1991 the Israeli government appointed the first commission of specialists—the Sadan Commission—to propose structural reforms for the army. In 1993, Tzahal was the first army in the world to adopt the so-called Total Quality Management (TQM) strategy. This management theory, which is widely adopted in private companies, particularly in Japan, is based on 14 principles, the main ones being product quality management and customer satisfaction as a guarantee of a company's sustainable profit (Deming 1986). The conclusions of the Sadan Commission were presented in June 1994 and suggested "redistributing tasks between the army and the private sector according to criteria of competition and market forces." The IDF[6] thus had to professionalize the military tasks considered its "core competencies" and to open up its non-core functions to competition from the private sector, opening the way for outsourcing to third parties (Liber 1999; Levi 2010).

6 The Israel Defense Force.

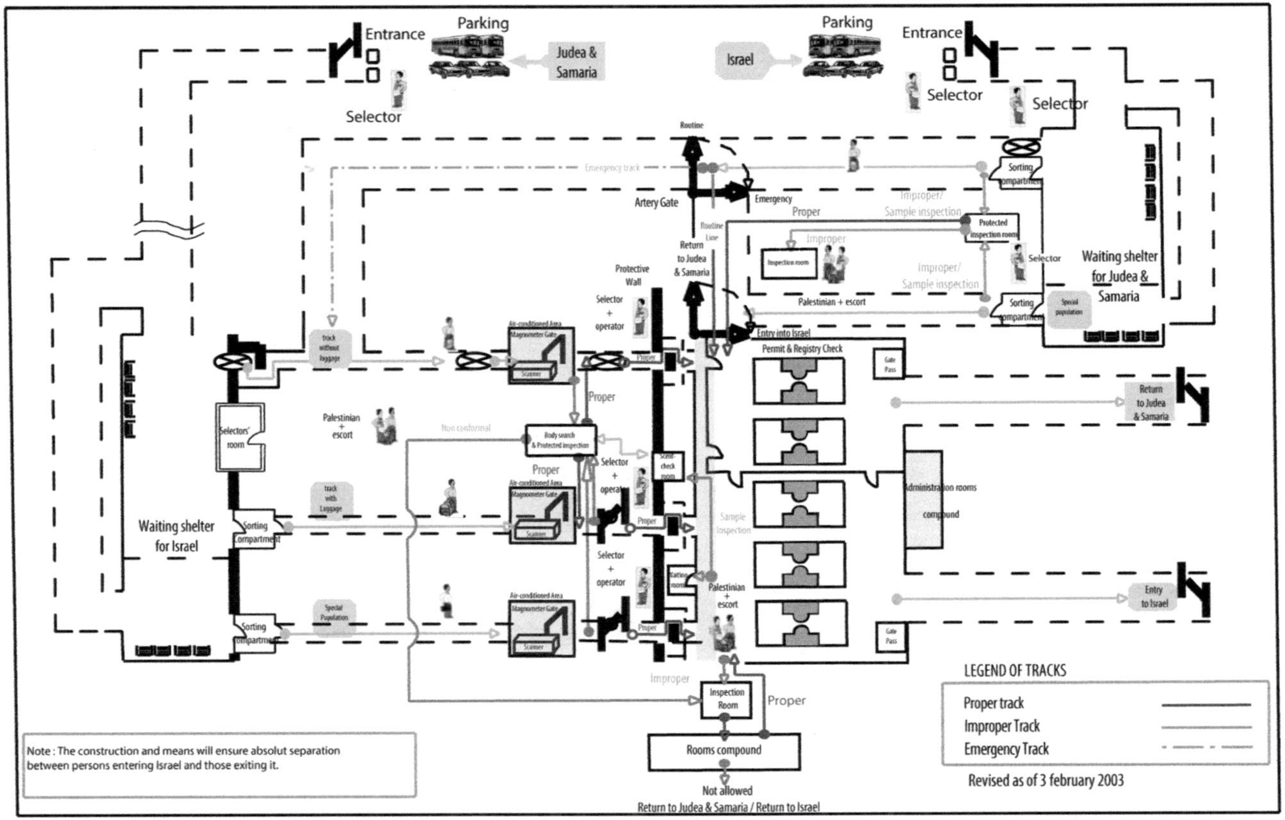

Figure 1.1 Sketch of a crossing. This sketch was originally published in an official call for proposal for the management of the checkpoints. Israeli Ministry of Defense, May 15, 2005.

The outsourcing of the checkpoints occurred at a time when the practices of the Israeli army in the OPTs in general and management of the checkpoints in particular were subject to intensified criticism. During the First Intifada (1987–1993) the criticism focused mainly on the brutality and violence of military practices. In the 1990s the criticism from within Israeli society changed focus and the IDF was largely blamed for neglecting its main task—combat—while concentrating on the everyday policing tasks of the occupation (Shelah 2003). At the beginning of the Second Intifada (2000–2005) the critics targeted the poor management of checkpoints, which became a major field of the daily confrontation between Palestinians and the occupying forces. In 2004 the Spiegel Commission criticized the lack of regularity and consistency in applying the regulations, problems "of discipline, behavior, ethics and morale," insufficient military training, lack of personnel, infrastructures and control technologies. It concluded that poor management "was detrimental to the image, trust and credibility of Tzahal in the eyes of the international community and of foreigners on site" (NRG 2004). During the most violent period of the Second Intifada, from 2002 to 2003, Israeli and international "civil society" stepped up their criticism of army practices at the checkpoints, supported by images, eyewitness accounts, and information diffused by the media.[7] Checkpoint management involved a great number of soldiers and was broadly criticized both within and beyond Israeli society, thus becoming an increasingly costly task, both materially and politically, for the Israeli army. In this context, outsourcing checkpoint management provided the army with the means to offload the task and, to a certain extent, avoid responsibility for it.

Privatization and Separation

The outsourcing of the checkpoints should also be seen in the context of recent transformations of the Israeli occupation regime, i.e. the implementation of a policy demanding separation since the early 1990s. The term "separation" takes on a special meaning here: it does not refer to a simple territorial distinction between Israeli and Palestinian spaces, but rather to the application of a control and domination strategy in a territory that remains entirely under Israeli control (Benvenisti 1988; Weizman 2007; Gordon 2008; Ophir and Azoulay 2008; Ophir, Givoni, and Hanafi 2009).

Historically, separation and annexation have been rival political projects for solving the supposedly temporary situation of military occupation. They first envisaged the creation of a Palestinian state alongside the state of Israel, while the second foresaw a single political system governing the entire Israeli-Palestinian state, granting Palestinians access to civic rights. Since the initial discussions on the future of the OPTs in 1967 and until the present day, both programs have remained marginal. The extension and entrenchment of the occupation have transformed a temporary situation into a permanent state. The rationale that the Israeli policy finally followed was based on treating the population and the territory as distinct entities. This policy aimed to maximize the territory under Israeli control without integrating the Palestinian population into the political community, thus maintaining a

7 See the creation of the women's movement Machsom ("checkpoint") Watch in 2001, the broadcast of the film *Machsomim* ("Checkpoints") by filmmaker Yoav Shamir on Israeli television in 2003, and reports by various human rights organizations: B'Tselem (Israeli Information Center for Human Rights in the OPT), Hamoked (Center for the Defence of the Individual), and Gisha (Legal Center for the Freedom of Movement), and international organizations such as OCHA, Human Rights Watch, Amnesty International, etc.

Jewish demographic majority. The resultant political dynamic is based on a dialectic that sets out two strategies of domination: separation and appropriation. All Israeli political processes, strategies and projects in the OPTs are marked by this double logic: behind each appropriation initiative lies one of separation and every separation initiative is doubled by strategies of appropriation (Azoulay and Ophir 2008, 2009).

The balance between the two dynamics has, however, changed over time. In the first 20 years of the occupation the state of Israel focused on the strategy of appropriation, allowing the Palestinians to cross the Green Line in order to work in Israel. In 1972 the army granted Palestinians in the OPTs a general entry permit and by the year 1987, 39 percent of the labor force in the OPTs was working in Israel (Kemp and Raijman 2008). This strategy of integrating or appropriating the Palestinian population of the OPTs into the Israeli economy did not mean integration into the political community; on the contrary, it was based on the fundamental distinction made between Israeli citizens, who had social protection and civic rights, and Palestinian non-citizens, who had none. The appropriation policy was one of the main factors that created a growing dependency of the Palestinian economy on the Israeli economy. At the same time, it created a large back yard for the Israeli economy that could therefore rely on this source of cheap, easily exploitable labor. It was not until the early 1990s, i.e., a few years after the First Intifada had erupted, that this policy began to change. In January 1991 the general entry permit was cancelled, a measure that began to be effectively applied from March 1993. Gradually, Palestinians who wished to enter Israel needed a travel permit, which was issued individually and according to criteria that became increasingly strict (B'Tselem 2007; Kemp and Raijman 2008; Handel 2009a; Barda 2012). The 1990s therefore marked a change in the balance between the separation and appropriation strategies, in favor of the first. The general entry permit was cancelled at the same time as the peace negotiations between the state of Israel and Palestinian political leaders began. However, these negotiations and the various agreements which resulted did not lead to the founding of a Palestinian state. Instead, declaring a transition into an intermediary state that could lead to the end of the occupation (Azoulay and Ophir 2008) marked a reorganization of the power apparatus based increasingly on the separation of populations. In particular this reorganization resulted in the construction of a "security fence" (*gader bitahon*) around the Gaza Strip in 1994 and, from 1995 onwards, increasing numbers of checkpoints along the "Seam Zone" in the West Bank. The situation cannot, however, be fully summed up in terms of this separation dynamic. During this period the appropriation process continued in the form of increasing numbers of settlements, the annexation of land and the redefinition of spaces and levels of control. Thus in 2000, just before the Second Intifada, Israeli sovereignty in terms of security and movement, covered most of the OPTs.

Since the beginning of the Second Intifada the state of Israel has applied three measures that reinforce the strategy of unilateral separation: (1) the construction of the Wall in the West Bank, beginning in 2002 and being the largest construction project ever initiated by the state of Israel (Arieli and Sfard 2008); (2) the withdrawal from the Gaza Strip in 2005, followed by strict control on entry and, since 2007, a continuous blockade on the territory; (3) reinforcing the control and restrictions on the movement of the Palestinians by the mean of an elaborate checkpoint system from the 1990s onwards. In order for Israeli society and the international community to consider these measures legitimate, they were presented as the means of implementing a separation policy (Rabinowitz 2003), a separation that was now considered the best solution for governing the OPTs.

Despite being less visible and attracting less media attention than the construction of the Wall or the withdrawal from the Gaza Strip, the checkpoint system is the main mechanism used to implement the policy of separating Israelis and Palestinians. In June 2012, the United Nations office counted 542 physical obstacles of which 61 were permanent checkpoints, 25 partial checkpoints, and 436 trenches, road barriers, mounds, concrete blocks, etc. in the West Bank. (OCHA September 2012).[8] These high figures suggest a power apparatus in its own right, which imposes control over the movement of Palestinians and fragments and divides the Palestinian territories. In this context, the "border" checkpoints in the "Scam Zone" play a particularly important role—together with the Dividing Wall, they embody the separation itself.

The implementation of this policy falls into the model analyzed above according to which separation does not divide two territories or two sovereign entities but, on the contrary, maintains Israeli control over the entire geopolitical area between the Mediterranean and the River Jordan. In addition to the separation policy, the Israeli governments have reinforced the appropriation strategies. In the West Bank the army has multiplied the number of obstacles and control systems to the east of the Wall, the settlements have been expanded as well as road infrastructures designated to their inhabitants. In this situation, it has become impossible to distinguish Israeli from Palestinian territories, as the areas in which the two populations live became closely interlocked (Benvenisti 1988; Handel 2009a; Parizot 2009c).

However, even though the implementation of the separation policy may not produce territorial separation, it creates a false image of two separate entities. The withdrawal from the Gaza Strip is a spectacular demonstration of this policy's capacity to act on two levels: it has enabled the reinforcement of Israeli control and, at the same time, the use of the concept of separation as a discursive tool, allowing Israel to abdicate its responsibility. In Gaza, this responsibility referred first to the life and wellbeing of the occupied population in a bio-political sense; and then for the life and death of the inhabitants as enemies (Azoulay and Ophir 2008). The discourse of separation has concealed the fact that the Gaza Strip was under siege and justified Israeli military operations.

We can observe a parallel process in the West Bank, as regulations restricting the movement of Palestinians do not create a divide between areas and the populations that live in them. Rather they create a filter system that grades populations according to their right to mobility in a space which is appropriated and separated by the same movement. The checkpoints, particularly the "border" checkpoints, embody the fundamental ambivalence of the occupation: they are intended to represent an international border for Palestinians and "resemble terminals like in every country in the world." Yet they do not represent an international border either for the Israeli army or for Israeli settlers who cross, and for whom security is no more demanding than a motorway toll or a shopping center entrance (Parizot 2009c). As in shopping centers, security screening varies according to ethnic and racial profiling.

Privatization and Restructuring of Power

Outsourcing the management of the checkpoints does not imply withdrawal of the state from border controls between the OPTs and Israel. It is rather a question of redeploying its

8 Ariel Handel points out that one of the main characteristics of documents such as reports published by OCHA is that at the time of publication the maps of checkpoints and figures given are already out of date (Handel 2009a).

administrative and institutional procedures implying a redefinition of the relations between public and private, military and civil, economic and political. An analysis of the changes in architecture, security screening, agents and discourse concerning privatization gives a concrete image of the reorganization of the power apparatus which "reduces friction" by displacing certain repressive procedures in an attempt to gloss over the control and domination procedures.

Political "Economies" of Privatization

Outsourcing the checkpoints means transferring one of the core functions of the state, i.e. security, to third parties. This process therefore involved setting up measures intended to define clearly the context and scope of the transfer. During parliamentary commissions the ministries' legal advisors identified certain problems posed by authorizing private security guards to resort to violence. These concerned the legality of such authorization, the legal responsibility of the security guards and the possibility of legal protection if complaints were made against them (Knesset Commission on Internal Affairs and the Environment 2005a, 2005b).

Two types of measures were taken to overcome these problems. In 2007 the law on the defense of public security was amended to extend the authority of civil security guards assigned to checkpoints. The amendment authorized them to search the personal belongings of people crossing the checkpoints, to conduct body searches and, under certain circumstances, to hold people until the police arrived (The official announcement gazette, 2007, p.2, 626). This legal amendment entitled private security guards to resort to force and violence but posed spatial and temporal limits on their interventions. The security guards' mandate also depends on the continuous presence of state representatives at the checkpoints, which is ensured by two types of agents: an army or police representative (according to the area) who is present at the checkpoints alongside the private security guards, and a civil servant from the CPA. Through their presence, they represent state sovereignty and may delegate part of their power and authority to the employees of private firms.

This type of organization is very economical because the CPA only has overall responsibility for management of the checkpoints while the work is delegated to the inspectors and security guards. The state also subcontracts to private firms that supply the staff, thereby abdicating its responsibility as employer: as the security guards and inspectors are employed by the private security firm, the Ministry of Defense is not responsible for hiring them, for their employment conditions, or for dealing with their demands. According to a well-known neoliberal approach, the administration—including the army, the sovereign institution par excellence—has integrated the idea that the state and its civil servants are less efficient at management tasks than the private sector and has adopted the standards of flexibility and downsizing characteristic of the private sector. As the director of the CPA Betzalel Traiber, despite being a civil servant himself, says to justify reducing the number of civil servants employed in his department to a strict minimum, "Civil servants are like nails without heads—they go in and you can't get them out." The neoliberal approach can also be seen in the increasingly popular notion of professionalization and expertise: it is in this way that the "civilianization" of checkpoints is also justified, presented as being a means of ensuring competent, efficient service using the mechanisms of the market. It is also used to depoliticize control by considering it an expression of knowledge and know-how that young recruits do not possess, something that belongs to the professionals.[9] By using this type of organization the state abdicates its legal responsibility. Unlike soldiers and police officers, private inspectors and security guards are not governed by any internal legal body. If any complaints are filed against them they are now judged as individuals in a criminal court (Breiner 2009). The political consequences of this abdication of responsibility are fundamental: instead of being considered the structural result of the occupation, violations of human rights become criminal cases committed by particular individuals. The state is therefore no longer obliged to respond to the increasing criticisms regarding the legitimacy of its use of violence. One of the functional effects of this outsourcing appears clearly: if, according to the Weberian formula, state sovereignty is based on its monopoly of control over legitimate violence, when legitimacy is not ensured, responsibility may be diluted by transferring that violence to third parties (Gordon 2002, 2006).

Outsourcing and Restructuring Control Practices: De-politicization as a Political Art

In addition to reorganizing the way the state operates and the principles of responsibility, the outsourcing of the checkpoints has led to many other changes: in the checkpoints' architecture, in the security screening procedures, in the positions of the agents who operate them and in the vocabulary used to describe how they are run and their purpose. The various changes contribute to giving these infrastructures a surface gloss and displace the forms of domination, rendering the violence less visible.

The architecture of the checkpoints is the first visible expression of these radical changes. While in the 1990s the checkpoints were improvised structures, built of concrete blocks and military accessories, privatization gave rise to permanent, complex buildings that gave the checkpoints a more stable, elaborate form (Braverman 2011a). One of the objectives of the new layout was to "reduce friction" by limiting direct contact between the Palestinians and the Israeli personnel. In the new checkpoints the security agents speak to Palestinians through tinted glass via loudspeakers or intercoms. Areas designated for Palestinians are monitored by cameras and Palestinians are prohibited from entering areas for personnel only. The checkpoints are not only called "terminals": their architecture and the remote control make them comparable to high-security locations such as airports or public institutions such as embassies, police stations or prisons.

These changes alter the forms of violence and the domination procedures used in the interaction between occupants and occupied. First, the imposing sizes of the buildings as well as the construction materials mark the institutionalization and sedentary nature of the system. The new architecture also imposes new discipline on the Palestinians who must follow set paths, stop at each station and wait until they are told to move forward (Mansbach 2009). However, unlike the disciplinary apparatus described by Foucault, the discipline imposed here cannot be fully internalized and is not intended to create docile subjects. It is rather enforced to contain and control the mass of non-subjects who are seen as a threat. Finally, the layout of the area makes the violence less visible. If we adopt the typology of violence proposed by Azoulay and Ophir (2008, 2009), these locations are marked by "contained" violence: even though it is evident, it hardly ever explodes. The effort to "contain" and reduce direct demonstrations of violence constitutes one of the main

9 Traiber has the following to say about checkpoint control: "I have always believed this is not a job for soldiers. We need professionals to cope with the Palestinians. Civil guards will earn a wage and it will be in their own interest to work as efficiently as possible" (Greenberg 2006).

means of legitimizing the checkpoint system, not in the eyes of the Palestinians whose perceptions do not count, but in the eyes of the Israeli public and international actors.

Moreover, limiting the opportunities for direct contact changes the interactions between Palestinians and Israeli personnel, reinforcing their unilateral character. The security guards and Israeli soldiers maintain their role as omnipotent agents while the Palestinians are deprived of any means of expression. Given that the only contact is through the technological remote control infrastructure, Palestinians are no longer able to negotiate the content of the interactions. Human Rights organizations and individuals who wish to protest against the repressive nature of these locations are also rendered dumb. As it has become very difficult for them to enter the checkpoints, they can no longer see what happens inside nor base their criticisms on direct observations. Moreover, because the authorities present the "civilianizing" of the checkpoints as a means of "humanizing" them, it has become increasingly difficult to denounce the "terminals" on the basis of human rights. Finally, the language is also changing: the Israeli authorities are redefining relations between the control agents and Palestinians according to neoliberal terminology, combining service and "humanitarian" vocabulary. This new approach was illustrated by the statements of Betzalel Traiber when he presented setting up the new private checkpoints as "a new phase in the service provided to passengers: the terminals have now toilets, drinking water, air conditioning, surveillance equipment and a technological infrastructure" (Knesset Commission on Internal Affairs and the Environment 2005c). This language therefore contributes to silencing human rights groups and political movements facing this new apparatus (Wigoder 2007; Masbach 2007, 2009; Kotef and Amir 2007; Braverman 2011a), the "contained" violence of which leaves traces of destruction without visible signs (Azoulay and Ophir 2008, 2009).

The security screening procedures have also changed. They have been formalized, restructured, and standardized in order to optimize efficiency: they operate like a production line according to a Taylorian working process in which each stage of the inspection has its place. Checkpoint control operates like a series of security inspection "stations" separated by obstacles intended to examine the movement from one "station" to another. The procession along the route may be shown in a flow chart (see diagram). The first station is where the "passengers" and their belongings are scanned separately. Depending on the inspectors' decision, the "passengers" are either sent to the next control station where their documents and magnetic strip cards are checked using a computerized identification system that records their passage, or they are sent to further inspection stations. These include an explosives detector, a room in which body searches are conducted and an investigation room. People enter and leave the "terminals" and move between "stations" via electronic turnstiles operated via control booths (Ministry of Defence 2005). The inspectors and security guards have defined responsibilities such that each person controls one fragment of the process.

Contrary to what is claimed in the official discourse, the changes in security screening procedures have not eliminated the arbitrary treatment that is characteristic of checkpoints, but have changed it. Certain types of arbitrary treatment that characterized the old checkpoints have disappeared. It is no longer the soldiers who decide, according to ambiguous and constantly changing orders (Gazit 2009), whether people may or may not cross. Inspectors now follow formalized regulations that standardize certain aspects of the decision. Similarly, that which the Spiegel commission deplored in 2004 as problems "of discipline, behavior, ethics and morale" has been reduced. The formalization of regulations as well as the close supervision of the security agents who may easily lose their jobs have prevented (or limited) certain types of brutality. Nevertheless, other types of arbitrary treatment have appeared: checkpoints may be closed from one day to the next on a unilateral political decision, as occurs in the West Bank when a blockade is ordered and in the Gaza Strip since the blockade. Security firms can also impose their own rules, which may differ from one company to another and from one checkpoint to another. Such an incident was observed recently: security guards from the Modi'in Ezrahi company that manages the Sha'ar Efrayim checkpoint to the south of Tul Karem, prohibited Palestinians who work in Israel from carrying quantities of food that the private company considered to be greater than their daily requirements (Hass 2009). Arbitrary treatment also arises due to technical difficulties, such as problems with the correct functioning of the electronic security system or the technical skill of the operators. Thus frequent scanner breakdowns slow down the crossings and cause long disruptions. Similarly, the demagnetization of cards often means that Palestinians are prevented from crossing (Sylvia Piterman and Machsom Watch 2009). Last but not least, the distribution of permits remains just as arbitrary and is performed using unclear criteria and for reasons that are withheld from the Palestinians (Barda 2012; (Sylvia Piterman and Machsom Watch 2009).

The changes are not limited to the general, collective level: they occur at the individual level of agents as well. The outsourcing of the checkpoint has consequences on the status of the personnel, which is now that of employees of manpower companies and not that of soldiers. It also affects the selection and training procedures, which are now in line with the human resources management procedures in private firms. The people, however, often remain the same. When security guards and inspectors are recruited they must firstly satisfy a certain number of criteria: Israeli citizenship, 12 years of schooling, military service for inspectors or military service as fighters for armed security guards. In order to qualify they then have to take a "psychotechnical" exam (outsourced to a private evaluation center), which assesses their skills and determines whether they are suitable for the work. Their training is run by private companies and includes military combat training. Each year they must also attend a two-day training course on "passenger service." In reality, more often than not the selection and training processes, which are called "professional," result in employing people who have already managed the checkpoints during their military service. Even though it is impossible, in so far as I am aware, to obtain statistics on the social profile of the security guards and inspectors hired by private companies, the conditions of employment may give an idea: renewable short term contracts, relatively low salary for work that does not carry much prestige and which is relatively risky, unsocial working hours, inflexible, severe discipline. This type of work can only therefore attract people who do not have a real alternative, particularly young people from the peripheries who have already done the same work during their military service (Havkin 2011). This is also the profile described by the head of a team of inspectors and security guards in the Mikud company: "As a supervisor, my ideal employee is someone who finished his military service a month ago, who lives near the checkpoint and wants to work for a couple of years to save money for his studies or a trip".[10] Therefore, sociologist Yagil Levi's analysis of policing tasks also applies to the employees of private companies: the job of managing the occupation has gradually been offloaded onto marginalized populations, working classes, geographical peripheries, "orientals" (*mizrahim*—i.e. Jews from Arabic countries) and Druzes, new immigrants, particularly from the former Soviet Union and Ethiopia (Levi 2007, 2009).

10 Interview with a supervisor in the Mikud private company at the Shaar Efrayim checkpoint, January 2008.

This process of checkpoint de-politicization and redefinition of control procedures is completed by a linguistic change. The military term "checkpoint," the Hebrew word for which, *machsom*, literally means "blockage," has been replaced by *ma'avar* meaning "passage"; the control and inspection procedure is referred to as a "service," the "Palestinians" (or "Arabs," which is the term the army prefers) have become "population," "passengers," or even "customers." As Commandant Oren Julian explained in an interview: "The population who use the crossing points have to be given a service. These places must be welcoming, places that provide service, not dark places" (Shikler 2005). At first glance, the new vocabulary euphemizes the control practices and the violence between occupiers and occupied. If, however, the starting point is that discourse and terminology influence the way reality is represented, then the discursive change contributes to how knowledge is structured and therefore changes the way reality is perceived and interpreted. Thus it also participates in redefining the scope of possible action (Foucault 1969).

Conclusion

The outsourcing of the checkpoints implies a reality that is much more complex than a mere outsourcing of state control to private firms. It results in the restructuring of space, the bureaucratization and standardization of procedures and the adoption of a management discourse that rebuilds the effectiveness of the occupation. The new "terminals," which are aimed at optimizing "efficiency" and "quality of service," have written regulations and documentation of events and produce new assessment criteria, similar to banks, business firms or telemarketing companies. The scientific organization of work based on "rational" and "objective" criteria of profitability, is at the basis of contemporary management theories. Since its inception, the doctrine of management has been presented as an ideologically neutral system, apolitical par excellence, and even claimed to announce the end of politics (Taylor 1985; Shenav 1995). The new privatized checkpoints reproduce, through mimetic isomorphism, the managerial rationale and strategies considered valid for all problems and spheres of action, whether public or private (Dardot and Laval 2009). In doing so, the privatization of checkpoints highlights changes in the modalities of Israel's occupation of the West Bank and Gaza Strip, combining occupation rationalities with the dynamics of neoliberal globalization, by using increasingly sophisticated technological resources and limiting direct contact between Israelis and Palestinians.

The new "terminals" do not replace the other types of checkpoints. Only the 33 "border" checkpoints have been privatized or are in the process of outsourcing compared to a total of 64 permanent checkpoints and 542 physical obstacles counted in July 2012 (OCHA 2012). Trenches, earth mounds, concrete blocks, road barriers, and flying checkpoints still appear on the roads in the West Bank, limiting and controlling the movement of Palestinian men and women. The brutality and the arbitrary nature of these encounters remain as explicit as before. Therefore, we are witnessing a spatial redistribution of the various forms of violence and repression in the Occupied Territories: the power apparatus applied to the "border" adopts a more "civil" and "civilized" form, while visible violence has been moved further away from the border, out of view of (un)concerned agents, such as NGOs and the Israeli and foreign press. It appears in checkpoints located further from the Green Line, in military operations and offensives and probably in closed areas within the "terminals" themselves.

Finally, in order to understand the political significance of this transformation, we have to consider the overall project in which these checkpoints participate. In the occupation regime relying on the dialectic between the dynamics of separation and the dynamics of appropriation, every attempt to reduce friction through separation brings parallel developments: annexation of territory, the expansion of settlements and the de facto or *de jure* creation of new points of friction and control (Azoulay and Ophir 2007). This strategy of domination, which has been developed over 40 years of occupation, is based on the unilateral nature of the decisions made. The new "terminals," like the Separation Wall and the withdrawal from the Gaza Strip, operate like a demarcation line that marks the spatial distribution of forms of domination over a territory and population, rather than as a border separating two disparate entities. Their "normal" appearance, "like terminals in every country in the world," conceals and pacifies both the political and symbolic threat that the borderless reality of the Israeli-Palestinian spaces materializes and the violence required to maintain it.

At the checkpoint between Ramallah and Jerusalem an electric sign flashes: "Welcome to Atarot Terminal, Have a Pleasant Stay."

Chapter 2
Denial of Borders: The *Prison Web* and the Management of Palestinian Political Prisoners after the Oslo Accords (1993–2013)

Stéphanie Latte Abdallah

The imprint of prison on the everyday life of the Palestinians has been strong since the occupation of the West Bank and the Gaza Strip in 1967. Every family has undergone this experience, particularly since the massive waves of arrests that marked the years of the First Intifada (1987–1993), which were renewed during the Second Intifada (2000–2004). Since 1967, different sources estimate that approximately 40 percent of the male population has been jailed, while the percentage of women is far lower.[1]

They are usually called prisoners of war (*asra*), or political prisoners, by the Palestinians and depicted as "security detainees" by the penitentiary administration, the Israel Prison Service (Shabas). This qualification has no legal reality: defining prisoners as "security prisoners" has no basis in law but is a decision taken either by the army at the time of arrest, or by the Israel Security Agency (Shin Beth or Shabak) during interrogation or lately by the prison administration.[2] This category cannot be applied to Jewish Israelis but is used for Palestinians—irrespective of whether or not they are Israeli citizens—and more generally for Arabs. Its application is therefore *de facto* an ethnic one. This security category greatly weakens the status of prisoners: the conditions of their interrogations, their access to lawyers and their conditions of detention are much tougher than those of other prisoners. Nor is there any possibility of sentence adjustment or remission.[3] Furthermore, the conditions of imprisonment are not clearly defined but are continuously being re-evaluated according to changes in the political and security context in both the Occupied Territories (OT) and Israel (inside its 1948 borders).

Confinement is therefore not a marginal experience. It is a main control tool that has pervaded many aspects of daily life in the Occupied Territories (Cook, Hanieh, and Kay 2004). A policy that can be described as one of "mass imprisonment" or "governing through imprisonment," using the terminology used by Loïc Wacquant (1999, 2009) in the context of

1 There have never been more than about a hundred women in Israeli prisons at any given time. A total of a few thousand women have been imprisoned compared with an approximate total of seven hundred and fifty thousand Palestinian men.

2 Interview with Lila, M., Legal specialist, Association for Civil Rights in Israel, April 27, 2009.

3 The legal system sanctions security-related offences according to a doctrine inspired by the so-called "Justice Model" developed in the United States in the 1970s and based on a penal philosophy known as just deserts, which, in general, prevents remissions and condemns to morally codified sentences. This results, for example, in sentencing those found guilty to several life sentences depending on the number of Israeli deaths the offence caused directly or indirectly. Nevertheless, for heavy sentences, a remission of a third of the sentence can be asked for. It is rarely granted, except when the sentence is considered *a posteriori* as excessive.

American prisons from the 1980s onwards, was actually implemented with the mass arrests of the First Intifada: in November 1989, the prison population reached the maximum number of about 13,000 inmates (Amnesty International 1990; B'Tselem 1999). At that time, the incarceration rate in the Occupied Palestinian Territories (OPT) was the highest worldwide (Hajjar 2005): it was estimated at 750 for 100,000 persons (Human Rights Watch 1991).

If incarceration has been a mode of government of the Palestinian territories for the Israeli state, this experience has been re-appropriated by jailed Palestinians and in relation with the PLO as a key national phenomenon, as evidenced by the creation in the 1980s of a political entity: the movement of Palestinian political prisoners. Political parties and the PLO took charge of the prisoners financially. Prisoners had a political and symbolic role, as an important part of the national movement. They were considered soldiers at the front, and a dedicated department took care of families of martyrs and prisoners (Nashif 2008a).

With the Oslo Accords, the majority of the detainees were released. Under the Cairo agreement, 5,000 people were released from prison first, followed by another 10,000. In 1997, not a single woman was imprisoned, and there were 350 men left behind bars. Former inmates resumed or acquired positions in the PLO and the Palestinian National Authority (PNA) administration, which was formed in 1994. Many of them have become high-ranking civil servants of the new institutions.

But arrests did not come to a complete halt during the years of the peace process[4] and during the Second Intifada they once more became considerable. The prison population rose steeply to reach more than 8,000 people in 2006–2008. According to B'Tselem (2014),[5] there were 4,999 "security prisoners" on March 31, 2014 and 5,265 on April 1 according to Addameer (2014).[6] In such a context, the Palestinian national responsibility towards detainees and dedicated budgets were increased. In 1998, a ministry of prisoners of war and ex-detainees was created to deal with this issue, which, quite unexpectedly, did not disappear with the Oslo peace process.

Since 1967, the practices of political imprisonment wove what I called a "*Prison Web*" in the Occupied Territories, which is a reality and virtual, i.e., the possibility of detaining a large number of people, men, and women from the age of 12. The prison system has indeed developed not only to punish proven offenses defined by law. Linked to the military justice system, it participates in the suspension of the international law of war and occupation (the Fourth Geneva Convention) that the State of Israel does not consider globally applicable in the Occupied Territories. Thus, "security detainees" are not considered by the Israeli authorities as prisoners of war, nor as civilians detained in a context of occupation.

The way Israeli authorities have been dealing with Palestinian political prisoners has changed greatly in 40 years. I will not give a detailed historical analysis of these changes between 1967 and 1993, but rather concentrate on the major transformations in prison management since the Oslo Accords. I am therefore interested in understanding

---

4 Mainly administrative detentions of people suspected of being opposed to the peace process (Bertrand 2006).

5 B'Tselem takes into account all Palestinian detainees from the Occupied Territories (irrespective of whether they have been convicted or are detained for the time of examination or trial or whether they are administrative detainees or, since 2002, prisoners classified as "Unlawful Combatants"). Since August 2008, these statistics also include residents of East Jerusalem. These figures, however, do not include Palestinian citizens of Israel in 1948, and common law prisoners. B'Tselem statistics are those provided by the Israel Prison Service.

6 The difference between these two figures is mainly due to the inclusion of the Palestinian citizens of Israel classified as "security detainees."

the implications and reasons for choosing to renew mass imprisonment policies and "government by prison" during this period. Furthermore, I consider the practices of incarceration in a broader perspective. Such practices contribute to the redeployment arrangements of Israeli occupation throughout the OT: Israeli authorities have indeed gradually installed a population management system based on the control of its movements and on various forms of confinement, including incarceration.

Some works have discussed the increasing carceralization of West Bank and Gaza Strip spaces through three main devices: prisons, checkpoints, and walls (Bornstein 2008). The OPT have thus been described as a "carceral society" (Bornstein 2008) or a "carceral archipelago" (Gregory 2004; Parsons 2010). Though, none of them have focused on imprisonment policies in such a context or analyzed the in-depth spatial, political, and social implications of this double confinement.

Thus, I consider this prison government as part of the Israeli policy of "separation" which in fact blurred borders and boundaries and enabled the occupation to be recomposed and maintained. This policy is based, on the one hand, on a variety of other limitations and separations which do not follow the contours of national borders. On the other hand, it is anchored to a state of uncertainty created by the perpetuation of an emergency state where law and fact merge, and everything becomes possible (Agamben 1997).

Researchers Ben Naftali, Gross, and Michaeli (2009) have rightly pointed out that the lack of clearly defined territorial boundaries had as a corollary—if not a goal—of erasing a whole set of distinctions, especially between occupation and non-occupation, annexation and non-annexation, temporary and indefinite, and thus between the rule and the exception, which created a space temporarily and legally undetermined beyond the Green Line. This uncertainty has deprived the Palestinians in the territories, on the one hand of the protection of international law of occupation and on the other, of the rights of Israeli or Palestinian citizens to the benefit of the government by a series of regulations, military orders, and procedures. This state of emergency has de facto enabled an unlawful occupation to be sustained.

This chapter analyses the various detention practices and regulations which turned imprisonment into a population management technique creating and expanding a *Prison Web* over the Occupied Territories. Being both real and virtual, this *Prison Web* participated in the creation of a space which is kept suspended and therefore indeterminate.

While the *Prison Web* and the blurring of distinctions have been gradually established since the beginning of the occupation, after the Oslo Accords two major changes occurred. Firstly, political prisons located in the OT which were military facilities were relocated to Israel (in its 1948 borders) and institutionally incorporated into the Israeli civil prison system. I will show how this integration enacted the absence of borders between Israeli and Palestinian areas. Secondly, with this integration process, a new management of political prisoners was introduced aimed at separating and fragmenting the prisoners' community, and by extension Palestinian society as a whole. The new management has relied on both a liberal approach seeking profitability and on humanitarian arguments (improvement of prison conditions, modernization of prison buildings).

These same dynamics are found in other security areas (see Havkin in this volume). The new arrangements of the occupation were indeed intended to standardize and normalize its procedures in order to sustainably manage the conflict (Weizman 2009). Thus, the occupation has been perpetuated while transforming its main devices: its operating tools have become less visible, less openly violent, less costly and therefore more politically acceptable.

Beyond documentation and ethnographic observation, this chapter is based on oral sources: 125 interviews were conducted between 2008 and 2012 with lawyers, people working for various NGOs working on imprisonment in Palestine and Israel, and 73 ex-prisoners, men, and women (35 women). The names of most of my respondents have been changed except representatives of NGOs or individuals holding a public role.

The *Prison Web*: Suspended and Limitless

Since 1967 the terms of political imprisonment for "security"[7] reasons have deployed a web over the Occupied Territories. The prison system was not only set up to sanction proven security offences. It is linked to the military justice system, which applies along racial or ethnic lines in the Occupied Territories only to Palestinians,[8] and more widely to so-called security offences committed by Palestinians with Israeli citizenship (Palestinians of 1948) or by other Arab nationals from Lebanon, Jordan, Syrians from the Golan Heights, Saudi Arabia,[9] etc.[10]

An Evidence Based System which Increases the Number of Indictments and is a Major Source of Intelligence

This judicial system operates out of the intelligence services (Shabak). Together with the army (Israel Defense Forces) or police, the intelligence services arrest suspects and are responsible for conducting the interrogations—*takhqiq* (Arabic)—before imprisonment and judgment.

The predominant role of intelligence services involves, firstly, that in some cases the charges may be kept secret and not divulged to the defendants' lawyers (incommunicado detention) and, secondly, that a special evidence system is used. This system is not based on any investigation or evidence discussed at trial but on confessions of the defendants or on statements and denunciations of third parties on their alleged activities—known in Hebrew as the *Tamir* practice. Interrogators therefore have to obtain such confessions at any price. Hence, since the beginning of the occupation and until 1999, physical and psychological violence—similar to torture—was regularly used during interrogations. The 1987 Landau report[11] even recommended such practices. Since the 1999 Supreme Court decision, heavy psychological pressures have replaced, in most cases, physical and sexual abuse (Latte Abdallah 2010b).

This system of confession-based proof appears particularly significant considering 95 percent of cases never come to trial but are settled through a system of plea bargaining—*safqa* (Arabic)—between lawyers and judges. This procedure requires a confession of guilt from the defendant (Hajjar 2005). Plea bargaining is advantageous for the military legal authorities who settle cases faster by reducing the number of trials. Defendants are particularly

7 Opposed to detention for common law offenses.

8 Not anymore though in the Gaza Strip since the unilateral Israeli disengagement of 2005. The Palestinians living in Gaza since then have been subject to specific regulations.

9 In April 2009, among the Arabs imprisoned for security reasons, there were 13 Syrians from the Golan Heights, 19 Jordanians, and 1 Saudi. All the Lebanese were freed after the last exchange deal of prisoners with Hezbollah in summer 2008.

10 It is worth stressing that nationals of non-Arabic countries who are of Arabic origin are categorized in the same way, e.g. the French national Salah Hammouri. In some cases, foreigners may also be categorized "security prisoners," e.g. the Ukrainian woman Irina Polishchuk Sarahneh who is married to a Palestinian, who is himself imprisoned.

11 This report institutionalized the use of physical and psychological pressure.

encouraged to plea bargain as negotiation usually results in shorter sentences. In contrast, those who take the risk of going to trial receive heavier sentences and the proceedings, often adjourned, are endless. Lawyers and families of detainees usually opt for plea bargaining because at the individual defendant level, the results are better. Some detainees state they prefer plea bargaining to trial for political reasons: choosing plea bargaining rather than a trial is then presented as non-recognition of Military justice (Hajjar 2005).

Whatever the reasons for this choice, this procedure significantly extends the grip of the *Prison Web*. It increases, at a low cost, the number of imprisonments and convictions. Almost all defendants were declared guilty of all or part of the charges against them while less than 1percent were acquitted (Halevi, Shlonsky and Machsom Watch 2008, p. 11). More broadly, this proof system based on confessions or denunciations greatly strengthens the grip of the intelligence services over the Palestinian population. Hence, they benefit from this significant source of information and compile files on the defendants, and members of their social and political networks.

The confession system has individual and collective psychological effects on defendants (Dacca 2009/2011) and, on a larger scale, on Palestinian society in which such confessions are silenced. The possibility that individual people have confessed or denounced others' actions instills doubt and mistrust.

The courts are constantly trying to increase the use of *safqa* not only to save time and money but also to justify the involvement of the intelligence services in military courts and to give such military justice national and international legitimacy. "When I attended hearings," a member of Court Watch[12] told me, "I understood that they were aiming at a 99 percent statistic."[13] This confession-based proof system and the practice of plea bargaining provide some kind of justification to the mass arrests and to the objectionable functioning of military justice in Israeli and international public opinion, through convictions for crimes that are recognized by their supposed authors or by third parties but remain unproven.[14]

A Vague, Timeless, Virtual Definition of Offences

These military legal practices make offences virtual and increase the deployment of the *Prison Web* through a lack of clear distinctions.

In 2007 offences related to security and assigned by the army to the category of "Hostile Terrorist Activities"—*fakhaï* (Hebrew)—accounted for 47 percent of indictments. But most of the people were not accused of "terrorist acts" resulting in or attempting to bring about death, but of simply belonging to or having activities inside an "illegal organization" (Halevi, Shlonsky and Machsom Watch 2008).

Membership of an illegal organization is defined in vague, general terms and includes an entire spectrum of relations summed up in the expression "having links" to an illegal

---

12 Court Watch was established in 2005 by activists from the NGO Machsom Watch (formed in 2000 by a group of women to monitor soldiers' practices at checkpoints). More engaged yet, members of Court Watch attend military court hearings and report on court practices.

13 Interview with Roni H., Jerusalem, April 8, 2010.

14 The Franco-Palestinian Salah Hammouri was convicted for intending to murder the Shass leader Ovadia Yousef and convicted by plea bargain to seven years in prison in 2008, after spending three years in prison waiting to go on trial. He only confessed his activism as a youth in the Popular Front for the Liberation of Palestine (PFLP), a party considered illegal, as are all Palestinian parties. Once the judgment was obtained, the French authorities have hardly mobilized as they were bounded by the respect of a court's decision.

organization irrespective of their nature. Moreover, publicly expressing political views can be enough for an indictment as is mere presence at a demonstration. Merely expressing political opinions or the Israeli authorities learning about such opinions can thus be enough to justify a suspended prison sentence.[15] Incidentally, these detainees are known as "ideological criminals" (Halevi, Shlonsky and Machsom Watch 2008).

The sense of time is blurred. Such "links" may indeed have been forged in the past, at a time when the said organizations had not yet been classified as illegal (Halevi, Shlonsky and Machsom Watch 2008).

The list of illegal organizations has gradually expanded to include more and more social and civil organizations such as non-profits and NGOs, etc. During trials, blurring the distinction between civilian organizations (parties, associations, etc.) and military (armed groups) has become increasingly common. Since the beginning of the occupation of the West Bank and the Gaza Strip, none of these classified as illegal organizations have been removed from the list. Hence, all Palestinian parties are on the list. Fatah, which was declared a "terrorist organization" in 1986 is still on the list despite its role in the Oslo Accords and the fact that it constitutes the basis of the Palestinian Authority (PA) (Halevi, Shlonsky and Machsom Watch 2008).

Clearly a distinction is made between memberships of these different organizations. Although few people nowadays are arrested merely for belonging to Fatah, the option is still there. The virtual nature of this classification has created a diffuse web capable of affecting every Palestinian. This *Prison Web*, which may be defined as the possibility of imprisonment, i.e., a prison reality and virtuality, can be activated according to the situation and the needs of the intelligence services. The arrests of members of Fatah were so widespread during the period of the Second Intifada (2000–2006) that Fatah members still form the large majority of detainees. Today, activists engaged in the peaceful protests of the Popular Resistance Committees (in Na'lin, Bil'in, Nabi Saleh, etc.) are frequently imprisoned.

The establishment of a *Prison Web* creates a diffuse system of control and in-depth knowledge of the Palestinian population and political, social, and daily life in the Palestinian enclaves. Through widespread arrests, the intelligence services gather information, recruit informers, and infiltrate the entire society.

At the beginning of the First Intifada, an average of 25,000 people per year were arrested. Between December 1987 and September 1991, of the 79,000 arrests, approximately 3,000 were women. In more than half the cases the sole aim of the arrest was to gather the maximum amount of information before release (Thornhill 1992). This was, and is, particularly true for women. They were arrested on account of their activities or their social or political relations, or simply because their family or friends were wanted—*matloubeen* (Arabic). By arresting wives, sisters or mothers, the intelligence services put pressure on people they want to imprison or about whose whereabouts they seek information. These women, relatives or friends were sometimes held under administrative detention for varying periods of time.

Genealogies, individual and family histories are indeed used to put pressure on people under interrogation or negotiate specific services. The use of emotional bonds and family ties by intelligence officers to exercise psychological blackmail is a constant in my interviewees' accounts of interrogations or in the ones found in prisoners' memoirs and written sources (Latte Abdallah 2010b, 2013, 2014a). Such ties have been severely exploited

15 Suspended prison sentences result in a series of restrictions on movement that significantly affect people's lives. These may include prohibitions on entering Israel (including East Jerusalem), holding a work permit, etc.

to make defendants, men and women, talk. For example, Theresa Thornhill (1992) reported that in 1987, according to her lawyer, the interrogators of Na'ila A. threatened to fetch her husband, mother, and brothers to the Moscobiya interrogation center and rape both her and her mother in front of them.

The *Prison Web* is both a reality and a virtuality, i.e., the possibility of imprisoning a large number of people, men and women, from the age of 12 onwards, unlike under Israeli civilian legal provisions which do not allow the arrest of juveniles below the age of 14. Until autumn 2011, military justice was treating juveniles in prison from the age of 16 onwards as adults in contrast with Israeli civil law and international law for which the age of majority is 18.[16] On April 1, 2014, 202 juveniles were in prison, 24 of which were under 16 years old (Addameer 2014).

Not even members of parliament are immune; in 2006, shortly after the kidnapping of Gilad Shalit, the Israeli authorities arrested several Palestinian members of parliament and ministers from different parties: On April 1, 2014, there were still 11 MPs behind bars (Addameer 2014).

A Prison Web *that Denies Temporality, Boundaries, and Limits*

The *Prison Web* breaks temporality. People can be arrested and detained according to their present, and even past and future, family, social, and political networks. Such ties are used to control and are objectivized as potential threats. As a preventive measure for acts not yet committed or for future and virtual acts, the Israeli authorities have resorted to numerous arrests.

The provisions of administrative detention,[17] which allow the holding a person in detention for renewable six month periods without charge,[18] greatly expand the *Prison Web* and the temporal boundaries of the prison system. The purpose of such detention is indeed presented as a way to avoid "future activity" that would threaten security (Cavanaugh 2007).

To the administrative detention's provisions, the Israeli authorities have added the category of "Unlawful Enemy Combatants." As for administrative detention, it is intended to apply to potential future actions of detainees, but has been applied to foreign nationals. The category of "Unlawful Combatants" was indeed created in Israel in 2002 in the international context of the redefinition of terrorism by the Bush administration after September 11, 2001. It was originally intended to detain Lebanese Hezbollah fighters to keep them as bargaining tools.[19] Then, with the Israeli withdrawal from Gaza in 2005, it was applied to the Gaza residents. According to the lawyer of the Association for Civil Rights in Israel, it was a "type of administrative detention for 'foreigners,' especially for the Gazans."[20] This implies that administrative detention concerns "non-aliens" that are Arab citizens of Israel (Palestinians of 1948), Palestinians in Jerusalem, and the inhabitants of the West Bank. The inclusion of the inhabitants of Jerusalem within the national borders of Israel is hardly surprising, but the inclusion of the residents of the West Bank is much more so, given the displayed separation between Israel and the OT.

---

16 Since then, some changes have been made concerning juvenile detention.

17 As of the April 1, 2014, 186 people were held under administrative detention, including nine MPs (Addameer 2014).

18 They are usually kept secret and at the discretion of the Shabak.

19 Since the 2008 exchange between Hezbollah and the Israeli authorities, no more Lebanese national is imprisoned in Israel.

20 Interview with Lila M., Jerusalem, April 27, 2009.

This blurring of distinctions in terms of citizenship is also at work in the way Palestinians with Israeli citizenships, the Syrian prisoners from the (Israeli annexed) Golan and Palestinian residents of Jerusalem[21] arrested for political reasons are treated by the prison system. The provisions applied to these prisoners are indeed becoming increasingly similar, thus confirming Israel's annexation policy of Jerusalem as a whole.

Lastly, Palestinians from Israel and Jerusalemites are held under conditions similar to those applied to the Palestinians from the OT and to Arab foreign nationals. Unlike Jewish prisoners detained for political offences,[22] they are treated as security prisoners. Therefore, they have little prospect of reduced sentence or remission. In addition, they are more heavily punished as enemies from within. For the State of Israel, as citizens or residents (regarding Jerusalemites), they are in general excluded from political negotiations for their release and from prisoner exchange deals. This is why the release of 48 Jerusalemites and 7 Palestinians of 1948 at the end of 2011 was considered a success of Hamas in the Shalit deal.[23] The legal boundaries of their citizenship, and even of their nationality, are blurred since such limits are flexible and subject to change: they are Israelis when out of prison or in a state of virtual liberty, but Palestinians in jails. This dual status greatly increases the time they spend in prison and the harshness of their treatment: they indeed constitute the majority of the longest-serving prisoners.

The Integration of Prisons: The End of Borders

Since the Oslo Accords, Israeli prisons have been relocated from the OT to inside the Green Line (inside Israel's 48 borders), the only exception being the Ofer prison which is located in Zone C.[24] These transfers involved mainly male prisons, women having been imprisoned for a longer period inside Israel.

Since 2003, this gradual relocation of prisons within Israel has been accompanied by the integration of all such facilities—previously managed by the military authorities—in the Israeli civil prison system, therefore under the sole authority of the Prison Service (under the supervision of the Ministry of Internal Security). This national integration of the prisons and security prisoners from the OPT was completed in 2007.[25] It contributes to blurring the borders between Israel and the OPT. The management of "security prisoners" by the national Prison Service denies their inherent rights as detainees in the context of an occupation, in breach of the Fourth Geneva Convention which stipulates that occupied populations must be imprisoned on their own territory, thus making the legal status of Palestinian detainees

21 On April 1, 2014, 279 Palestinians from Jerusalem and 234 Palestinians of 1948 were behind bars (Addameer 2014).

22 Ygal Amir who assassinated Itzak Rabin, for example.

23 In October 2011, this agreement provided for the release of the soldier Gilad Shalit held since June 2006 by Hamas in the Gaza Strip in exchange for the release of 1,027 Palestinian prisoners.

24 The Oslo Accords divided the West Bank and the Gaza Strip into three zones: Zone A, where the Palestinian Authority runs the police and civil administration, Zone B, where it is responsible for civil administration and manages security in coordination with Israel, and Zone C, which remain under Israeli sovereignty. The redeployment of the occupation and checkpoints, and the incursions by the Israeli army into virtually every part of the West Bank since 2000 cast considerable doubt on Palestinian sovereignty over these zones.

25 According to B'Tselem's statistics, starting in 2005 a gradually higher number of prisoners were incarcerated by the Israel Prison Service rather than the army. In 2009, all prisoners came under the authority of *Shabas*.

undetermined. This integration of political detention within the Israel Prison Service also increased the (strictly Palestinian or Arab) "ethnic" dimension of the category of "security detainees" and beyond, of military justice. At present, the military justice system tends to become the legal system applied to Palestinians in the West Bank regardless of the offenses considered. A major impact of this relocation on the lives of prisoners is the restriction of family visits which now require a permit to enter Israeli territory, often denied for "security reasons" (Latte Abdallah 2014a).

The integration of military prisons under the authority of the Prison Service has been justified on grounds of professionalism, economic viability (the possibility of reducing the cost of detention), and on humanitarian grounds (improvements in the conditions of detention including modernization, renovation, and construction of prisons). It has been argued that Shabas was the most competent and able to manage such a large number of prisoners. But other considerations were at work: the intelligence services played a key part in this integration process which was decided when the head of the Shin Beth, Avi Ditcher, was minister of internal security. Since, as officio members of the board of Shabas, they participate in decisions and greatly inspired the new prison management initiated in 2003.

Reducing the Cost of "Prison Government" and Increasing the Financial Coverage of Detention by the Palestinian Authority

The State of Israel reduced the costs of detention by using more sophisticated technology and increasingly neoliberal prison management: human resource management (interim jobs, trained and paid less), privatization of certain services such as prisons' stores (canteen) where prisoners find food, hygiene products, clothing, etc. But the reduction of cost is also the result of the presence of the PA, which, since 2003, has become a de facto financial and administrative intermediary in the Israeli prison system. With the resumption of mass incarceration from the Second Intifada, the role of the Ministry of prisoners of war and ex-detainees and its financial investment increased.

To help families and strengthen its national role, the PA has provided legal assistance and granted a monthly sum to each Palestinian security inmate—whether from the OT, Jerusalem, or from Israel—and for every Arab prisoner, irrespective of their political affiliation. The amount depends on the time spent in prison, family status and place of residence.[26] In addition, The PA pays the fees of any prisoner who aims to study in jail, i.e. at the Open University of Tel Aviv.[27] In total, the Palestinian Authority transfers

---

26 In 2011, they were revised upwards. The minimum is 1,400 NIS—310 euros—for a single person of the Territories incarcerated less than three years. And the maximum is 12,000 NIS—2700 euros—for an effective sentence of more than 30 years. Added to these are 300 NIS—70 euros—if the person is married and 50 NIS—10 euros—per child. Prisoners from East Jerusalem receive an additional 300 shekels, and those of 1948, 500—110 euros—due to higher costs of living. Interview with Saad Nimr, chief of staff of the minister of prisoners and ex-detainees Issa Qarake, Ramallah, April 27, 2011.

27 Prisoners serving long sentences have been allowed to study, but only at the Open University of Tel Aviv. Paid access to such courses is a privilege that is only granted if security officers certify that an applicant's behavior deserves it. Only certain disciplines are open to prisoners such as literary subjects, political science, sociology, economics, psychology and management. The study of hard sciences and information technology is not allowed, nor is any other subject considered as constituting a security threat. See Military Order No. 04.48.00, dated 8 January 2004. Following Gilad Shalit's abduction, studies in jail have been suspended for "security prisoners" and did not resume after his release. Their authorization is highly dependent on the political situation in Israel and the OPT.

approximately 20 to 25 million shekels (4.5 to 5.5 million euros) to the Israeli authorities each month for the prisoners.[28]

Since the 1970s, parties or families started sending money to prisoners on a joint account in the name of one or more leaders of political factions. This amount allowed prisoners to purchase goods or commodities (for canteen) which the leaders redistributed. A few years ago the PA was compelled to pay such sums individually, thereby breaking some of the collective organization. In addition, to reduce the cost of the mass imprisonment government policy, Shabas took advantage of the failure of the hunger strike organized by prisoners in August 2004 to significantly reduce food allowances (fruit and vegetables) as well as hygiene products and basic necessities provided to inmates. Collective and families' donations were banned. Multiple daily purchases at the prison shop, where privatization has increased prices, have thus become necessary.

At the same time, Shabas introduced a lucrative system of fines for any breach of the internal prison regulations (approximately 400 shekels, or 90 euros) in addition to the usual punishment (solitary confinement, prohibition of visits, etc.). These fines are deducted directly from the amount allocated to each prisoner by the PA. In addition, the amounts paid by defendants during their trial as a part of their sentence—which has become routine in recent years—have sharply risen.[29] To prevent this inflation continuing and to avoid supporting the cost of military justice, the PA has limited its support to 4,000 shekels per prisoner to pay court fines.

Since the Second Intifada to the short-lived resumption of Palestinian-Israeli negotiations in 2012–2013, the PA has not been the interlocutor with Israel in the negotiations for the release of prisoners, to the benefit, since 2006, of Hamas which was holding Gilad Shalit. However, the PA continued to play a national role vis-à-vis prisoners by increasing its financial investment and assistance to all Palestinian and Arab detainees. Hence, the PA (and the European and international contributors to its budget) are paying a large proportion of the cost of detention in Israel.

Furthermore, since the split between Hamas and Fatah in 2007, Hamas militants have been imprisoned in the PA's detention centers in the West Bank for their political or suspected armed activities considered as illegal.[30] This is part of the security coordination with Israel. The development of a Palestinian prison system has de facto reduced detention in Israel. In 2009, unofficial data from the International Committee of the Red Cross (ICRC) estimated that approximately 2,000 people were being held in Palestinian prisons run by the PA in the West Bank and by Hamas in the Gaza Strip. Since 2007, the number of security prisoners in Israel was concomitantly reduced of approximately 2,000 people.[31] The cooperation agreements between the PA and the United States, set up for the training of police and of the security services, responsible for the military prisons in the West Bank where political prisoners are held, have facilitated the detention of Hamas prisoners in PA facilities.[32]

Since 2000, the PA has rarely been considered by Israel as a political partner for negotiations on prisoners and has mainly assumed a financial, management, and

---

28 Interview with the president of the Prisoners' Club (*Nadi al-asir*), Ramallah, November 3, 2010.

29 Data from the legal system of Judea and Samaria (West Bank) shows that 12 million shekels (2.7 million euros) were collected in 2006 and 9 million (2 million euros) in 2007 (Halevi, Shlonsky and Machsom Watch 2008).

30 In Gaza, a number of activists, particularly from Fatah, have also been imprisoned by the Hamas authorities.

31 Interview with an ICRC member, Tel Aviv, October 30, 2009.

32 These accords were first implemented by General Dayton.

administrative role. Hence, it has developed other forms of action, including the filing of complaints with the Israel Supreme Court. This legal activism enshrines the absence of borders between the Israeli and Palestinian spaces and the non-recognition by Israel of the PA as a political interlocutor. The objective of these legal proceedings is to improve individual and collective conditions of imprisonment. The cases have been monitored by an Israeli lawyer, Avigdor Feldman, who has long had a commitment to political prisoners. The PA is sometimes associated with collective petitions made by Palestinian and/or Israeli NGOs.[33] Recently, the former minister of prisoners—in charge until 2014—, Issa Qarake, has adopted a more political legal strategy. He developed the project of posing the problem of security detainees, this time in terms of citizenship and in the framework of international law: by asking the International Court of Justice to settle the thorny issue of the legal status of Palestinian prisoners. They are indeed deprived of any clear status since they are not considered either prisoners of war or prisoners under occupation.

The Humanitarian Argument: Sustaining the "Prison Government"

The shift from military prisons to the Israeli prison system has also been publicly legitimized by humanitarian considerations. Arguments in favor of this process have included the expertise of the Israel Prison Service, an improvement in the conditions of detention, modernization, renovation, and construction of new prisons, putting an end to the exception due to a military prison system in favor of a civil one, etc. It created an unexpected connection and an objective convergence of interests between the security-based management of the intelligence services and humanitarian arguments. The Israeli NGO Adalah recently filed a complaint with the Supreme Court contesting the distancing of prisoners from their relatives in breach of the Fourth Geneva Convention and the resulting increased difficulty in visiting. Yet, despite Israel's non-compliance with international law, the Supreme Court accepted the argument made by the Israeli authorities, stressing on the contrary the improved treatment of prisoners in facilities run by Shabas and the positive effects of integration regarding respect for human dignity of prisoners.[34]

The humanitarian arguments accompanying this new prison policy is part of a broader normalization of Israeli control and occupation in order to make it sustainable by humanizing it, at least in appearance (see also Havkin, in this volume). This approach, known as "humanitarianism" or "strategic humanitarianism" has developed considerably since 2000 in the military and intelligence services around programs such as "Another Life"[35] (Weizman 2009). The control and occupation devices have been made "softer," more invisible; they were routinized and in this way more easily perpetuated. Faced with the legal activism of some Israeli NGOs calling for rights for Palestinians and contesting occupation practices in the OT, the army and intelligence services have integrated their demands, some of their practices[36] and even their discourse. In doing so, they also aimed at integrating the Israeli opposition to the occupation.

---

33 Interview with Ashraf el-Ajami, minister of prisoners and ex-detainees, Ramallah, April 26, 2009.

34 Interview with Rachela M., professor of law at the Hebrew University, Jerusalem, April 7, 2010.

35 Launched in 2003 by the army, this program aimed to limit the damage caused by the occupation on Palestinian society in order to prevent a humanitarian crisis in the OT that could have compelled Israel to take responsibility for certain services (Weizman 2009, p. 562).

36 For example, Eyal Weizman notes that a humanitarian officer was allocated to the check-points in 2000 (Weizman 2009, p. 562). This was at the same time as the NGO Machsom Watch had

A New Faceless Management of Prisons: Multiplying Separations, Fragmenting the Community of Prisoners

The integration of security facilities under the Prison Service was presented as economically advantageous and aimed to improve the conditions of detention, thus presenting advantages in political terms.

However, the relocation of prisons in Israel and the integration of military prisons under the supervision of the Israeli civil prison administration paved the way for a new prisoners' management policy, and more broadly of Palestinian society. This new policy was inspired by the intelligence services. The administration of security prisoners by the Prison Service enabled the intelligence services to participate directly in decision making, management and regulations of security prisons, as members of the Board of Shabas.[37] Since 2003, the new prison management was largely inspired by these intelligence services. It is anchored on the extension of the *Prison Web* and on a series of techniques intended to manage the prisoners' society, and more broadly the Palestinian population, using confinement and fragmentation processes. This new prison policy multiplied separations in prison, modeled on isolation techniques implemented outside jails, between the Palestinian enclaves in the OT, in an increasing parallel between inside and outside.

Spatially Separate, Promoting Differences and Disagreements between Groups of Prisoners

The effects of this new prison management began to be felt after the failure of the long hunger strike observed by inmates in August 2004. In all prisons, detainees went on hunger strike for around three weeks depending on the prisons and inmates. The strike was then broken by the prison administration and ended erratically. The strike had been sparked in part, no doubt intentionally, by a series of arbitrary and coercive measures taken by Shabas starting in 2003 (Dacca 2009/2011). The strike was a failure and resulted in a step backwards: major rights previously obtained by the struggles of the Movement of Political Prisoners were lost. The Movement of Political Prisoners, tied to political parties, had strongly structured, ideologically, culturally, physically, and psychologically the lives of inmates until the Oslo Accords.

Following the 2004 hunger strike, the security detainees lost the responsibility of the prison kitchen in favor of common law inmates. The information flow which was spread by kitchen assistants and collective organization were temporarily disrupted. Israeli common law detainees were brought back into "security" prisons and prisons' sections though the separation between Israeli common law and Palestinian political prisoners had been achieved through struggles in the 1980s. The failure of the strike already reflected the weakening of the Prisoners' Movement and the first effects of the sudden change in prison management. The failure of the hunger strike marked consciences and the new prison management was then more easily and systematically implemented.

At first, prison authorities aimed to widen Palestinian political divisions, especially the Hamas/Fatah split which occurred in 2007. Even though the disagreements in the national movement and the split between Hamas and Fatah affected the struggles in prison

---

just been set up and started its monitoring of the check-points.

37 Interview with Rachela M. and Leslie S., legal specialists at the Hebrew University, Jerusalem, April 7, 2010.

and weakened the Prisoners' Movement, ideological division had less impact inside the prisons than on the outside. Above all, it never involved open conflict or violence between inmates. The imprisoned leaders (representatives of Fatah, Hamas, PFLP, Islamic Jihad, the PLO, and the Democratic Front for the Liberation of Palestine: Marwan Barghouti, Abdul Khalid al-Natsheh, Ahmad Saadat, etc.) have, on the contrary, reacted quickly to the divisions by drafting the 2006 prisoners' document, which called for national unity and put forward a specific unitary prisoner's identity which aimed to oppose the splits. Though, the prison administration intended to institutionalize divisions by bringing together in separate sections the PLO secular parties—Fatah, PFLP, DFLP, and Communists—on the one hand and religious parties, Hamas and Islamic Jihad, on the other. Women were even jailed in separate prisons (Hasharon and Damoun) according to their political affiliations.

These separations between groups of prisoners established by Shabas were not limited to partisan differences. The regulations of the prison administration were designed to separate inmates according to their status and their geographical origin, and to individualize daily life behind bars. These divisions have been reflecting the spatial separations and fragmentation implemented in the OT by the construction of the Wall, the proliferation of checkpoints, and the creation of enclaves, the isolation of towns, villages, etc. Indeed, detainees were separated according to their citizenship (Palestinian of 1948-Israeli citizens, Palestinians of Jerusalem, Palestinians of the West Bank, or Gaza Palestinians) and their geographical origin or status (refugees, residents of cities, villages, or camps). The separation between the different quarters and sections is all the more keenly felt as kitchen assistants are now not Palestinian political prisoners and restriction of movement between the different prison areas has sharpened. Moreover, Shabas resorted more frequently to long periods of solitary confinement, punitive measures targeting small groups or individuals.

According to the testimony of Walid Dacca (2009/2011), imprisoned in Israel for over 20 years, in some prisons, Shabas multiplied separations and procedures aimed at isolating groups (in separate cells, sections, or quarters) according to geographical or statutory distinctions which exist in the OT: city residents have been separated from camp dwellers, those from a particular town separated from those from another town, etc. It increases the separation between cities (Ramallah, Nablus, Jenin, Hebron, etc.) which are already isolated from each other by the Israeli system of movement control.

This distinction between groups of inmates reactivates the primacy of local or primary networks and solidarity between refugees, inhabitants of camps, city dwellers, or villagers, as well as family ties against which the national movement was built. This new spatial organization is rooted in a series of techniques, which are applied differently depending on the prison. They intend to break collective organization and the Palestinian party and national leadership in favor of social and political representations modeled on geographical, status or family solidarity. In some cases, the prison administration tried to replace the unique prison representative of all the inmates—*moumathel al-mou'taqaleen* (Arabic)—by a spokesperson for each region or city. He was supposed to be selected among those proposed by the detainees and would only be able to channel the personal demands and concerns of the inmates of their quarters. This mode of selection and attributions contrasts greatly with the prisoners' representative who has long been elected by the political factions according to their respective influence and is the only interlocutor of the Prison Service. In most prisons, transfers of prisoners are regularly orchestrated to influence the results of internal prison elections and reinforce geographical, statutory or family solidarity (Dacca, 2009/2011).

Recently, Palestinians of Israel and Jerusalemites were treated in the same way and were removed from others[38]. In male prisons, they were often physically separated from other inmates. Until now, women who are far fewer in number have managed to oppose such separation.[39] The specific provisions applied to these inmates are part of the Israeli objective of the total annexation of Jerusalem. Since 2007, Palestinian Israeli citizens or residents of Jerusalem can no longer receive direct payments by the Palestinian Authority. The amount is sent to their families who forward it, sometimes late, sometimes not fully, thus widening the gap between the living conditions in prison. In addition, as "enemies from within," they receive heavier sentences than other convicts and they are subjected, once released, to repressive measures that also affect their families, such as being prohibited from working or experiencing difficulties in finding employment.

As these distinctions became more pronounced, some inmates made differentiated demands according to their geographical origin and status to the administrative body that "governs" them, i.e., not to Shabas but to the PA (which provides them with a monthly allowance and assistance in a number of areas related to their incarceration). Thus, Jerusalemite prisoners or Palestinians of 1948 inmates have requested—and obtained—from the PA, larger allowances and assistance when in prison and once released, due to the cost of living which is much higher in Jerusalem or in Israel than in the OT and because these former prisoners experience greater problems of reintegration. Each city or region was long represented by its own committee of prisoners. Today, these committees tend to empower and make independent claims.

These requests of prisoners or former prisoners have become in some cases more aggressive vis-à-vis the PA. A group of former women prisoners founded in 2010[40] has recently threatened to use the form of protestation inherited from prison experience, i.e. hunger strikes, to demand that their studies be fully financed, and that they then be recruited into the administration[41] or receive financial aid until they find a job. These emerging claims show a form of professionalization of the status of prisoner and increased expectations vis-à-vis the Palestinian "state" authority.

Individualizing Prison Life, Isolating Prisoners

The redeployment of prisons within the Green Line has limited family visits (restricted since 1996 to first degree relatives: parents, children, siblings) because an entry permit to Israel has to be obtained. In some cases kinship is contested by the Israeli Authorities and has to be proven by families. Such permits are rarely granted to politically active people or former prisoners who are frequently refused entry for "security reasons." Men between 16 and 45 can only receive security permits (valid once for 45 days) which take a long time to be renewed, and are frequently refused. Thus, it is mainly women who visit the inmates and, increasingly, unaccompanied children because the other members of the family are refused entry. Due to long administrative procedures, distance, and transportation regulations (organized by the ICRC), visiting prisoners became quite an ordeal for families (Latte Abdallah 2014a). The increasing grip of the prison system on the Palestinian population,

38 Prisoners from the Golan are also grouped in the same facilities.
39 They were 20 as of April 1, 2014 (Addameer 2014).
40 *Lajneh al-asirat al muhakhareen.*
41 As is often the case for men.

and consequently the rising number of families of prisoners (Latte Abdallah 2010b), contributes to the separation of prisoners from their families.

Since the 2004 hunger strike physical distance has become the rule in the visiting room: visits take place through a windowpane and via telephone. Only children under eight may approach prisoners and touch, kiss or hug their mothers or fathers. Moreover, families from Gaza were collectively banned from visiting their relatives between 2007 (takeover by Hamas) and July 2012.[42] After that, they could gradually resume visiting following the agreement that was reached by the Movement of Prisoners after the hunger strike in spring 2012.[43] For Gazans, the Israeli government had previously proposed virtual visiting rooms using video conferencing. Prisoners rejected this proposal.

Policies aimed at isolating and separating prisoners have used both a humanitarian discourse and approach and a more managerial logic of promoting the material comfort of inmates.

Ensuring material comfort was initially directed at imprisoned political leaders such as deputies, leaders such as Marwan Bargouti (Fatah) or the Hamas leadership. Rapid improvements took place in certain prison facilities (where over 10 security prisoners used to be crammed into a room sleeping on mattresses on the floor) as a result of their integration into the Shabas administrative system. This was particularly the case of Ofer and Ksiot prisons, of the new American-style wing in Hadarim prison, and of the Rimonin building recently constructed in Telmond for juveniles who were previously held in deplorable conditions above the women's prison in Asharon.[44] The buildings of the American-style wing of Hadarim prison, where some leaders are held, are new, light, divided into reasonably large rooms and provided with every comfort (TV, DVD, and CD players, electric fans, hot and cold running water). Certain quarters include a kitchen and even a washing machine. These places are designed to enable inmates to organize their daily lives.[45] Leaders first, then other security detainees, have benefited here from a higher level of comfort than common law prisoners, except for the lack of food and basic commodities that they need to buy at the prison shop (canteen*)*. In doing this, the authorities have aimed to influence the imprisoned leadership, reform or form a new one which would be more in line with the objectives of the intelligence services. A leadership who could, in turn, influence Palestinian society.

Walid Dacca (2009/2011) analyzes these transformations as a normalization by material comfort implemented in some prisons only to model the conscience of inmates and mold a new generation of Palestinians. He examines the psychological discomfort caused by the new style of management, particularly the use of material comforts that renders the violence of imprisonment and domination apparently softer. This discomfort is caused on the one hand by the difference between the unchanged committed image of political prisoners and the current reality of their daily lives in certain prisons. On the other hand, it is the result of the difficult adjustment to a new faceless management of prisons.

Indeed, the modernization of some prisons has been based on new prison technologies which render control and deprivation of freedom less visible by confronting detainees with faceless space and prisons systems rather than guards. Inmates close their own doors

42 In October 2008, 900 Gaza prisoners were cut off from their immediate family. As of April 1, 2014, there were 377 prisoners from Gaza in Israeli jails (Addameer 2008, 2014).

43 The Strike for Dignity which lasted from April 17 to May 14, 2012.

44 Interview, Rachela M., legal specialist at the Hebrew University, Jerusalem, April 7, 2010.

45 Ibid.

before the guard on duty activates the central locking system of 100 cells. Others live in an "independent" area without seeing any warders, washing their clothes and going about "freely" in their daily lives. This modernization and new technologies have offered the advantage of reducing the costs of detention in a neoliberal economic logic. These measures considerably reduced prison staff, while enabling the employment of less educated and less skilled people on fixed-term contracts. A younger and more feminine staff has hence been contracted which contributes in return to euphemize violence and domination and disorient political prisoners.

Walid Dacca observed that some inmates withdraw into themselves and focus on activities that are far from national concerns: sports, numerous hours spent in front of TV shows that have partly replaced political and cultural training activities and reading which had hitherto been the central axis of political prisoners' socialization.[46] According to him, the most widely read books are those on astrology, personal development, or novels. He noted that more and more inmates follow university courses in prison and interprets their motivation as mainly personal or professional. Lastly, he commented on the powerlessness created by the new management describing the terrible episode of the Gaza war in 2009 which the inmates watched on the Al Jazeera TV channel, exceptionally authorized for the occasion, and did not arouse any large-scale mobilization. The Israeli authorities symbolically ended this war by raising the Israeli flag in the courtyard of his prison, which had never previously occurred (Dacca 2009/2011).

Testimonies of prisoners as well as interpretations do not converge, and the recent collective mobilizations of detainees require a more nuanced analysis. Moreover, these modernization and comfort policies are currently far from being applied uniformly to all prisoners and to all prisons. Unequal treatment has been another active factor of dissent which creates suspicions among inmates, or vis-à-vis some leaders who are consciously distinguished by the prison administration.

Conclusion: Inside/Outside, Parallels and Effects of the Prison Processes

To the old political techniques of playing on partisan divisions and creating statutory inequalities are superimposed geographical and citizenship divisions modeled on the territorial fragmentation in force in the Occupied Territories. In addition, managerial and neoliberal practices of isolation, material gain, and comfort have been introduced to encourage more individualistic, powerless, and passive subjectivities in sharp contrast with political prisoners' culture. Indeed, political prisoners have long constituted and represented a strongly committed and nationalist collective body.

In this chapter, I focused on the changes in modes of incarceration in the perspective of a discussion on borders and boundaries between groups rather than on the experiences and actions of prisoners. Hence, I cannot come to conclusions about the effects of such a prison management, nor on its intentions, such as those suggested by Walid Dacca (2009/2011): to use the prison to model the consciences of a new Palestinian generation according to Israeli

46 Ismail Nashif (2008a, ch. 4) has shown how in the 1970s a "Palestinian revolutionary education" was created and organized in men's prisons, turning culture, reading, and writing into a way of overcoming the effects of imprisonment and of existing as Palestinians and as a political group. For women, too, prison was used as a school, a Palestinian university (*academieh falestinyeh*) where, for a time, feminist ideas were developed (Latte Abdallah 2010b, 2013).

national interests. Nevertheless, the growing similarities and parallels between the spatial organization and the management of prisons on the one hand and territorial fragmentation and control policies in the OT on the other is a main effective process.

The Movement of Political Prisoners and the Palestinian collective body have been weakened in the post-Oslo period, especially after the Second Intifada and the imprisonment of a whole new generation, as a result of several factors: the changes in prison population management since 2003, the wider territorial and institutional reformulation of the occupation, and political developments in the Territories, with the split between Fatah and Hamas (2007).

However, reconstructions are at work. The "hunger strike for Dignity," which was widely undertaken by inmates for nearly a month in the spring of 2012, resulted in an agreement which was favorable to most of prisoners' requests. The strike was widely publicized and relayed around the world, especially through social networks and contributed to reviving the fight against the occupation inside and outside prisons, as well as Palestinian unity. Most of the social and political forces, the Popular Resistance Committees, Stop the Wall and BDS (Boycott, Divestment and Sanctions) groups, and the population largely supported the peaceful protest movement of the prisoners contesting the terms and conditions of detention. Making reference to the Arab Spring, the communiqués of the leading strike committee (where all the parties were represented) stated that the struggle was exceeding the conditions of detention and was more broadly directed against Israeli occupation.

In addition, the banality of the prison experience, the number of comings and goings between inside and outside by many Palestinians, have over time created an increasing porosity between prison life and life outside. Inmates intend to live beyond and despite the prison, to continue their individual itinerary: witness the five-fold increase of those following university studies in prison in the post-Oslo period and encouragement to marry during the prison period (Latte Abdallah 2013). Social networks that are powered by associations of ex-prisoners and NGOs, the web, and numerous radio broadcasts (establishing communication between families and inmates beyond bars) participate to forge a real and virtual community of prisoners. A web of links is being rebuilt with the support of technologies that can also serve to maintain ties in a spatially fragmented context. These technologies, notably new ICT, are recreating a web of links against the fragmentation induced by the *Prison Web*: from 2002 to 2003, mobile phones came into some male prisons and facilitated dialogue with the outside world[47]. The digital presence of inmates through the net and Facebook profiles supplied by relatives or NGOs have enabled them to exist virtually outside prison and has given new collective resonances to the mobilizations of inmates.

47 While introducing a large black market contrary to previous values of political prisoners and a suitable tool for monitoring by the prison authorities.

Chapter 3
Constitutionalism in Colonial Context: The Palestinian Basic Law as a Metaphoric Representation of Palestinian Politics (1993–2007)

Emilio Dabed

Introduction

Trying to understand social and political transformations through the law can be enlightening because legal and institutional processes are historical products that bear the traces of the social and political context in which they take place. In other words, legal structures are social creations reflecting the role played by the actors, the relative positions that they occupied in different periods of time, and the strategies that they deployed. I argue that the Palestinian constitutional process (1993–2007)—and the institutional structures that it implied—was a central battlefield for the actors of the Israeli–Palestinian conflict. In the course of this process, significant political dimensions—i.e., the nature of the actors' power relations, their conception of the Oslo process, the role that they attributed to the Palestinian Authority—were spelled out in legal-institutional terms and mirrored in the political structure of the Palestinian Interim Self-Government Authority (PA). It is in this sense that the constitutional order can be understood as a sort of metaphoric representation of Palestinian politics.

With the "institution building process" that Oslo launched, the PA structures—which in practical terms replaced the Palestinian Liberation Organization (PLO) as the head of the Palestinian national movement—became both its main product and the anchor point of a potential settlement to the Israeli–Palestinian conflict. I will argue that with the PA as the major Palestinian player, the legal and political forms of the new institutions became crucial. Specifically, the Palestinian nationalist expectations, the evolution and outcome of the "indirect rule" regime that Israel intended to establish, and the international interest in settling the conflict in a "two state solution" framework depended greatly on the legal form that the new authority would take. Thus, the constitutional structure of the PA was at the very center of this process and quickly became a main arena of negotiation and battle.

Effectively, the new main actors in the Israeli–Palestinian conflict (which—ever since Oslo—are the PLO and its leadership, the internal Palestinian leadership of Gaza and West Bank, the occupying power Israel, and the "international community," namely, donor countries) were aware of the fact that the structures of legal authority and the creation of political hierarchies and dynamics are inextricably intertwined. Hence, the legal structures were one locus point of struggle precisely because they implied a reshaping of the conflict, and the power relations of the actors in the Palestinian-Israeli chessboard. Consequently, divergences on the conception of the Oslo process and its expected outcome, when represented in legal form, turn into a conflict over the nature and structure of the political authority. For instance, when, during Oslo, Palestinian and Israeli negotiators diverged on the name of the new authority ("Palestinian Authority" or "Palestinian National Authority"), on the title of its head ("president" or "chairman"), on the designation of its legislative body ("parliament" or "council"), or on the jurisdictional power of the PA, they knew that they were not just struggling for symbolic markers, but over the very articulation of a new structure of power relations. Given the centrality that the PA's institutional design acquired, and the strong involvement and influences of Palestinian, Israeli and international actors on the constitutional process, the constitutional making itself strongly challenges the idea of a real "border of sovereignty," so to speak, between the Palestinian territory and Israel. Rather, I argue that the PA's constitutional structures and practices are strongly influenced by the porous lines and walls separating Palestinians from Israelis, and other international actors—influences framed by asymmetrical relations of power.

Accordingly, my argument is that the views and strategies of the multiple actors were systematically echoed in the legal and institutional design of the PA, prompting a phenomenon of gradual "constitutionalisation" of Palestinian politics inside the Occupied Territory. Indeed, each one of the four periods, in which the constitutional process is divided here, represents a historical turning point in the evolution of the conflict, implying changes in the actor's "relative positions" within the Palestinian-Israeli political space and, consequently, changes in political strategy. Changes in political strategy were often articulated in the form of a new legal approach and implied not only textual legal changes but also changes in legal practice and discourse, especially in the constitutional field.

Indeed, since the beginning of the constitutional drafting in the early 1990s and until the suspension of the constitutional regime in 2007, the envisioned constitutional structure, reflecting the complex interactions between the Palestinian and non-Palestinian actors and their influences on the constitutional making, passed through different stages, with each period establishing a different political system. It evolved from a strong presidential regime establishing the PLO as a background power (1993 to December 1995); to a presidential system with some parliamentarian trends while increasingly sidelining the PLO's role (January 1996 to July 2000); to a semi-parliamentarian regime in which important executive powers were transferred from the presidency to the cabinet (August 2000 to November 2004); to a progressive re-concentration of power in the president's hands (December 2004 to June 2007, and onwards).

To make sense of the seeming incongruousness of the constitutional process, this chapter advances what I would call a historical "deconstructive" analysis intending to make explicit one way in which the process acquires meaning and can be understood. To do so, I will focus on the issues that have generally been ignored by previous works: namely, the colonial context in which the constitutional process took place, the non-statist character of the Palestinian authority, the changes in the structures of distribution of power in the Palestinian social space, and the different relative positions that the actors occupied in the above mentioned four major periods of the constitutional process. From this perspective, the constitutional process will be understood as the outcome of choices articulating the actors' positions, strategies and actions within the frame of a colonial conflict and its power relation structures.

Drafting a Constitution without a State: Between Liberation and State-Building

Eugene Cotran has stated that "the main issue to keep in mind to understand the constitutional drafting is the fact that it was done in a situation of occupation."[1] I will argue though that the occupation is only one dimension of a much larger context. The PA's constitutional and institutional frame indeed represents the product of an ongoing settler colonial conflict. The settler colonial character of the Zionist enterprise, its objective to conquer land based on political, economic, and religious reasons, to settle a sizeable foreign population in it, and to displace or dominate the indigenous population, created a very particular structure of power in the territory of Mandate Palestine. This context was reinforced after 1967 with the occupation of the West Bank and Gaza, and its power structures continued being implemented throughout Oslo, and until today. This structure of power has been framed by some classic colonial patterns and practices seen in many other colonial projects, which include the rejection of real sovereignty for the indigenous population, establishment of an authoritarian political regime, and the division of the colonized people to facilitate their domination.

Recognizing that Oslo is the product of this colonial context—and not the end of it—becomes crucial to understanding the PA's institutional/constitutional structures, and dynamics. Rather than a process that lead Palestinians to liberation/self-determination, the constitutional evolution came to rearticulate the colonial power relation in a model of "indirect rule regime," freeing Israel from occupation costs while inhibiting the coalescence of a Palestinian resistance into a wide challenge to the reconfigured colonial order. It does not mean that colonial forces acted as something to which Palestinian institutional and legal processes can be causally attributed but, rather, as an inescapable context of analysis within which they can be intelligible. I argue that, as an expression of this context, the PA's regime tended to reproduce these same colonial patterns—i.e., the Palestinian non-sovereign status, authoritarian forms of domination and the division of the colonized/occupied population in social groups with different legal status and often antagonistic interests.

The first of these features, the non-sovereign status of the PLO was reproduced in the non-statist character of the PA, and primarily shaped the political field within which the constitutional process took place. The Interim Agreement on the West Bank and Gaza Strip signed in Washington on September 28, 1995 (the Oslo II agreement) came to consecrate what previous documents of the Oslo process already announced: the new authority as a non-sovereign entity. For instance, the Paris Protocol of 1994 deprives the PA of real economic or monetary independence vis-à-vis Israel; the PA does not have military power nor control of natural resources or borders; its jurisdiction is limited to areas A and B, excluding issues that "will be negotiated in the permanent status negotiation" and the powers not transferred to the Council (Art. XVII-1-a-b, Oslo II agreement); it does not have powers in the domain of foreign relations (Art. IX-5, Oslo II agreement); the Palestinian Legislative Council (PLC) is given the power to adopt legislation (Art. XVIII-2, Oslo II) but all legislation must be communicated to the Israeli side which can refer any law—that it considered may violate the Agreements—to a "joint Palestinian-Israeli Legal Committee" (Art. XVIII-5-6, Oslo II) in which the Israeli side enjoys a veto power.[2]

1 Eugene Cotran was one of the drafters of the first versions of the Palestinian Basic Law between 1993 and 1995. Interview, London, July 2008.

2 The influence of Israel on the PA's legal production through this joint legal committee and other mechanisms is certain. For example, see footnote 4. In the constitutional field, the Palestinian Legislative Council archives confirm that—as early as October 19, 1994—the Palestinian negotiators

The fact is that the Basic Law (BL) has not been the outcome of a process by which a "state gives to itself a constitution"; but, on the contrary, it is a constitution drafted by and for a "non-state," the PA. Effectively, the Oslo Agreements envisioned a phased solution to the Israeli–Palestinian conflict, concluding in a permanent settlement at the end of the "interim period" in 1999. The Basic Law was thus to be drafted and applied during this interim period, when the conflict was still unresolved, and had to regulate the functioning of a provisory authority deprived of the features which normally belong to modern states. The recent UN admission of Palestine as a non-member observer State has not changed these facts on the ground, and with the exception of headings on ministry letter heads declaring a Palestinian State, the Palestinian Basic Law continues operating up to the present as the constitution of a non-sovereign entity, the PA.

Lacking the attributes of sovereignty and enduring extreme political and economic dependency vis-à-vis external actors, the Palestinian Authority has been prey to diverse and strong influences that crisscross its institutional design and, in particular, the drafting of the Palestinian constitutional text and its past and current constitutional practices. On the one hand, the traits of the PLO as a national armed liberation movement—turned later into a non-state authority—heavily molded the constitutional process. On the other hand, the non-statist nature of the PA has permitted great interference by external actors, namely, the occupying power Israel and the "international community" embodied by donor countries.

A more detailed examination of the context and phases of the constitutional process illuminate these dynamics.

Guerrillas Drafting Constitutions: Building Institutions to Concentrate Power

The first stage of the constitutional drafting encompasses the period from 1993 to 1995. It begins with the first constitutional drafts prepared in 1993,[3] and it ends in December 1995 just before the election of the PLC, which marked the first major change within the PA's structures and specifically in the constitutional drafting process.

This first period was dominated by the PLO leadership whose internal power dynamics and objectives under the Oslo frame strongly shaped the constitutional projects and the PA's institutional building process thereby re-actualizing a second colonial pattern, namely the authoritarian form of domination. The reproduction of authoritarian rule—which was characteristic of the Israeli occupation regime—into the PA's structures, was facilitated by the complementarities between the imperatives of Israeli colonial design and the PLO's aspirations for political hegemony in the Occupied Palestinian territory (OPT). The subordination of the Oslo process to the Israeli imperatives of security and territorial expansion was clear from the wording of the agreements and the Israeli interpretation of them. To this end Israel agreed to and allowed the PLO leadership to create a political

---

submitted a draft of the Basic Law to the Israeli delegation, sharing with their Israeli homologues comments and observations on the constitutional text. "Notes on the project and suggestion on the drafting of the Basic Law for the Palestinian Council that the Palestinian delegation sent to the Israeli delegation in Cairo on October 19th, 1994" (PLC Records, Volume I, document 1). These records are the written tracks left in the PLC regarding the constitutional drafting. The author separated them in three volumes corresponding to the three last periods of the constitutional drafting identified in this research.

3 The first draft Basic Law prepared by the Legal Committee of the Palestinian National Council was made public on January 1, 1994.

regime in the OPT characterized by concentration of power and the organization of a strong security sector.

In its turn, the PLO primarily conceived the new Palestinian Authority as an instrument to consolidate its power inside Palestine, and to this end legal structures were given priority.[4] In the constitutional arena, the PLO's organs tried as much as possible to consolidate the existing PLO's leadership and dynamics of power within the PA's constitutional frame. Indeed, the rationale of PLO hegemonic power was adopted in the first constitutional drafts proposed by the legal committee of the Palestinian National Council (PNC) (1993–1995). Despite the efforts of those who drafted the initial versions of the BL—Anis Al-Qasem and Eugene Cotran—to create the basis for a Palestinian democratic regime, the frame of any possible constitutional structure had to consider—in Eugene Cotran's words—"what Arafat would accept."[5] It turned to be nothing less than total control of the new authority by the PLO-Fatah and, more specifically, his own leadership.[6]

For instance, in the third "Draft Basic Law for the National Authority in the Transitional Period" issued in May 1994[7] and despite the fact that the second Declaration of Independence of 1988 mandated a parliamentarian form of government for Palestine, a strong presidential regime was established. The chairman of the Executive Committee of the Palestinian Liberation Organization (PLO-EC), Arafat himself, was designated as "the President and the head of the National Authority" and given extensive powers.[8] Regarding the institutional relations between the PLO and the PA, the PLO's leadership maintained a strong hold on the new authority which was to be "appointed by the Executive Committee of the PLO" while the PLO-EC was to be considered the PA's reference authority (Art. 58-1-2).

After public discussions, a fourth and last "Draft Basic Law for the National Authority in the Transitional Period" was prepared by the Legal Committee of the PNC and rendered public in December 1995 (PNC Legal Committee 1995). This draft sought to harmonize the constitutional structures with the provisions of the signed Oslo II Agreement which clarified the status of the PLC and the issue of the elections of the PA authorities.

In this fourth draft, the president was no longer the head of the PLO-EC and, rather, general elections for the presidency were established (Art. 68); the president's powers were limited to those prescribed "by this Basic Law" (Art. 75); all the indications of the PLO-

4 Even before arriving in Gaza in 1994, and in an act of independence vis-a-vis Israel, Yasser Arafat issued a resolution (No. 1) stating that all laws in force prior to June 5, 1967, in the WB&G shall remain valid (Palestinian Official Gazette, November 20, 1994). For various reasons (fundamentally the opposition of Israel) this resolution was modified on April 17, 1995, when Arafat issued two new laws which, with the above mentioned, constituted the legal framework for the PA until the election of the PLC in 1996. Law 5 of 1995 (Palestinian Official Gazette, issue four, May 6, 1995) transferred to the new authority all the prerogatives provided for in legal acts effective in West Bank and Gaza Strip before May 19, 1994, which included all the Israeli military orders since 1967. Law 4 of 1995 (Palestinian Official Gazette, issue four, May 6, 1995) on the legislative procedure left in the hands of the executive branch (Arafat himself) the whole legislative process. This improved its internal position giving to the PA an extraordinary power of internal control.

5 Eugene Contran. Interview, London, summer 2008.

6 Anis Al-Qasem, was the head of the Legal Committee of the Palestinian National Council of the PLO. Interview, London, fall 2009.

7 (Third) Draft Basic Law for the National Authority in the Transitional Period. PNC Legal Committee: May 1, 1994, Jerusalem Media and Communication Centre, documents, available at: http://www.jmcc.org/politics/pna/basic1.htm, accessed March 3, 2014.

8 The president enjoyed a great deal of legal power (Art. 50, 78, 107), legislative (Art. 55), judicial (56), executive (Art. 63-3), security (Art. 54), and institutional (Art. 96) prerogatives.

EC as reference authority are deleted; the Legislative Council is not appointed by PLO-EC but elected in general elections (Art. 48); and the ministers are no longer responsible to the PLO-EC but to the president and the PLC (Art. 83). Despite these changes, and the fact that the fourth draft Basic law (Art. 2) proclaimed a parliamentarian regime, a presidential system and concentration of power with the executive were the main characteristics of this constitutional text.[9]

At the time of the fourth "Draft Basic Law" (December 1995), the new Palestinian Authority had been operating for almost a year and a half and the concentration of power and authoritarianism were already a reality. Without the Basic Law having been approved or a parliament elected and given the supremacy of his leadership, Arafat was ruling the PA without any real challenge. In the legal domain, he monopolized the legislative process and within the political structures he occupied at the same time the offices of head of Fatah, Chairman of the PLO, President of the State of Palestine (since the declaration of independence in 1988), and President of the PA. Moreover, this fourth draft fulfilled the Tunis-PLO leadership's strategy of maximizing political power, consecrating it in constitutional terms. Nevertheless this draft, which secured Arafat's political control, was never approved by the executive, showing Arafat's unwillingness to accept any limits to his authority.

The authoritarian feature of the PA's regime was underpinned among other reasons by the fact that the PLO's lack of territoriality was rearticulated in the PA's institutional design. Indeed, the PA was given a weak territorial jurisdiction mostly limited to 'area A and B' (approximately 28 percent of WB&G territory). Loyalty and identification with the PA order was undermined because the authority was unable to provide to its population protection against occupation measures, personal and social security, and to reach any real nationalistic objective. As in the PLO before, the PA leadership resorted to neo-patrimonial and authoritarian rule, co-optation, corruption, and allocation of material benefits to reinforce stability, and compliance to the PA's order. These trends would prevail not only during Arafat's period but after his death as well (2004), becoming characteristic of PA's rule.

The Election of the PA Legislative Council: Arafat is no Longer Alone

The second period of the constitutional process includes the time from the election of the PLC in January 1996 until the failure of the Camp David talks in July 2000. After July 2000 there was a sweeping change in Arafat's position in the political field, and in the constitutional strategies of the actors.

The election of the legislative body on January 20, 1996, entailed the first major political adjustment within the PA. Namely, the influence of the local Palestinian leadership in the decision making process was institutionalized in the PLC with a large majority of its members being local leaders. According to the Oslo II Agreement, the Council was

9 The Executive Authority is exercised by the president and a cabinet of ministers (Art. 67); the president is the head of the executive (Art. 68, 77); the cabinet only assists him in the performance of the executive authority (Art. 78-1); the president maintains the control of security services (Art. 72); he can initiate or propose laws and issue secondary legislation (Art. 73-2) and promulgates the laws (Art. 73-1); the president appoints the head of the Judicial Authority and president of the Supreme Court (Art. 110); he can establish "specialized public bodies ... to assist the government" (Art. 94); the executive exercises the powers and duties provided for its predecessors in legislation in force before this Basic Law (Art. 92).

called to adopt the Basic Law for the PA (Art. III-7). Though the elected parliament was widely dominated by the Fatah party, there was eagerness, especially among local leaders, to restrict arbitrariness. Reacting to the authoritarian political practices of successive occupying forces in Palestine and the PA itself since 1994, the PLC sought to de-concentrate power, increase accountability and fortify the protection of civil rights. During this period the constitutional draft—approved by the PLC in its third reading on October 2, 1997 under the name of "Palestinian Basic Law"[10]—was modified along these principles: it emphasized the separation of power (Art. 2) and the rule of law (Art. 6) as constitutional principles, and the responsibility of the government to the PLC (Art. 5, 64, 68-2, 71); it improved the catalogue of rights (Art. 9–33); it strengthened the prerogatives of the PLC (Art. 34–49) and its participation in the nomination of key public officials (Art. 84, 87, 98); it widened the prerogatives of the Constitutional Court (Art. 94); finally, it defined the legal frame for a declaration of a state of emergency giving to the PLC the right to review all the measures "implemented during the emergency state" (Art. 101) while restraining the prerogatives of the executive during the emergency (Art. 102, 103, 104).

Though these initiatives were taken, the approved BL failed to challenge the strong presidential characteristics of the regime operating since 1994. Indeed, despite the fact that the BL stipulated that "the governing system in Palestine shall be a democratic parliamentary system" (Art. 5) and that some features of parliamentarianism were established,[11] the presidential regime remained in place while the president held strong symbolic and political power: he was to be elected in general and direct elections (Art. 5, 51), and was responsible for the Executive Authority and its head (Art. 50, 62).[12]

The text approved by the PLC though consolidating a presidential political system, outlined the legal and institutional basis for a potential unprecedented democratic regime in the Middle East. Indeed, it established direct elections for the President (Art. 5) and the PLC (Art. 34); a specific term for them (the interim period); and the elections were to take place in a legal context of political and party pluralism (Art. 5). The democratic expectations were, nevertheless, to be frustrated by the political and legal practices of the Executive Authority, helped by the dynamics of what seemed to be a one-party (Fatah) regime.

Many important battles took place between the PLC and the PA executive throughout this period, but two issues are especially pertinent here: on the one hand, the conflict over the legislative initiative that Arafat deemed reserved to the executive and, on the other hand, the question regarding the promulgation of the constitutional text and its application. Concerning the first question, President Arafat managed to weaken the role of the parliament by leaving numerous laws passed by the PLC without promulgation, with the

---

10 See Palestinian Legislative Council, Volume I, document 24.

11 The appointment of the Council of Ministers (COM) requires the vote of confidence of the PLC (Art. 64); the COM is responsible to the Legislative Council (Art. 5); which can withdraw confidence from the COM prompting the termination of its mandate (Art. 44).

12 In this task he is 'assisted' by the COM (Art. 63); the government is accountable to the president (Art. 5 and 68); the president appoints the ministers and presides over their meetings (Art. 62); He maintains control of the security forces (Art. 55); he can declare the state of emergency (Art. 101-1-2); he monopolizes the authority's diplomatic prerogatives (Art. 56); he has the right—in exceptional cases—to issue decrees that have the power of law (Art. 60); he can propose laws, issue regulations, and take the necessary actions to execute them (Art. 58); he promulgates laws and can exercise a legislative veto power (Art. 57); has the right to pardon or commute sentences (Art. 59); the execution of a death sentence requires his endorsement (Art. 100); and he appoints key authorities (Art. 87, 98).

Basic Law being the most important one. Rather than minimizing the significance that the legal framework had during Arafat's regime, his resistance only reinforces the idea that the law and especially the constitutional process was a political battle field. As regards the second question, given his central role and leverage on the political process Arafat avoided the promulgation of the BL for many years (1997–2002) without being subject to any significant political pressure. This state of affairs changed after the failure of the Camp David Talks in July 2000, leading to the promulgation of the BL in 2002 and far-reaching constitutional reforms in 2003.

Forcefully "Democratizing" the PA? The Influence of International Actors

The third period of the constitutional drafting (August 2000 to November 2004) was defined by the decline of "Arafat's reign," which meant a new political adjustment inside the national movement and the PA. Until then, Arafat was the personification of the "Palestinian revolution" and, as such, an autonomous source of political legitimacy. He enjoyed a political stranglehold much larger than the entire ruling leadership and his personal position placed him above all groups—allowing him an extreme discretion in ruling the PA. But Arafat's rejection of the "generous" Israeli offer in Camp David in 2000 made clear that he was willing to accept only a part of Israel's conditions for a final settlement to the conflict. Despite the propaganda campaign launched to fabricate a consensus in international public opinion that Arafat did not want peace (Dary and Sieffert 2002), the facts of Ehud Barak's offer are well known today, and essentially entailed a Palestinian State without East Jerusalem, no real territorial continuity, a fictional sovereignty, and no resolution regarding the millions of Palestinian refugees (Malley and Agha 2001). However, Arafat's refusal rendered clear to Israel and its allies that the Palestinian leader was willing to accept only a part of Israel's conditions for a final settlement to the conflict. The political response of Israel, therefore, was to attempt to replace Arafat, by isolating him diplomatically and debilitating him politically. A significant number of donor countries led by the Unites States followed and supported this strategy.

With Arafat's marginalization, the influence of the occupying power Israel and international actors prevailed. Yet, for the Israeli leadership the main role of the PA was to deliver security for Israel itself. The international actors, while financing the whole PA institutional development plan, seemed also to prioritize the security of Israel at the expense of a more democratic state-building process. Indeed, during the first and second phases, they (Israel included) accepted the Arafat-PLO leadership's attempts to concentrate power based on the assumption that executive centralization could help to deliver security and to reach a final settlement to the conflict under Israel's preferred terms. Military actions, especially the suicide attacks inside Israel, carried out by resistance groups strongly eroded this argument. The outbreak of the Second Intifada and its militarization along with the "war on terror" launched after the 9/11 attacks also reduced the range of Arafat's maneuver. In this battle, international actors and internal challengers privileged the constitutional field. Under internal and international pressure, the BL of 1997 was belatedly promulgated by Arafat in 2002. However, as Eugene Cotran puts it, "the constitutional drafts begun by being a set of hopes and promises which would be betrayed by the authority's legal practices."[13]

13 Cotran Eugene, Interview, London, July 2008.

When the constitutional text proved to be insufficient in weakening Arafat and retained the concentration of political power in his hands, Israel and donor countries pushed for constitutional reforms. To do so, they united with a section of the Palestinian leadership (from outside and inside the PLO) that had been demanding these reforms for a long time. Indeed, in 2003, the PLC undertook significant constitutional amendments introducing a strong dose of parliamentarianism in the Palestinian constitutional system. These changes sought to divide the political power—until then concentrated in the president's hands—and to invest the newly created post of prime minister and the Council of Ministers (COM) with the bulk of executive prerogatives.

Persuaded that the reforms introduced were being used by Israel and the donor countries with the objective of politically weakening him rather than improving the Palestinian political system, Arafat paid lip service to the motion but used all his remaining political leverage to maintain his power. The result of this battle was the "Amended Basic Law" (ABL) of March 18, 2003[14] consecrating a constitutional regime which seemed to have two heads, the president and the prime minister. Two general changes were consecrated in the 2003 reforms: first, the president's position was no longer in the chapter pertaining to the executive authority, as was the case in the Basic Law of 2002. Thus, the prime minister became the head of the executive (Art. 68). Second, regarding the executive prerogatives the general principle established by the ABL is that the Council of Ministers is the highest executive and administrative instrument.[15] Except for the executive powers of the president "executive and administrative powers shall be within the competence of the COM" (Art. 63). In addition, the formation and continuation of the COM—as in the BL of 2002—remained being based on parliamentarian confidence (Art. 74-3, 77-1, 78-2).

Nevertheless, the political independence and executive prerogatives of the prime minister vis- à-vis the president are weakened by certain provisions of the ABL which give to the president strong political status and powers to intervene in the political process: first, the political legitimacy of the president is not less than that of the government since he is elected in a general and direct election (Art. 34) and he "shall appoint the Prime Minister and authorize the latter to constitute his government".[16] Second, the president has his own executive, diplomatic (Art. 40), legislative (Art. 41-1, 43), judicial (Art. 42, 109), institutional (Art. 92-2, 96-3, 107-1), and security (Art. 39, 110-1) prerogatives for which "the Council of Ministers shall assist the President" (Art. 46). Indeed, under international and local pressure, the reforms were rapidly approved without foreseeing the problems arising from such an intricate division of the executive and symbolic power.[17] The constitutional system became the bearer of institutional dissonances and tensions whose implications were to be unveiled during the period of the first Palestinian Prime Minister, Mahmoud Abbas, leading to his hasty resignation,[18] and became much more acute after the electoral victory of Hamas in 2006.

---

14 Palestinian Legislative Council, volume II, document 27.

15 Some of the most sensitive powers attributed to the president in the 2002 Basic Law are now within the jurisdiction of the COM (Art. 69 and 70).

16 The president can dismiss the PM, and accept his resignation (Art. 45); The PM is accountable to the president (Art. 74-1); the president can refer the PM for investigation as a result of crimes (Art. 75-1).

17 Arafat accepted the amendments on February 12, 2003, and in less than a month, on March 10 the constitutional changes were approved by the PLC.

18 "These were precisely the core of the conflict between Arafat and Abu Mazen … in a matter of months Arafat succeeded in knocking out Abu Mazen who ended up resigning. Later Arafat

During this third period, the core of the constitutional drafting was concluded. The next and last stage of the constitutional process (December 2004 to June 2007) does not include any significant constitutional changes with the exception of a very minor one in 2005 establishing the term of the PLC and the President as four years (instead of the interim period). Nevertheless, between 2005 and 2007 the most acute inter-Palestinian political crisis since the inception of the PA took place, having important constitutional dimensions and far reaching social and political consequences still present today. In what follows we turn to this issue.

The End of a "Colonial Democracy" (2004–2007): Toward an Authoritarian Non-State

This last period of the constitutional process (December 2004 to June 2007) was marked by Arafat's death in November 2004 which had "psycho-political" effects: the figure of the "father" of the Palestinian movement, providing his personal legitimacy to the entire PA structure along with relative unity and stability, disappeared. In his absence, after 10 years of Fatah rule, increasing perception of corruption and more than 8 years without elections, the instrumental and ethical limits of the neo-patrimonial rule of Arafat-Fatah leadership was exposed. The result was growing Fatah internal division and rivalry while the fall of Fatah's political legitimacy accelerated.

In this context the legal and constitutional field was again at the forefront of the political scene. Indeed, seeking new forms and sources of legitimacy among a dispersed constituency, the new PA leadership emphasized reference, albeit rhetorical, to the law and the democratic process. In the institutional succession of Arafat, for instance, the PA closely followed the constitutional framework. As per the articles, during an interim period, the speaker of the PLC replaced the deceased president and organized new presidential elections on January 9, 2005, in which Mahmoud Abbas (Abu Mazen) was elected. His democratic election was the main political credential that Abu Mazen could brandish to revitalize his leadership.

Strengthened by his electoral victory, the new president and his closest partisans sought to provide the PA with a renewed democratic legitimacy, something that had also been increasingly demanded in domestic and international circles. To this end, the PA's leadership scheduled two rounds of elections: local elections in 2005 and general legislative elections in 2006. Nevertheless, these elections ultimately did not deliver the expected results for the Fatah-PA leadership. The positive performance of Hamas in the local elections in 2005 was a first warning for the PA rulers, which prompted a return to institutional-constitutional skirmishes.

Foreseeing a potential victory for Hamas in the legislative elections in 2006, Fatah leaders and MPs introduced a motion for constitutional amendments in fall 2005.[19] The proposition had several components, but it basically tried to strengthen the Fatah-president's constitutional prerogatives while stepping back from the reforms of 2003 by explicitly recognizing a presidential character for the Palestinian political system (Art. 5) and giving

nominated Abu Ala as PM. But Abu Ala had learned the lesson: he (PM) cannot do business without the approval of Arafat. At the end of the day we succeeded in putting the amendments on paper but we could not implement the reforms … the PM continued working as though it was another office of the presidency." Azmi Shuaiby (PLC member 1996–2006) interview, Ramallah, November 2007.

19 Draft amending Basic Law 2003, (2005). PLC Records, Volume III, document 5.

a central role to the president within the institutional system as "the head of the PNA" (New Art. 55). It gave the president the right to call for a general referendum regarding issues that concern the "higher interests of the homeland" (New Art.); the power to call early elections—dissolving the PLC—under several scenarios (New Art.) many of which strikingly resemble situations that transpired after the election of Hamas; it strengthened the president's legislative (Art. 63) and security prerogatives (as the head of a National Security Council, New Art.); finally, a new post of vice president (New Art.) would allow Fatah to avoid losing control of the presidential office in case of Hamas' victory and vacancy of the post of president.

The amendment was not approved, perhaps reflecting the over-confidence that Fatah still had at the time regarding the electoral result. A few days before the 2006 legislative elections, the motion was reintroduced but, on two occasions, the parliamentary quorum was not reached. Apparently Fatah members were too absorbed in their campaigns but to no avail. In the January 25, 2006 legislative elections Hamas obtained a sweeping victory winning a majority of 74 out of 132 seats in the parliament and, therefore, the mandate to form a new government.

Despite the fact that Fatah officials made public statements about abiding by the popular vote, the PA's leadership undertook actions that exposed their distress after losing political power and its material privileges. Having failed in the constitutional field, the outgoing Fatah government and parliament started issuing a series of legal measures (especially in the domains of public finance,[20] security[21] and institutional jurisdiction[22]) that reverted a significant part of the 2003 constitutional-institutional reforms and obstructed the performance of the new Hamas government. The internal Palestinian conflict and the constitutional crisis that followed were reinforced basically by three factors.

First, apart from the denial of Palestinian sovereignty and the authoritarian form of domination, the legal-institutional design of the PA rearticulated a third colonial pattern, namely, the division of Palestinians and the national movement. It is in this sense that I argue that if ever the Palestinian political regime was a democracy it was a "colonial democracy," that is, one which tends to reproduce colonial categories and practices. The fact is that Oslo's institutional and legal structures contributed to redefine identities and reshape subjectivity while entailing political, economic and symbolic incentives, which broke the unity of purpose of the Palestinian national movement culminating in the political crisis of the present. Indeed, in addition to the pre-Oslo divisions imposed on the Palestinian people, after Oslo they were progressively re-divided along new lines:

---

20 On February 4, 2006, the Palestinian Investment Fund's board was modified and PIF oversight transferred under the president's office thereby limiting the access of the new government to PA resources (Presidential Decree No. 4, 2006, Palestinian Official Gazette, issue 63, April 27, 2006). The board of the Palestinian Monetary Authority was modified to prevent the new government from resorting to it (Presidential Decree No. 25, 2006, Palestinian Official Gazette, issue 63, April 27, 2006).

21 The presidency reinforced its prerogatives and security powers—undoing most of the 2003 security sector reforms—so as to prevent Hamas from having control over the security apparatus. In response, the Hamas government created a new security agency, the Executive Force, in April 2006.

22 On February 13, 2006 the law of the Constitutional Court was passed with some last minute amendments increasing the president's powers to nominate its members (Palestinian Official Gazette, issue 62, March 25, 2006). In the last meetings, the Fatah-led outgoing PLC passed amendments creating new administrative posts in the PLC and appointed Fatah people to them; the ongoing government appointed Fatah loyalists to the head of several governmental bodies; and on February 12, 2006, the oversight of the main official Palestinian media was transferred to the president's office (Resolutions 29 and 30 of 2006, Palestinian Official Gazette, issue 63, April 27, 2006).

for instance, by creating new groups (i.e., bureaucracy, new economic and political elite), excluding more than a half of Palestinian people (refugees and diaspora) from the political process thereby symbolically reducing Palestinians to the population of West Bank and Gaza, and by labeling Palestinians as moderates (pro-Oslo) and extremists (against Oslo). The fates reserved to these different groups not only diverge but are very often antagonist, consecrating a new cognitive grid for social meaning and political action. It generated new forms of interaction between Palestinian actors and transformed alliances, whereby former "liberation fighters," for instance, could begin to see each other as enemies and indeed some could begin to treat the occupying power as an ally. For instance, during the PA-Israel meetings for security coordination, Fatah officers have expressed their conviction that they (meaning the PA and Israel) are fighting a "common enemy" (Elad 2008).

Second, extreme pressures were exerted on the political and constitutional system. The non-state character of the PA and its extreme vulnerability to Israeli power and dependency on foreign actors (fundamentally donor countries), allowed them a broader ability for unsettling, often insidious, intervention. Even though the inter-Palestinian conflict has internal dynamics and cannot only be attributed to foreign intervention, it undoubtedly contributed to accentuating the disparities and political rivalry. Fatah party—under international influence—declined to participate in a national unity government. In turn, the donor countries made financial aid contingent on the fulfilment by the new Hamas government of certain conditions (i.e., to recognize Israel and its "right to exist," to renounce violence, and to accept past PLO-Israel agreements) that Hamas rejected. Subsequently, all official international aid to the Hamas government was suspended. In addition, Israel froze the clearance revenues that it collected on the PA's behalf, as stipulated in the Paris Economic Protocol (1994) increasing economic chaos. Later—June 2006—Israel started a campaign of arrests against members of the government and more than 40 Hamas deputies, thereby completely paralyzing the Palestinian parliament which, in the context of a parliamentarian system, is to say the virtual paralysis of the political regime. With the parliament unable to fulfill its role, the constitutional hermeneutic about the jurisdictional prerogatives of the presidency and the prime minister and cabinet seemed to dominate the public political discourse.

Third, the strongly contradictory influences introduced in the institutional and constitutional Palestinian system in the preceding stages seemed to contribute to the escalation. Indeed, different political objectives promoted by the actors of the conflict at different moments, were inscribed—without a clear political vision—in the constitutional system which became the bearer of institutional dissonances and tension. In many senses the regime seemed to have two heads, the president and the prime minister. It permitted confusion and interpretative fighting about their prerogatives, and eroded the respective institutional authorities and aggravated the political situation.

Thus during 2006–2007, the constitutional text continued being a central issue in the political struggle. In May 2006, a constitutional interpretative battle invaded the public discourse when the PA (Fatah) president claimed to have the right to call for a referendum to put an end to the divide. After acknowledging the evident fact that the Basic Law did not foresee this procedure, the presidency gave up this idea. Later, in December 2006, a new "constitutional" way out of the deadlock was put forth by the PA presidency. The measure threatened to dissolve the PLC and call early legislative elections. However, the constitutional basis for such an action was no stronger than that for a referendum.[23] The initiative found a wide opposition and was never put in action.

In early 2007, the fragmentation of the Palestinian national movement had reached such levels that Palestinian society was confronting a major political crisis with strong Fatah-Hamas armed clashes taking place. Rumors of an imminent Fatah "push" with international support further poisoned the situation (Al-Majd 2007; Perry and Woodward 2007). In May 2007, these fears took root with the arrival of 500 new Egyptian-trained recruits loyal to the Palestinian president (Issacharoff, 2007). There were also rumors that Fatah was mobilizing troops towards Gaza. *The Washington Post* wrote on May 15, 2007 a report titled: "Fatah Troops Enter Gaza with Israeli Assent: Hundreds were trained in Egypt Under US-Backed Program to Counter Hamas." On June 14, 2007 Hamas undertook military action resulting in a total takeover of all PA security and political apparatus in Gaza.

The action was labeled by Fatah as a "Hamas coup d'état." President Abbas declared a state of emergency and dismissed Prime Minister Haniyyeh (Hamas) and his cabinet.[24] In his place, the president appointed a new prime minister (Salam Fayyad) to form a government that completely excluded Hamas. Finally, after the takeover of power in Gaza by Hamas in June 2007 and the division of the PA into two governments, the application of some significant chapters of the constitutional text were legally and practically suspended.[25]

Since June 2007, an "Emergency Government" has been ruling the West Bank outside constitutional legality.[26] Nevertheless, Israel, the US, and the EU have given their total support, and international financial institutions and donors hastily resumed aid for this government. Meanwhile, the PA-Fatah leadership has been using a range of legal and extra-legal means to re-concentrate and consolidate his power and marginalize Hamas politically. Currently, a long list of ministerial and presidential decrees governs the West Bank, excluding Hamas from the PA, and repressing internal opposition and resistance to Israel. The situation is not better in Gaza. Despite participation in several rounds of meetings to solve the internal deadlock, both groups concomitantly oversaw a crackdown of its opponent. In all the major cities of the West Bank, the Abbas government has mobilized its forces; a "police quasi-state" has been set up, one which entails repression and the arrest of dissidents; Hamas[27] and all other armed militias[28] are outlawed; NGO's linked to Hamas are closed;[29] freedom of speech exists only in appearance; the state military courts have been reactivated after almost four years of inactivity and their jurisdiction extended;[30] some efforts

23 The amended article 47 (2005) gave the PLC a four year term, and did not foresee its dissolution.

24 Presidential Decree No. 8 and 9, 2007. Palestinian Official Gazette, issue 71, August 13, 2007.

25 By Presidential Decrees No. 11 and 12, 2007 (Palestinian Official Gazette, issue 71, August 13, 2007), the president suspended the application of some chapters of the ABL, especially those concerning the appointment of a government and the vote of confidence of the PLC.

26 The Palestinian Basic Law does not authorize any government to enter into office before receiving a vote of confidence from the PLC. The current government has never met this condition.

27 Decision No. 257, June 16, 2007, Palestinian Official Gazette, issue 71, August 13, 2007. Decision No.4, August 12, 2007, Palestinian Official Gazette, issue 73, September 13, 2007.

28 Presidential Decree No. 17, June 26, 2007. Palestinian Official Gazette, issue 71, August 13, 2007.

29 Presidential Decree No. 16, June 26, 2007. Palestinian Official Gazette, issue 71, August 13, 2007.

30 Presidential Decree No. 28 for the year 2007, on the jurisdictions of the military courts during the state of emergency, Official Gazette, issue number 73, September 13, 2007.

have been made to renew the PLO as a way to bypass the paralyzed PLC;[31] and the High Court—acting as a constitutional court since 2006—has not challenged this authoritarian paradigm.[32] Ironically, and despite the fact that between 2009 and 2011 the PA's regime has cancelled three calls for local elections and one call for legislative and presidential elections,[33] the discursive reference to the law and the democratic process—from all political actors—has only increased.

31 The efforts in this direction had gone so far that the PLO Central Committee has met to discuss the dissolution of the PLC.

32 See, for instance, the decisions issued by the high court acting as a constitutional court since 2006. Some of them are in http://muqtafi.birzeit.edu/en/courtjudgments/ShowDoc.aspx?ID=52112, accessed August 22, 2012.

33 Central Elections Commission, "Presidential Decrees and Cabinet Decisions." Available at: http://www.elections.ps/tabid/739/language/en-US/Default.aspx, accessed August 22, 2012.

Chapter 4
What Are We Talking about when We Talk about "Geographies of Occupation"?

Ariel Handel

Introduction

"The attitude to occupation," writes the Israeli geographer Elisha Efrat, "is mostly political, juristic, economic, demographic, or administrative, but rarely geographical" (Efrat 2006, p. 4). Before asking ourselves whether this claim is accurate,[1] or trying to undertake such research ourselves, we must ask what exactly do we mean when we talk about "geography of occupation." I would like to claim that there are several geographies of occupation, and that one must be aware of the different spatial languages in order to avoid confusion, which might lead to mistakes in the current situation analysis as well as in the suggestions for future solutions.

Roughly speaking, there are four spatial-geographical languages in the OPT: 1) that of classic modern geography, from geopolitics to rational planning; 2) that of the Israeli control system; 3) that of NGOs and human rights organizations; and 4) that of the Palestinian inhabitants. It is important to make clear that I do not talk about four different existing discourses—and my purpose is not to make an academic literature analysis, mapping who said what about the subject—but rather about different languages in the deep sense of "language." One can think only through one's own toolbox of words and logic. Language in this sense is a pair of glasses that designates the way in which one sees the world. Therefore, a spatial language is not only what a group of people say about the space in the Occupied Territories, but rather how they see it, analyze it and use it.

This chapter will deal with three of these languages.[2] As a matter of fact, it will criticize two of them—that of classic political geography and modern planning discourse, and that of the human rights' discourse—by using the fourth, namely the Palestinian phenomenological language. Although those three languages strongly condemn the Israeli occupation and thus should share the same view about the spatial problems and suggested solutions—this is not the case. The purpose here is to analyze critically the linguistic gap and to show why well-intentioned languages fall short of catching the situation as a whole. This happens due to built-in biases in their basic assumptions: the bias towards the mappable in political geography and the bias towards liberalism and neoliberalism in human-rights discourse. On the one hand, those seem to be complementary. While the first looks at the territory and the spatial representations of human activity, the second abandons the question of

1 Efrat's book is based on an earlier publication in Hebrew (Efrat 2002), to that date, his argument was stronger. For more recent reference on the topic, see Harker (2010), Amir (2011, 2013) and Handel (2013).

2 I have dealt in details with the language of the Israeli system control in other places (Handel 2007, 2009a, 2013). See also Weizman (2007); Kotef and Amir (2011).

territory altogether and concentrates on the individual's movement in an abstract space. However, both overlook the same thing: the actual people and the actual territory—not in their abstract representation but rather in their daily phenomenological life and perception. Harker (2010, pp. 203–4) argues that "to employ only a geopolitical epistemology to encounter Palestinian lives and spaces is to run the risk of abstracting these spaces and subjects in much the same way as the practices of the Israeli occupation do." The same thing can be said regarding neoliberal epistemology, turning the Palestinians into moving particles, without history and emotions.

This section will introduce the inherent bias of geography towards the mappable and expose that tendency's weaknesses in the case of the Occupied Territories. The second part will do the same for human rights' discourse, and mainly to "the right to the freedom of movement." I will argue that this right is meant to serve not the freedom of movement, but rather the freedom to arrive. Namely, that it sees human movement in space only in a rational and goal-oriented manner. In the third part I will introduce the phenomenological way of seeing the space through the Palestinian inhabitants' eyes; that of making a territory through walking and experiencing—in which territoriality and subjectivity go hand in hand. The main source is Raja Shehadeh's book *Palestinian Walks* (2008). The last part will criticize one of the future solutions suggested by the American RAND Corporation under the title "The Arc", and will attempt to show a possible consequence of unrecognized language gaps—creating problematic solutions. In order to solve a problem one must, firstly, define the question. When the problem is worded in different languages, the solution cannot be found. This inherent problem, as will be shown later, subverts all the suggested solutions, and limits the political imagination of all sides.

Political Geography and the Bias towards the Mappable

Richard Hartshorne, who is considered one of modern geography's founders, wrote in 1939 that "[s]o important, indeed, is the use of maps in geographic work, that … if (the) problem cannot be studied fundamentally by maps—usually by a comparison of several maps—then it is questionable whether or not it is within the field of geography" (Hartshorne 1939, p. 249). Although many years have passed since then, geography is still considered to be a visual discipline (Rose 2003; Braverman 2011b) and maps have an important role within it. Guntram Herb (2008) distinguishes between three main currents in political geography. The first is the "geopolitical," in which the state is the most important and influent agent. It focuses on the scale of the state upwards (e.g., the Middle East or the "global order") and presents dichotomies of "us and them," "east and west," etc. Generally it uses global or regional maps that focus on territories and borders. The second is the "professional" attitude, in which the state is taken for granted. This attitude sees itself as scientific, objective, neutral, and nonpolitical that concentrates in management, in improving the spatial administration, and the like. It is therefore a powerful administrative body that is deeply connected to the state's planning and control arrays. The maps it creates are delimited to the state, and show mainly spatialized statistical data such as income, voting patterns and so on. The third and most recent attitude is the "critical" approach towards the state and its actions. Its scale is flexible and varies from the gendered body in urban space to global networks. It recognizes that the political does not only concern the state and wishes to change power relations. This is the only attitude that usually does not use or create maps, as it recognizes that not only are maps biased and manipulated (cf. Harley 1989; Monmonier 1991), but more important that there are things that cannot be mapped. The criticism brought here relates exactly to this: that classic political geography looks for the mappable and the visual, while in the OPT especially these are far from telling the whole story.

Traditional geopolitics deals mainly with states and mappable territories, and, therefore, recognizes the occupation mainly with the violation of the territory's wholeness, and concentrates on the legal-political questions of mutual recognition, one-sided annexation and the like. Efrat (2006) widens the geography of occupation beyond those aspects, and explains the spatial problems of the occupation mainly from the point of view of rational and sustainable planning. In other words, Efrat's innovation is taking the occupation from the geopolitical domain into the field of "professional" geography, the one that looks not only at the scale of states and above, but also at the "objective" and "neutral" processes that occur in the territory.

For him, the occupation is a destructive act. In contrast with a territory that grows organically, in time and in conjunction with the human and the physical environment, the occupation is a fast act, when all the human and nature resources are abused forcefully by the occupying forces. The geography of occupation is defined by the creation of parallel geographical systems: settlements, roads and administrative centers, which do not make an integral part of the zone. The new system is meant to serve only the immediate needs of the occupiers, namely control over the territory and the population. The result is usually a suburban and marginal array that is far from any geographical logic, and its maintenance has unreasonable prices, both economically and environmentally.

And yet, Efrat is loyal to classic geography, in its deep meaning of sticking to the mappable. The geography of occupation in his eyes is only the damage that can be seen, that which can be marked on an aerial photo, and that which will remain after the decolonization. He realizes that the legal-territorial aspect is not the only one, but still uses the old geographical toolkit. It seems that for him, the main problem with the occupation is the lack of an organized planning process and the priority given to control and expansion aspirations over the professional geographic rational. His criticism of the geography of occupation (any occupation, not just the Israeli case) is mainly that it is unprofessional, therefore harming the physical and the human landscape. It is a modernist geographical-planning approach. While its criticism is right and important, it is inherently limited.

Here is one example. Irus Braverman (2008) analyzes the practice of aerial photography in the West Bank, which is part of the process of land confiscation by declaring it to be "state land." In order to declare a "state land" the authorities have to prove that the land is uncultivated for at least 10 consecutive years. Braverman details the different discriminations in this "objective" and "neutral" practice: high cost, inaccessibility of the Palestinian to the archives, the expertise that is needed to decipher the photos, etc. I want to emphasize here only the fact that the aerial photos neglect many things which can not be pictured. The aerial photo, says Braverman, "ignores things that 'don't exist': namely, human behavior" (Ibid., p. 469). Malka Offri, the state's expert of aerial photography, explained to Braverman that:

> *There is no detail in the area that escapes our eyes.* Every detail in the territory gets a code ... That doesn't include humans. ... You can't really map people. Trees, on the other hand, don't move. People move, but things stay in place ... Sometimes I spot goats in the aerial photo. It is amazing to see them there. Of course, I don't mark them into the map, because they move. [I only map] *existing things.* (Ibid., italics in original)

Human behavior and "moving things" do not exist because these cannot be mapped and analyzed. This is not the whole truth, however, as people's routes can be viewed, collected, analyzed and mapped. This is commonly done in Israel/Palestine as well as in many other places. And yet there are things that are unmapped, and even unmappable. It might be hard to tell whether a certain field would be considered as cultivated if it is used only for grazing from time to time—but still the authorities can count the days and the hours in which the place is used and to give it a title: cultivated/uncultivated. Fear, on the other hand, cannot really be mapped. The same goes for uncertainty, human emotions, daily experience and spatial "tactics" (see de Certeau 1984). All of these direct human spatial use and behavior on a daily basis, much more so than the mapped issues. One can feel free to cross a private field in Europe, while transgressing a private property sign in the United States might summon an angry man with a rifle. The title of the land or the demarcation of borders means less than the human experience and the emotions of fear, joy, etc. The unmapped features of the Israeli occupation regime are discussed in the next section.

The Mobility Regime in the Occupied Territories: Ecology of Uncertainty

Since at least the mid-1990s, reaching its peak during the Second Intifada, a substantive part of the occupation and its evils is found in the restrictions on Palestinians' movement. The system of population control by policing and blocking movement nearly annihilated the possibility of mapping and absolute location and even ridiculed them. Azmi Bishara describes how people stopped asking their cell-phone interlocutors "where are you?"

> The question "Where are you?" ceased to exist as life in the shadow of the checkpoints turned it into a foolish, at times even taunting, question. For where else could one be under such circumstances? If a curfew was imposed, people would be at home, and if there was no curfew there was the choice of their immediate surroundings or the checkpoint, since moving from city to city had become nearly impossible. (Bishara 2006, p. 29)

Since the mid-1990s a regime prevails in the Occupied Territories that produces uncertainty as one of its main features. Generally speaking, technological dispositives are composed of three parts: tools, signs, and human actors. In this case, these are: *checkpoints* (manned, permanent or semi-permanent; mobile ("flying") checkpoints; and different kinds of physical roadblocks); *movement regulations* (more or less stable; most of them are not published. Some are revealed to the Palestinian's only when reaching the checkpoint, while others remain totally unknown, such as restricted "special security zones" near settlements, roads, and military bases); and *soldiers and private security companies' workers* (to whom the decentralized character of the control system gives huge and unproportional power, as for the number of the checkpoints as well as for their mode and pace of work (violent/non-violent, efficient/inefficient). These three factors are multiplied by each other and thus create a state of extreme spatial uncertainty. One cannot know how many checkpoints and roadblocks will be met on one's daily journey; one cannot know whether these will be closed or open, and what the daily orders and restrictions are; one cannot know how the soldiers or the private workers will behave: they can run the checkpoint efficiently or deliberately slow, they can be threatening and insulting or nice and respectful. That is exactly the problem: the uncertainty. On one day the road between two Palestinian cities can be quick and easy; on another day it can be slow and annoying; on a third day, one can find oneself detained and beaten. Not only does the space decompose into many tens of "land cells," the passage between which is hard and slow, but it also liquidizes when there are not any fixed distances between certain points (as the journey can take one hour or two days, depending on the above-mentioned factors), and the concepts of near-far, and in-out, nearly lose their meaning (Handel 2009a).

The spatial state of uncertainty as described here was mainly relevant in the years 2001–2008. Since 2009 there is a feasible relief of the movement restrictions due to the Intifada's decay and the pacification of the area, and as part of the Israeli Prime Minister Benjamin Netanyahu's ideas of "economic peace," which is based on economic growth and a rise in the quality of life, while rejecting the political questions of the control over the West Bank. And yet, that state of affairs is still relevant as 1) it exists as a potential that can be renewed and employed at any time: all the checkpoints remain in place and the gates that are now open can be easily closed. The same goes as for the physical roadblocks: the cement cubes are still there, the ditches can be dug in one hour, etc. It means that the institutional memory is there as well as all the necessary military plans, and that movement is still conditional. 2) Despite the apparent relief of the restrictions, an AIDA[3] report from June 2011 shows how the movement restrictions still affect their own work, increasing the costs of giving aid to the Palestinians. The estimations are an additional 4.5 million US dollars per annum. According to the report, the restrictions decrease the effectiveness and sustainability of aid operations. It denies the most vulnerable populations from vital services. Ninety two percent of the organizations had difficulties in obtaining permits for their staff—mainly to East Jerusalem and the Gaza Strip, but also to C areas, that encompass 60 percent of the West Bank. Many organizations had to hire international workers at higher costs (because they can move through the checkpoints) and to open new offices, sometimes just a few tens kilometers away (AIDA 2011).

The problem lies not only in the closures and the spatial separations, but mainly in their unpredictability. Permanent and "organized" checkpoints, even when they take some time to go through (but are relatively stable, and thus "mappable") still allow a schedule to be planned. Clearly those blocks will affect daily routines and reduce the possibilities of the way the space used, but they will not completely disrupt it. The state of affairs is different when the three elements of spatial control, namely the map of roadblocks, the passage regulations and their operation by the junior soldiers change a few times a day. Said Zeedani (2005), a professor from Al-Quds University, wrote:

> Because of the checkpoints and the arbitrary permit system or systems associated with them, most Palestinians in the OPT are more inclined to give up moving from one region to another, and from one city to another, unless it is so important or essential or urgent to do so … Checkpoints haunt people's minds all the time. You cannot make a plan, a promise, an appointment or a commitment without taking them seriously into account, of course if implementation requires crossing checkpoints … checkpoints mean that the rest of the country with its space and landscape recedes into the distance and the background. It becomes off limits, too remote and unreachable.

This is, in a deep sense, a nearly unusable space. Heidegger (1996) had developed a concept of "readiness to hand" (*zuhandenheit*). According to Heidegger, when a carpenter is using

3 The Association of International Development Agencies: a coordinating organization of 84 NGOs in the OPT.

his hammer, the tool gradually becomes transparent to him. The "ready to hand" tool allows a craftsman to concentrate on what he wants to do *with* the tool and to forget the tool itself when doing routine work. The hammer comes back to be an object of consciousness only when it brakes, or for some reason, does not fulfill its task. In the same manner, space normally becomes transparent to its users. People are usually not aware of the door handle, and just use it in order to get in or out. The handle is revealed only when it is stuck—namely, when it stops being transparent. When the handle is broken, or there is an oil stain on the way to the kitchen, the person becomes aware again of the space that lies between him and the cup of coffee he wishes to drink. In the same manner, nearly any intent in the OPT is mediated in the sense that the Palestinian user of space is always aware of the medium in which he moves, and that the medium may "revolt," not allowing him to fulfill his intentions. Basic human actions decompose in a situation like that. The wish to go to work, to study or to visit friends is not so simple anymore. Living in a "normal" space one can arrange a meeting on the other side of the city and turn one's mind to things other than the location and the direction on the way there. In the OPT, the space and the time stopped being transparent, exactly because they do not cooperate and demand attention to the movement itself.

The daily disruption of routine presents the *space itself* as an unobvious resource in a state of permanent scarcity. The control over the space, that uses the space itself as a tool of control, and makes the space opaque and always-present, makes present also the state of "being under occupation." It haunts people's minds all the time and creates friction and a sense of submission even when there is no representative of the control regime in sight.

"The Rights Discourse": The Right to the Freedom of Movement, or the Right to Arrive?

One could think, therefore, that the solution is to be found in the human rights discourse, which sets aside the question of actual space and time, and grants every man and woman with a personal and natural right to the freedom of movement. Whatever disturbs that right would then be considered a violation—no matter if it is the checkpoint itself or the way it is run by the soldiers. This is only partly true, first of all, because of its legal background, there always needs to be "proof" of the violation—and again, if you cannot map it or picture it, it does not exist. Second, it turns the Palestinian inhabitants into a moving particle, without emotions, wishes or history. Most of the references to the right to the freedom of movement do not refer to the movement in the space for itself, but mainly to the things that are affected by its absence. Namely, that the violation of the "right to the freedom of movement" is a problem, essentially because it offends other rights: to education, to health, to employment, etc. Following Raja Shehadeh's book *Palestinian Walks*, I would like to emphasize the right to walk per se, and maybe even the right to get lost.

In a previous book, *The Third Way*, Shehadeh (1982) describes his life as a Palestinian *sāmid*,[4] insisting on staying in the prison which is his own house. "Living like this," he writes, "you must constantly resist the twin temptations of either acquiescing in the jailer's plan in numb despair, or becoming crazed by consuming hatred for your jailer and yourself, the prisoner" (Shehadeh 1982, p. viii).

---

4 Literally, a steadfast person, one that sticks to his land.

In the contemporary Occupied Territories, the double temptation had become fourfold. The first is the temptation to give up and leave Palestine; the second is the active or passive cooperation with the occupation regime; the third is the hatred and the violent reaction; and the fourth, which is relatively new, is the temptation of misery, and the reduction of humaneness into a pattern of "humanitarian object." This "humanitarian object" is the subject of most of the human rights organizations' reports nowadays, and increasingly of the Israeli control system itself as well. According to this perception, the Palestinian subject is no more than a statistical detail, without name, dreams or history, whose biological needs must be supplied together with a basic set of "human rights."

The humanitarian discourse suffers from a neoliberal bias towards the rational and the goal-oriented imagined man. Therefore, it reduces the movement in space to practical needs only. That is how, for example, the right to the freedom of movement is conceptualized in B'tselem's report (2007):

> Israel's legal obligation to respect the freedom of movement of residents of the West Bank results first and foremost from the basic duty that international humanitarian law imposes on the military commander to ensure the needs of the civilian population in occupied territory. This obligation is important because every impediment to freedom of movement almost inevitably impairs the ability of the population under occupation to meet other vital needs, by denying access, for example, to medical-treatment facilities, job sites, commercial centers, and educational institutions.

It seems, therefore, that it is less the issue of the right to movement as it is the "the right to *purposeful* movement," or "*the right to arrive*."

More than movement itself, it is *access* that is consecrated by this approach. Following the neoliberal turn, an overriding importance is attached to access and to economic activity in the broad sense of the term, as rational, goal-directed human activity. Efficiency and productivity or output became the watchwords. Space was measured in relation to time (which became exchangeable with other economic activities: "time is money"). The time spent on movement between places is lost time, and therefore distance itself becomes a problem—a bothersome and superfluous factor that should be reduced and eliminated as much as possible.

Using Raja Shehadeh's *Palestinian Walks*, I would like to claim that there is a significant gap between the neoliberal spatial perception—that praises the "flow" and prefers the speed and the arrival more than the road itself and its experiences—and the aspiration to meaningful spaces. Let me explain it using a critic on functional contemporary architecture, by Michel Houellebecq (1997):

> Since contemporary architecture ... reaches its own optimum in the constitution of functional places so that they become invisible, contemporary architecture is a transparent architecture. To enable the rapid movement of people and goods, [contemporary architecture] tends to reduce space to its purely geometrical dimension. To be traversed by an uninterrupted succession of textual, visual and iconic messages, it must ensure maximum readability (only a perfectly transparent place is capable of providing a total conductivity of information) ... so it allows the individual ... to achieve its objective of movement while minimizing friction, uncertainty, and lost time.[5]

5 Translation by Cédric Parizot.

Houellebecq criticizes this type of functional architecture, because when speed and efficiency are the highest objectives, and the possibility of slowness and giving attention to detail is pushed aside, the result is that "They cannot grant an autonomous meaning, to evoke a particular atmosphere; they therefore cannot own any beauty or poetry, or more generally any proper character."[6] He feels that the functionality and the adoration of speed as the highest value reducs his human essence as a subject that is not only rational, economic, anl goal-oriented. It is an architecture that does not respect the person's subjectivity, and reduces him to a changeable unit which is measured only by its value rate.

The spatial language of the NGOs and the rights discourse also prefers functionality to walking per se. It seems that if the use of some kind of a science fiction, "*Star Trek*," launcher—one that can move people from a point to point in no time and no space—was possible, it would have been adopted both by the NGOs and the Israeli control system (as it would avoid friction and make the perfect separation).

It might also be that out of desperation, the launcher would also be adapted by some of the Palestinian inhabitants. It is important to emphasize that in the OPT the worst of all possibilities prevails. It is not the transparent architecture Houellebecq is writing about (as I showed above, it is meant to be inefficient and uncertain)—but at the same time it does not create its opposite, namely a meaningful space that encourages building subjectivity by slowing down, seeing and collecting experiences. Therefore, it is all too transparent (in Houellebecq's terms), and at the same time, all too opaque (in Heidegger's terms).

The Right to Walk as the Right to Space

It seems that the right which Shehadeh wishes to protect in his book is the right to walk. Just walk, without purpose and without justification. A given space only really belongs to someone, where one truly feels at home, in a place where when one can walk without planning in advance and without teleological justifications. "People are not robots, and their lives are not series of programmed actions arranged on a form," wrote petitioners against the "permits regime" in the "Seam Zone," incarcerated between the Green Line and the Separation Barrier.[7] Shehadeh's book is an excellent example for a non-robotized Palestinian, insisting on being a lively human being, living and experiencing life to the full, who is fighting against his flattening into a biological creature defined only by his medical, educational and agricultural needs, according to which the control apparatus' justifications are organized.

In that book, Shehadeh goes on six *sarhat*, the plural of *sarha*: "[a] man going on a sarha wanders aimlessly, not restricted by time and place, going where his spirit takes him" (Shehadeh 2008, p. 2). In the Israeli vocabulary, that option does not exist for a Palestinian for two reasons. The first reason, a material one, is that from the 1990s, and increasingly since the Second Intifada's eruption (2000), the Palestinians are subjected to a permits regime in which nearly every movement demands a permit and a justified purpose in one of the narrow categories listed by the Israeli control array. The second reason is older than the first and has deeper roots. It is that in the Israeli perception, the only spaces which are recognized as "Palestinian" are those already built on or that are cultivated for a certain

6 Translation by Cédric Parizot.

7 HCJ 639/04—The Association for Civil Rights in Israel et al. v. Commander of the IDF Forces in Judea and Samaria.

number of consecutive years. All the rest is conceived as "empty," and, therefore, is prone to be declared as "state lands" (the State of Israel, of course). A detailed description of the manipulative and tricky system of the land grab in the West Bank and the Gaza Strip—issues, by the way, in which Shehadeh is an expert, wearing his other hat as a lawyer and the founder of El-Haq—is beyond the scope of this essay (see, for example, B'Tselem 2002). I will only state that for that system the Palestinians are, in the best case, no more than discrete details with certain property rights—and not full human beings, who have rights and interests in the nature around them: for future development, for a picnic, or just for an aimless walk. The idea is manifested by the Israeli lawyer Michael Sfard, who wrote that

> The idea that Palestinians can use [their] land for recreational purposes, for a picnic, for a housing development, or even as open space is totally alien to the Israeli perception of the Palestinian … We [Jews], on the other hand, know how to make use of land. We know how to … construct a parking lot; [to] create a mall … But them? They don't need all these things. (quoted in Braverman 2008, pp. 459–60)[8]

However, Shehadeh, as aforementioned, does not accept these assumptions, and he resists with counter symbolic gestures. He describes one of his trips, in which

> It was as though the earth was exploding with beauty and color and had thrown from its bosom wonderful gifts without any human intervention. I wanted to cry out in celebration of this splendor. As I shouted "S-A-R-H-A!" … My cry of greeting echoed against one hill then another and another, returning to me fainter and fainter until I felt I had somehow touched the entire landscape. (Shehadeh 2008, p. 9)

The cry of greeting and freedom envelopes the whole landscape in a way that reminds those acquainted with the Israeli occupation's land laws of one of the articles in the 1858 Ottoman land law, by which most of the "state lands" were captured. According to the article, a state land is every uncultivated land, which is far enough away so that "the loudest noise made by a person in the closest place of settlement will not be heard."[9] The cry declares the landscape as owned by Shehadeh, as a human being and as a Palestinian. Nevertheless, in contrast with the Israeli land grab, which is exclusive and exclusionary—namely that this land is Israeli, and therefore, not Palestinian, restricted for building, cultivation or a journey—Shehadeh declares the land as open to walking and joy.

The book opens with two maps. The first, entitled "Ramallah and its surroundings," presents a hand-drawn map, which although having scale and familiar "factual" landmarks such as Palestinian villages and towns as well as Jewish settlements, doesn't really enable navigating or locating oneself. Not only are names of the rivers and fountains written in Arabic (which won't appear on any signpost or official map), but most of the landmarks are related to Shehadeh's own walks and to the intimate geography he creates in the space ("the place where Jonathan and Raja stopped for a conversation," for example). It is a map resembling a personal diary much more than an "objective" map. In the opposite page, is the "North West Bank" map, which is the familiar one, featuring the West Bank and Israel, the Green Line and the Separation Barrier's winding route.

8 Sfard is a lawyer, representing Palestinians in legal disputes with various Israeli agencies.
9 Section 6, Ottoman Land Law, in Planning, Building and Land Laws, p. 427.

In relation to those two maps, I would like to suggest a paraphrase of the book's title. "Palestinian Walks," with the addition of an "A" at the beginning becomes "A Palestinian Walks," namely, a Palestinian person who walks in his occupied lands. The two titles represent two distinct spatial languages. The first title assumes absolute values and a mappable space in which a given territory named "Palestine" exists; the second title focuses on the use values. In the West Bank, the potential use of the space for Israelis and Palestinians differ radically. While the settlers have fast and well-lit roads that connect the settlements to each other and to the Israeli cities west of the Green Line—the Palestinians are left with a slow, narrow, and block-filled road system (see Handel 2013). While the Israeli lands are saturated with history and heritage—the Palestinians have only empty and leftover spaces (in the Israeli perception, of course). While Jewish settlers' trip would be backed by guns and military forces, coupled by an arrogance of "belonging"—a Palestinian walk would be a scary journey beyond the boundaries of safety and certainty. The space's absolute values mark boundary lines and continuous spots on the map, signifying "ours" and "theirs;" A, B and C areas; "Israel" or "Palestine." The use values, on the other hand, ask what the actual meaning of these definitions is. Thus, "Palestinian Walks" assumes the existence of Palestine[10]—the ancient, the contemporary or the upcoming—while "A Palestinian Walks" simply describes the experiences of a Palestinian, trying to walk in the present.

Shehadeh revolts against the official, "objective" maps. He is disappointed when discovering that the paths he walks on are already marked on an old British map, and is deeply irritated when forced to use an updated map in order to find his way at a time he was lost, driving between new settlements and industry zones he didn't know:

> All the signposts pointed to Jewish settlements. I could find none of the features that used to guide me on my way … So I decided to consult a map of the hills. I had to … It was not a practice I would have chosen, for it implied submission to others, the makers of the maps, with their ideological biases. I would much rather have exercised the freedom of going by the map inside my head, signposted by historical memories and references. (Shehadeh 2008, p. 184)

The poetic geography in the head and the legs of a walking Palestinian are therefore the book's theme more than "the struggle for Palestine." Or in other words, the struggle is not only about certain borders and a national territory, but rather on the very right to use the space: to walk in it; and if one wants, also to get lost in it. Stuart Elden notes that there are two common uses to the term territory in geographical discourse. The first sees it as a closed container, under the control of a certain group of people, while the second sees it as a result of territoriality, which is a human behavior and strategy. In other words, the first emphasizes the absolute values, while the second—the use values. Shehadeh, who gives more weight to the space uses than to its official titles, shouts "S-A-R-H-A!" and not "P-A-L-E-S-T-I-N-E!," because he knows that there is no meaning to "Palestine" without being able to make a "sarha" in it. This is the difference between the juristic definitions of territory, and the territoriality which is made through the legs, the eyes and the mind. That is also the difference between the modern neoliberal perception, in which the space is just an indifferent container full of rational agents who only want to arrive, and a social perception

10 The Hebrew edition of the book (Tel Aviv: Am Oved 2009) is titled "Journeys in Palestine," making the abovementioned even more explicit.

which sees the man as more than just a rational particle, but actually full of aspirations, dreams and a rich background.

In a way, the spatial definitions and perception of the humanitarian logic are coupled by the concrete control regime to estrange the Palestinian from his land. It is a *cognitive occupation* and its effect on the occupied is not necessarily smaller than that of the physical.

The actual occupation blocks the Palestinian's way by the separation roads, by the settlements, by the fences, and mainly by fear: fear of the unknown and of running into soldiers or armed settlers. The West Bank is a territory in which everything is derived by ethnic origin. In Ramallah and its surroundings, there is not such a thing as a "walking man," as the walking itself, its possibilities and its experiences are derived straight from the question of ethnicity. When a "fire zone," a "nature reserve," or a settlement's "municipal area" is declared in a place which is already inhibited by Palestinians, those inhabitants turn automatically into "illegal" in their own houses, and their presence on the ground has no value at all.

The *cognitive* occupation is concentrated mainly in the deep estrangement of the Palestinian from the landscapes in which he lives and acts. In the introduction to the Hebrew edition, Shehadeh describes an encounter with an armed settler from Dolev, who tried to delay him and to summon military forces in order to arrest him. When asked by Shehadeh where he lives, the settler answered him arrogantly: "I live here. Not like you, I really live here."[11] And thus, newcomer settlers are creating an imagined Palestinians-free map, taking a huge 2,000-year leap from the Bible to the twentieth century, while the Palestinian is forced to write a travelogue in his own landscapes. What does writing a travelogue on a one's own country mean? What kind of estrangement does it require?

Allegedly, a real inhabitant should not write a book about walking, but simply walk. This privilege is, unfortunately, not given to a walking Palestinian such as Raja Shehadeh. Shehadeh is highly aware of the meaning of travelogues in general, and in Palestine in particular, and hopes he is not part of that tradition, which dresses a given land with a priori religious, historical and political schemes and expectations. However, the hand-drawn map and the human sensitivity to nature and to recent history prove Shehadeh's deep connection to the space. The space that is made his exactly by those walks and personal-political experiences.

That is why Shehadeh's insistence on walking in Palestine can be called resistance. The question, whether a resistance should always be aware of itself and purposeful, occupied many writers (cf. De Certeau 1984; Pile 1997; Giddens 1991; Bayat 2000). Seemingly, nearly any daily practice can be interpreted as resistance: staying in town or leaving it; moving in space or avoiding it; being seen or being unseen; retreating to the private space or going out in public—all these can be thought of as an act of defiance, subversion or protest. According to De Certeau (1984), every shortcut in the city is an act of resistance. He calls the alternative routes taken by pedestrians "everyday resistance," as they deviate from the planned-from-above routes. However, it is quite clear that in most of the cases, this act would not be thought of by the walker as "resistance to the existing order," nor will it challenge the authorities (who in most times would not care if someone took a shortcut through a building's backyard). So the question is what makes an everyday "small" action a resistance. Pile (1997) argues that the answer to this question lies in the action's context: namely, to its location in the dispositive of power and control. In other words, the more the action would be tied into the order and its logic, and act directly against control or

---

11 Tel Aviv: Am Oved (2009), Hebrew edition, p. 14.

surveillance practices, the more it would be "resisting." Therefore, not every shortcut or deviance would be resistance—and the question should be how much the spatial order is defined, organized and takes part in control systems. In the OPT, when the control is in the space and by the means of space management, walking is clearly a political resistance.

Daily movement—in spite of the difficulties, the risks and the explicit restrictions—gives back the meaning to the space and loads it with Palestinian use values. Rema Hammami (2004) writes that

> [T]he old nationalist ideology of *sumud* or steadfastness has re-emerged bearing new meaning. While in the 1970s, its meaning emphasized staying on the land and refusing to leave despite the hardships of the occupation, now it has a much more active connotation. In its new form it is about continuing with daily life and movement; the common refrain: "*al haya lazim yistamir*"— "life must go on." Thus, *sumud* has become about resisting immobility, the locking down of one's community, and refusing the impossibility of reaching one's school or job. (italics in original)

Shehadeh reconstructs his personal and political subjectivity through the walking. For him, the question of the destination or the narrow definitions of "purpose" (namely, the system's demand to define a priori one's reason to walk) is not important. He refuses to accept all the existing spatial languages: the "objective" maps, the Israeli logic of space management, and the reduction of movement as suggested by the legal language of rights—which directly reduces him into a neoliberal agent or a humanitarian case. His territory is made and not given (by the UN or by Israel)—and this territoriality is made through subjectivity. That is why the language of the space cannot be separated from the language of the subject. This is geography of resistance that is the mirror image of the geography of occupation—actual and cognitive.

The Lingual Gap and the Future Solutions

As was said in the beginning of this chapter, different languages also entail different definitions of the problem and its desired solution. I would like to finish by referring to one suggested solution, developed by RAND Corporation under the title "The Arc." The purpose would be to expose its basic assumptions, and to show why, although it has good intentions, it is still problematic. In other words, I would argue that The Arc suggests solutions to the problems as seen by both languages of the rational planners and the rights discourse—but it is still disconnected from that presented by Shehadeh in *Palestinian Walks*. Therefore, The Arc is unfortunately another case in which others plan for the Palestinians without being aware of their language and spatial perception in the old Orientalist approach, in which "they cannot represent themselves, they must be represented."[12]

The Arc is a utopian master-plan to the future Palestinian State, created by LA-based architect Doug Suisman for the RAND Corporation. The plan, called The Arc due to its shape, suggests a fast train that will move along the West Bank ridges from Jenin in the north to Hebron in the south. Then it will rush through the Negev (crossing Israeli territory) to the Gaza Strip. Suisman offered to build along the rail track a water canal, fiber-optic

12 The sentence appeared in Karl Marx's *The Eighteenth Brumaire of Louis Bonaparte*, and was quoted as an epigraph to Edward Said's *Orientalism*.

cables, electricity lines, a toll road, and green parks. Instead of a slow and disintegrated Palestine, the plan suggests a Palestine on the move, that all of its parts, including the Gaza Strip, are connected. The whole journey, from Jenin to the last stop at the Gazan airport, should take no more than 90 minutes. It should also be mentioned that the plan is highly ecological, keeping green open spaces and planning sustainable building that enables building for large numbers of refugees while saving the environment, and gives special attention to public transport instead to roads, private vehicles, and more.

The authors write that the conception of the plan is of an archipelago of medium-sized cities, connected by The Arc's infrastructure. The Arc's stations would be located 8 to 25 kilometers from the urban centers, and local bus lines would connect the cities to the main line. Two million returning refugees would be settled along the narrow corridors between the existing cities and the Arc. In the end, Palestine would look like this:

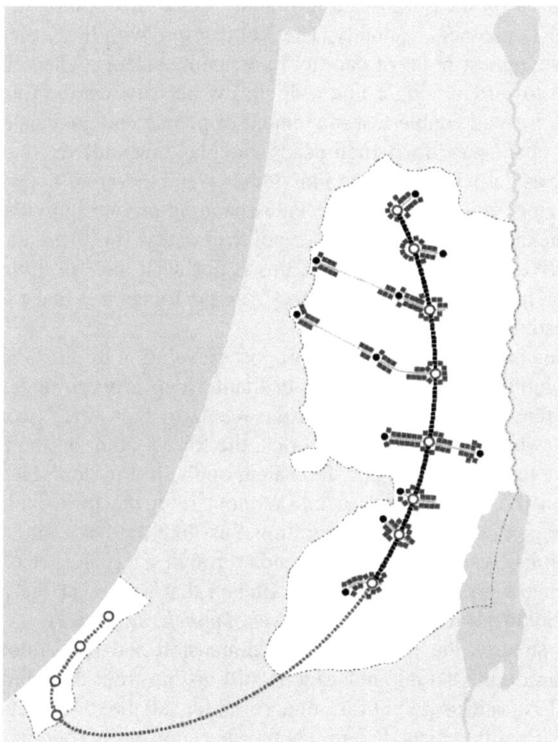

Figure 4.1 The Arc Project. Rand Corporation, 2005

Although it has clear advantages, I would like to point to some problematic issues, even beyond the question of its feasibility. Doug Suisman openly admits that he did not have any former acquaintance with Palestinian space. "His sense for Palestinians' nostalgia, for their attachment to the land, even for what their cities actually looked like—that would come much later" (Bennet 2005).

The Arc treats the state of Palestine as a line connecting six cities, in which all their inhabitants have free professions and enough money for frequent travel. It has no agriculture or traditional professions. It has no importance to politics in its wider sense. It is part of the neoliberal discourse that neglects local cultural and historical features in the name of "progress." The American planners and the Palestinians are talking in two different languages. Suisman, who is used to focusing on the flow of goods and people, "was surprised by the emotion with which Palestinians talked about flowers and olive trees" (Ibid.). Putting it explicitly, it seems that The Arc is trying to make Palestine into a Jewish settlement—namely, a detached non-place, with no actual connection to the ground, that is based only on the speed of movement. It answers the problems as defined by planners (by creating a sustainable, rational, and highly-connected system) as well as by the rights discourse (because it gives Palestine the ultimate "freedom of movement")—but it does not even think about "flowers and olive trees."

"We have chosen for the purposes of this study to set the question of Israeli settlement aside," write RAND planners. Actually, The Arc can live with the settlements' array as it is. It would be the highest point of parallel movement systems, when all the Palestinians will be canalized towards a single line (which, by the way, can be totally stopped with one single switch). It will enable fast movement of people and goods along defined lines, and leave all the open space for Israeli use.[13] The plan does not see the Palestinians as a society but rather as "efficient" private individuals—and therefore fits with the Israeli old aspirations of giving "autonomy" to the Palestinian as private individuals, but not as a society, without acknowledging any kind of political entity. It is clear, though, that at least from the perspective of *Palestinian Walks*, this is not what the Palestinians wish for their space and future. The "flowers and olive trees" are the things that make the territory—and the people—Palestinian.

The Palestinian artist Larissa Sansour proposes a solution for the "Palestine Problem." The proposed solution, as presented in a brilliant video artwork under the title *Nation Estate*, is a tall, fenced building in which the whole of Palestine is concentrated. From floor number −3 (where we find the Dead Sea, the lowest point on the face of the Earth) upward, the entire territory is laid out: Jerusalem on the fourth floor, Hebron on the tenth, an olive grove on the twelfth, and so on. Without referring to it explicitly, plans such as The Arc undergo a conceptual exaggeration. Just like the fast train, the rapid elevator transports Palestinian residents within seconds through every corner of their homeland. In the same manner time is nullified, in a kitchen that serves, at the push of a button, traditional Palestinian meals such as *mlukhiyeh*, *kubbeh*, and *tabuleh*. The ostentatiously futuristic design stresses the rapid temporal dimension and the smooth transportation. But not all is smooth: the actual landscape is still visible from the windows (the Temple Mount and the Al-Aqsa Mosque) but is not accessible; and the roots of the olive tree break through the floor. Reality resists. It reminds the spectator of the roughness, the delay, and the interferences in time and space that are an integral part of what it means to be human. As in Houellebecq's text, the exaggerated smoothness ironically produces remoteness. The building, where everything is close and accessible, leads to a boredom that is evident in the artist's face. She can get to every point in Palestine, but something is missing: the olive groves and the flowers, the food and the smell, and above all the ability to walk without a pre-given, rational, and efficient goal.

---

13 The map is amazingly similar to some of the right-wing Israeli plans for a future solution, which also give the Palestinians no more than a few corridors of the areas already built.

Figure 4.2 ***Nation Estate*—Poster. Paper print, 100 × 150cm. Larissa Sansour, 2012**

Figure 4.3 *Nation Estate*—**Olive Tree. C-print, 75 × 150cm. Larissa Sansour, 2012**

Figure 4.4 *Nation Estate*—**Main Lobby. C-print, 75 × 150cm. Larissa Sansour, 2012**

Conclusion

To sum up, there are several geographies of occupation. Although all three spatial languages presented here in detail (not including the Israeli system's perception) are critical of the occupation, not all of them share the same definitions of the problem and of the desired solutions. When the liberal language and the NGOs talk only about the importance of arrival and of "flow" as an abstract concept—or alternatively only of the legal problems of sovereignty—a real gap is created. The suggested solutions—whether it is the two-state, the one-state, or any other creative idea—are also victims of the different languages and expectations. That is why we must first ask: what are we talking about when we talk about "geographies of occupation"?

PART II
The Economy of Separation

Chapter 5
Porosity, Fragmentation, and Ignorance: Insights from a Study of Freight Traffic

Yaakov Garb

Introduction

In 2008, I was commissioned to conduct a study of all freight trucks between the southern West Bank and Israel, with a special emphasis on the newly constructed Tarqumiya crossing, which was intended to centralize all of this flow. In this chapter, I reflect on that study, and what it tells us about the functioning of borders under Israel's separation policy, and borders more generally.

In the years prior to my study Israel had begun directing the flow of freight between the West Bank and Israel to a limited number of designated and regularized passage points, and had recently built and begun to operate special back-to-back facilities in these locations, such as Tarqumiya. Trucks with freight originating in or destined for Palestinian locations would be required to use these facilities, at which they would not pass between Israeli and the Palestinian areas, but stop and transfer their goods through inspection facilities to counterpart trucks on the other side. Freight from or to "Israeli" sites in the Territories (settlements, army bases, or businesses) carried on yellow (Israeli) plated vehicles would go directly through a different lane adjacent to the back-to-back facility, with only a cursory check of the required documentation.

In November of 2007, the Israeli Civil Administration began a trial run of the Tarqumiya back-to-back facility, which was to be opened the following month as the only legitimate passage point for Palestinian freight to and from the southern West Bank. Against this background, I was asked to provide hard information on how many freight vehicles moved between Israel and the southern West Bank through all crossings (the overall flow) as well as patterns of flow through Tarqumiya itself. An important basic question on the table was: will the Tarqumiya back-to-back facility be able to serve freight to and from the Palestinian areas once the redirection of flow from other crossings to the official back-to-back crossing is complete? The Israeli authorities assumed most freight was already passing through Tarqumiya and that the back-to-back facility could handle any additional flows once the new freight regime was fully operational. Palestinian stakeholders were convinced otherwise. However, simply put, no one knew.

My report was commissioned by the Economic Cooperation Foundation (ECF), under contract to the US Agency for International Development (USAID), as part of their initiative to facilitate Palestinian trade. A range of stakeholders were interested in this question, ranging from the Israeli Ministry of Defense, on the one hand, on the other the Palestinian Shippers' Council and the mayor of Hebron (the economic engine of the southern West Bank). International agencies, such as various branches of the UN, were also supportive and keen to get the results. In this project, the ECF pieced together a delicate coalition of stakeholders which, despite their differing stances and social locations, all supported a survey to obtain this information. In parallel, our survey team pieced together the fine-grained logistics and understandings necessary to conduct the field survey without danger, discomfort, or disruption. This was important since our project, described below, entailed surveyors photographing and counting trucks, often around the clock, near all the guarded passage points Israel had erected on major routes between Israel and the southern West Bank as well as along internal corridors within the West Bank. The Israeli Defense Force and border personnel needed reassurance that we were not a security hazard, while Palestinians needed reassurance that we were not associated with Israeli security forces or compromising Palestinian interests. I came to this project after several years of work on the politics of mobility and with background in investigating traffic movement. My entry was through the technical skills needed for a good survey of movement, but I was also interested in the opportunity the project might offer to deepen my understanding of the surrounding issues related to mobility, its management, and borders. And, indeed, a series of unplanned encounters at the margins of the "official" study, and, in particular, the gap between my findings and the information available to the Israeli border administration, proved quite interesting. In this chapter I present both my findings about freight mobility and then reflections sparked by these broader contextual observations. In particular, I describe the relation or lack of relation between my findings on freight flows on the one hand, and the knowledge of these flows held by the bureaucracies that ostensibly monitored and managed them.

From the onset of the project, as I began to review what was already known, I was struck by the knowledge gaps regarding truck movement. Though the main contours for managing Palestinian freight had been in place for over a decade prior to the start of my study in early 2008, not much was known about the volumes and patterns of this travel. Israeli and Palestinian actors were willing, even eager, to give me any background materials they had, but there was not that much available. Of course, the emergence and effects of a regime of mobility management had been well documented for the movement of people, at least in qualitative terms (Hass 2002), and some fairly elaborate quantitative surveys (more or less equivalent to studies of workers) had been conducted (Portugali 1993; Farsakh 2002). Similarly, the movement of goods of different kinds (Van Arkadie 1977; Arnon 1997) and the partly corresponding transfers of capital (balance of trade) are, of course, documented by standard statistics, which have been collected on an ongoing basis. But data about how these goods moved (i.e., data on freight vehicles), seemed mostly lacking. This may be due to the fact that the constraints on goods movement were more partial and appeared later than those regarding people. Also, in most cases the management of freight movement took the form of redirection and impediment, rather than an outright ban. Thus, if buying and selling still took place and the goods ultimately arrived, the how and where of truck travel may have seemed less critical to those gathering statistical information. Whatever the reasons, basic information on truck travel was not available in any systematic way that I could discover. A freight survey was rumored to have been conducted by the IDF or Ministry of Defense in the early 2000s, but to have been lost, so that the data was not available to planners in these organizations or to me.

This absence of basic information on truck flows was striking given the large cost and consequence of the crossings infrastructures and arrangements being readied for freight in the years preceding my 2008 report. The location, nature, and capacities of these crossings would shape travel times, distances, and frequencies of the trucks that—quite literally—carried the Palestinian economy on their backs. Trade flows constitute 85 percent of the Palestinian GDP, with an estimated 85–90 percent of this trade with Israel, which is immediately adjacent, politically dominant, with a population twice as large, and an average per capita income five times higher (World Bank 2009). Almost all of this involves freight vehicles moving to and from Israel, and even Palestinian trade with the world at large has to cross by truck through Israel to the Israeli ports of Ashdod and Haifa. Yet I was told by informed individuals that the decisions about these crossings were made without reference to data about freight volumes or typical origin/destinations. In particular, despite the massive costs of its planning, construction, equipping and operation (a single X-ray scanner for trucks costs over $10 million), the construction of the Tarqumiya facility seems to have relied on a shaky estimate of how many trucks would pass through it. Before beginning the survey I spoke with people in the Passages Authority and was reassured that the Tarqumiya crossing, which had been operating for some months, was handling 300 trucks a day in an orderly way, so that as the five other crossings became increasingly closed to freight movement, the greater flow (as high as 600 trucks a day if needed) could readily be managed by the facility through moving to double shifts. The estimate was based, it seems, on an assumption that Tarqumiya was already carrying the bulk of all freight in and out of the southern West bank, and that the vast majority of Tarqumiya traffic was already going through the back-to-back. As I will discuss below, these assumptions were wrong.

In addition to the question of flows across the Tarqumiya crossing was the question of how trucks were traveling within the Palestinian areas to and from this crossing. The Israeli authorities assumed that most of this traffic approached Tarqumiya on the main highway, Route 35. Yet people in the area knew that villages off the highway close to the Tarqumiya crossing, such as Idhna and the village of Tarqumiya, had heavy and disruptive freight flows through them (see Figure 5.1). As described below, my study showed that, here too, actual flows were quite different than imagined by the Israeli authorities, with very few trucks traveling, as expected, along the full length of Route 35. This seems to have been due in large part to the fact that many of the trucks allowed to bypass the back-to-back facility—that is, those officially traveling between Israel and Israeli locations within the West Bank—had loads, drivers, and trajectories of, let us call it, a somewhat less official nature, which allowed or required them to reach the crossing through various back-road alternatives to Route 35.

Of course Israel has the capacity, in some overall potential sense, to stop and inspect every vehicle passing near Tarqumiya, and, probably, to track these in real time to source or destination. Some Palestinian collaborators could surely have described what most of the local villagers knew—the extent to which trucks were continuing to use non-Tarqumiya crossings or the direct Tarqumiya lane rather than the formal back-to-back facility, and explain the less-than-formal aspects of the trans-border economy, which, in part, shaped these choices. Indeed, it would not take any fancy or clandestine measures to know that most trucks were not using the Route 35 approach to Tarqumiya. The inhabitants of the villages off the main road found the large volume of trucks going daily through the small winding streets in these to be an overwhelming nuisance, and one would need to be blind not to notice the preponderance of trucks with yellow (Israeli) license plates in these villages and in downtown Hebron. In fact, if soldiers or officials at the Tarqumiya checkpoint and crossing as well as the soldiers in the constantly manned armored pillbox literally overlooking the turnoffs from the main section of Route 35 were to count trucks, simple subtraction would have been enough to shake up the authorities' understanding of what was going where.

Yet as far as I know, the Ministry of Defense did not assemble the kind of basic reliable information on truck flows produced by our small and moderately mobile team in a few days of focused survey. In fact, our results underscored that the authorities were operating on seriously wrong assumptions about the volumes and patterns of these flows. How could basic features of the massive and obvious traffic flows escape the knowledge of the apparently powerful and technically sophisticated state institutions planning for and administering them? This chapter is an attempt to explain this blindness—my first attempt at a sociology of ignorance.

I will suggest two non-exclusive avenues for interpreting the apparent gap between state knowledge and "shadow" practices. One is to loosen certain prevalent conceptions about Israeli authorities' capacities for surveillance, understanding and control of movement (as well as their own freedom to move in the Palestinian space), and, indeed, assumptions about a unitary integrated Israeli state power. A second and possibly related avenue for interpreting these apparent lapses in knowledge/oversight regarding informal passages through this formal checkpoint is to loosen the formal/informal distinction and suggest a greater coexistence (complicity) between them than is usually conceived. In other words, the formal apparatus (the "state") is both less powerful over and more entangled with the informal under-life that evades its notice, forgoing or even preferring to avoid certain kinds of readily available knowledge. Some frameworks for these lines of thought are given briefly below.

Many academic and activist descriptions of the nature and operation of borders and separation in the Israeli-Palestinian space seem to locate power within a monolithic and integrated Israeli state/army apparatus, possessing exceptional abilities to move freely around Palestinian space and to know and control Palestinian movement (also monolithically conceived) on the other. Separation lines and crossing points are seen as places where Israeli surveillance, control, and spatial demarcation are particularly heightened. Weizman's (2007) work on the architecture of occupation and the politics of verticality; many of the contributions to Sorkin's anthologies on Jerusalem (2002) or the separation barrier (2005); Zureik's (2001) writing on surveillance, and on border control in particular; Parsons and Salter's (2008) description of Israeli biopolitics and mobility regulation; Gordon's (2008) book on the mechanisms of occupation; and the work of many others, together, give a broad and often sophisticated account of the mechanisms and infrastructures of Israeli control. But, at the same time, they leave their readers with the cumulative impression that this control originates from a unitary and increasingly omniscient and omnipotent Israeli center. And, correspondingly, that what we see in border practices on the ground are either expressions of this control, or negligible mishaps in or resistances to it.

Alternative conceptions are few and far between, and with the exception of Parizot's (2009b, 2009c, 2010) more multilayered accounts of border practices, these have scarcely been brought to bear on the actual and concrete functioning of borders and mobility. Starting points for this might be, for example, Fischer's (2006) more poly-vocal accounts of the conflict and its landscapes, Hammami and Tamari's (2008) recognition that "while the complex web of [Israel's] control strategies works to achieve territorial ambitions, they do not form a seamless whole, in step with each other," and work that dissects cleavages and heterogeneity within the military subsystem and its complex relations to civic subsystems (Levy 1998, 2003, 2006, 2007a; Peri 2006; Barak and Shefer 2007; Michael 2007).

Let us move, now, from loosening a mistaken conception of the unity and near omniscience of Israeli control, to a second conceptual loosening: of the boundary between the formal and informal spheres. The flows through and around the formal arrangements of this and other border crossings were on the whole part of economic activities (the transport of workers and goods such as used furniture, scrap metal, quarried stone, or heating fuel) only partially in the formal realm, that is, part of registered transactions conducted by registered entities. While the informal sector (also termed "underground, " "irregular," or "shadow economy") has historically been considered almost by definition as being marginal and outside the domain of institutionalized state regulation, these views have changed. In recent years (Maloney 2004; Portes, Castells and Benton 1989; Williams 2007;), this perception has been challenged by less dualistic conceptions and studies that demonstrate the scale of these activities (60 percent of the workforce and almost 40 percent of GDP in developing countries; 17 percent and 14 percent respectively in OECD countries) (Schneider and Enste 2000), the cross-cutting social location and motivations of the actors and the impressive entrepreneurship they exhibit, as well as the tight, complex, even fundamental coupling of informal and formal sectors.

The latter dependence of formal and informal is particularly interesting to follow in the Israeli-Palestinian case. Noting Roy's (2009, p. 84) conjecture in her study of the Calcutta metropolitan region that "informality exists at the very heart of the state and is an integral part of the territorial practices of state power." or Elyachar's (2005) description of the deep interpenetration of informal state, and international organization practices in Cairo, it seems to me that the relation of Israel to the occupied Palestinian territories and the creeping (some would say galloping) de facto absorption of these over the last several decades is not only accompanied by but, in some ways, made possible by a play back and forth between informal and formal. The shifting and unsettled semi-formality of the legal, economic, and institutional structures and arrangements affords Israel tremendous flexibility to have domains and practices that are simultaneously off-record when needed, and on-record in other circumstances. The informal is, thus, a flexible holding pattern or reserve, allowing desired but unallowable things to coexist until such time when they can be made otherwise.

My own field experiences, and work in conducting this freight survey made clearer to me the need for more such nuanced accounts of control and informality. In particular, I was surprised at how patchy and fragmented Israeli power and knowledge sometimes were, at the differential (and sometimes extremely limited) abilities of different Israelis to move through Palestinian space, and at the degree to which motivations and knowledge of different "Israeli" actors did not necessarily align neatly. Officers and non-officers were differently positioned, as were enlisted army officers versus reserve duty soldiers, the army versus the Crossings Directorate or the police, and all of these were differently positioned than settlers. Palestinians, too, have differential stakes, abilities, and knowledges regarding borders and movement: East Jerusalem Palestinian truck drivers versus the Palestinian truck owners, traders versus Palestinian tax authorities, etc.

Thus, movement within "Palestinian" space and between it and Israel seemed organized by multiple and patchy fields of force. Indeed, Israeli and Palestinian lives, spaces, and economies were sometimes seen to be intertwined to an extent that blurs categories, rules and borders, and frustrates unilateral or comprehensive control. For sure, Israeli institutions wield a massive amount of far-reaching power. But the nature and operation of this power (and, in particular its disruptions and blockages) are better understood through breaking open the unitary "Israeli" category, and considering the operation of multiple players in an increasingly internationalized terrain. When it comes to cross-border practices, especially informal ones, the state is not the only—nor even the most useful—unit of analysis. I will return to these claims in a concluding section, after presenting the contexts of freight control and a more detailed account of our study's methods and findings.

The Emergence of a West Bank Freight Passage Control System

Labor and goods vehicles moved relatively freely in both direction between 1967 and 1987, a period during which Israel unilaterally and incompletely integrated Palestinian areas into Israel's economy in 1967. Yet through the implementation of its separation policy, at the beginning of the 1990s, and especially since 2000, labor flows have been drastically reduced and increasingly formalized, and there has been an effort to channel goods flows through a few formal and technologically monitored crossing points. While still unilateral in many ways, the operation of these crossing points now takes place in the crowded arena of observation and intervention by governmental and other bodies from outside the region, as the international community has become increasingly critical of the Israeli occupation and its constraints on Palestinian mobility in particular. Here I describe the emergence and functioning of the freight crossings Israel operates, and some of the stakeholders looking over its shoulders as it does so.

The Hardening of Freight Control 2000–2010

After the occupation of the Gaza Strip and the West Bank in 1967, the border between Israel and these areas was eliminated. From this date until the early nineties, during a period of "imposed incomplete integration" of the two economies (Cobham 2001; Arnon and Weinblatt 2001), the movement of freight (and labor) were unrestricted—indeed encouraged—to a degree that is difficult for the current generation of Israelis to imagine. A gradual repeal of these freedoms of movement began in the late eighties, and intensified in the mid-nineties, as rare closures became periodic and, eventually, permanent. Yet, until the Second Intifada (September 2000), goods still moved relatively freely between the West Bank and Israel, utilizing a variety of formal and informal locations for crossing, and with little if any inspection.

With the outbreak of the Second Intifada, however, all forms of movement within the West bank and, also, between the West Bank and Israel were much more forcefully hampered by an array of obstacles that prevented movement along some routes and fixed and "flying" check points that inspected and filtered flows on others. The construction of the separation barrier (2002) along and beyond the Green Line became the centerpiece of this new mobility management regime. As part of this, and especially from 2003 onwards, a more explicit policy emerged for centralizing people and freight movement to and from Israel to a few formal crossings. In July 2005, a Passages Authority was established within the Israeli Ministry of Defense, and several government "non-papers" describing border arrangements were released. One of these, entitled "Israeli Assistance Steps and Humanitarian Measures toward the Palestinians," declared that:

> Israeli security forces will transfer the bulk of their monitoring and control efforts from checkpoints inside the West Bank and the Gaza Strip to crossing points along the revised route of the security fence. This will mean a sharp reduction in the number of roadblocks and barriers within the Gaza Strip and the West Bank, alongside the construction of new terminals and crossing points between Palestinian-controlled areas and Israel. (Israeli Ministry of Foreign Affairs, May 2005)

The policy that emerged limits freight to five designated crossings (four existing and one planned), designated in Map 5.1. These were characterized by a greater reliance on civilian

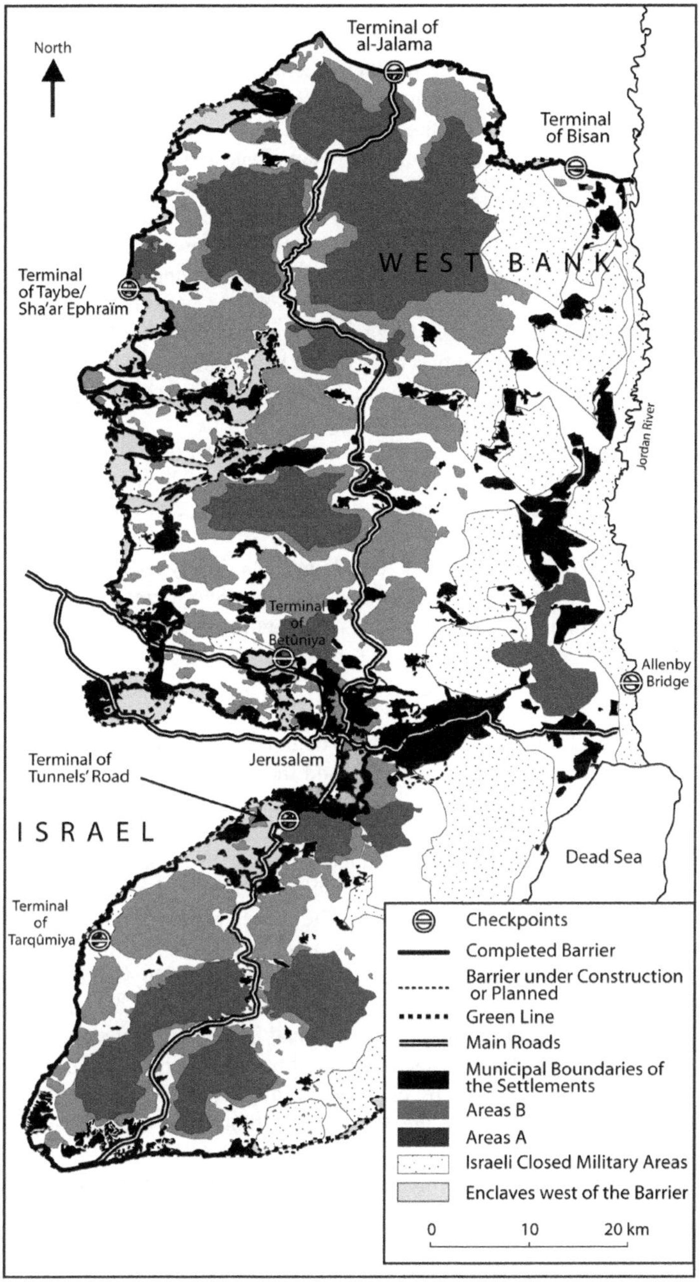

Map 5.1 Main crossings for the treatment of goods between Israel, the West Bank, and Jordan

Map realized by M.Barazani (CRFJ) and M. Coulon (LAMES).

Source: OCHA Information Management Unit; Map produced June 2009; Data Base and Statistics: OCHA, PA, MoP

staffing and technological measures for their operation (Havkin, this volume). Three of the existing crossings serve the northern West Bank: Al Jalame (northeast of the Jenin area), Al-Taybeh/Sha'ar Efraim (Tulkarm/Qalqiliya area), which is a main passage for the northern West Bank, and Beituniya, close to Ramallah, and serving the East Jerusalem area. In addition, the Bisan Terminal is intended exclusively for export of agricultural produce grown in the Jordan Valley. For the southern West Bank, Tarqumiya (Hebron area) is currently the dominant terminal (other terminals allow passage of certain kinds of load, such as quarrying materials). Tarqumiya is of critical importance to the economy as the closest to the Port of Ashdod, used for overseas goods, as well as being an important link in travel from the West Bank to Gaza -playing a role in some of the "safe passage" routes discussed for linking these two parts of the Palestinian Authority. It is also the site of a planned but stalled "Tarqumiya Industrial Estate." Planned, but not yet operational, is Mazmuriya, in the southern Jerusalem area.

The northern–southern West Bank distinction is a critical one, since freight movement between these is officially limited to the Wadi El Nar road. This makes the passage from Bethlehem to Ramallah areas through an extensive detour eastward of East Jerusalem, on a steep contorted road through the desert (Garb 2010). The difficult passage can be unpredictably delayed if the "Container" (see Map 5.2) military checkpoint midway delays trucks. Thus, southern West Bank trucks often prefer to use a southern West Bank crossing (Tarqumiya) even if their origin/destination is in the northern part of Israel.

Tarqumiya Crossing

The operation of the Tarqumiya crossing is through two alternative routes: one a back-to-back transfer station for Palestinian freight, and the other a direct crossing for "Israeli" vehicles. The former is mandated for freight originating in or destined for Palestinian locations. In this back-to-back arrangement, trucks themselves do not cross, only their goods. For example, a truck with a load from the southern West Bank would approach the Tarqumiya crossing, turning left toward a waiting area near the back-to-back terminal, some 150 meters from the direct crossing. The driver registers with the terminal, and when the counterpart truck is ready on the Israeli side, the Palestinian truck is called to enter the gated crossing compound. The driver enters the facility, and leaving his truck doors and engine cover open undergoes a physical security check. Then the truck passes through the X-ray scanner and, sometimes, a manual inspection of the goods as well, after which the goods are loaded onto a second, Israeli truck, which continues with these into Israel. The crossings are operated by civilian guarding companies under the supervision of security officers employed by the Passages Authority in the Ministry of Defense (Havkin, this volume).

Freight from or to Israeli settlements or army bases or businesses carried by yellow-plated (Israeli) vehicles is exempt from this back-to-back arrangement, and continues directly along the main road for brief inspection and continued travel into Israel. In a procedure usually taking a matter of minutes, shipping papers are examined to ascertain if the truck is, indeed, intended for or coming from "Israeli" locations. Detailed data on the number of trucks going through the back-to-back passage or rebuffed from the direct crossing were collected by the Passages Authority, but regarded as classified, and thus not available to me. I was, however, able to count trucks approaching and departing from both crossings through the methods described below.

The crossings and Tarqumiya in particular, can, on first appearances, seem to be international border crossings, but, in fact, perform complex filtering functions within Palestinian-Israeli spaces. Most are not located on the Green Line, but push the limits of Israeli geographical extent somewhat further into the West Bank. They are located unilaterally (i.e., without the consent of the Palestinians) and operated by the Israeli army or private companies controlled within C areas, where Palestinian police and customs police do not have any access. This undermines Palestinian officials' ability to monitor or control the goods passing through them (Natsheh and Parizot, this volume). For "Israeli" vehicles, Tarqumiya constitutes an internal inspection along a highway, whereas for Palestinian vehicles it is a dead end at which they transfer goods to or from another vehicle on the Israeli side.

On closer inspection, Israel's sovereignty at these crossings is actually made up of several actors: not just the Israel Defense Forces, but the Crossings Directorate (sometimes called the Passages Authority), established within the Ministry of Defense in 2005 to guide the demilitarization and upgrading of the border crossings, the private security companies to which portions of the screening and guarding work has been outsourced, as well as the police or Border Guard in the Jerusalem "envelope" crossings, and the Airports Authority in the Gaza crossing. These bump up against other actors that indirectly shape the crossings regime. This is because Palestinian mobility and its relationship to both human rights and economic viability have been a growing concern for various Israeli, Palestinian, and international actors from the mid-nineties until the present. The question of freight has come into sharper focus as the new crossings regime took shape. Thus while Israel makes unilateral decisions regarding the functioning of border flows, it does so in the context of considerable scrutiny and oversight from a range of stakeholders, which can exert diplomatic and normative pressures.

The Multiplicity of Stakeholders

The issue of freight mobility has been studied and advocated by international organizations including the World Bank, the European Commission, the Quartet, and the UN—especially UNOCHA,[1] UNWRA,[2] UNCTAD,[3] and the World Food Program (WFP). It has also been the focus of studies and advocacy by governments, notably the US through USAID, the Japanese Government, and the Norwegian Ministry of Foreign Affairs. The Turkish government, as well, has seen the "Tarqumiya Industrial Park" adjacent to the crossing as a commercial opportunity as well as a way of extending its "soft power" in the region (Altunisik 2008). Two Israeli non-governmental organizations involved in informal (Track II) diplomatic efforts and the economic dimensions of peacemaking, the Peres Center for Peace and the Economic Cooperation Foundation, also devoted considerable efforts to the topic. Palestinian organizations, such as PalTrade (a national non-profit trade promotion center and business membership organization, established in 1998) and the Palestinian Shippers' Council (a national organization established in 2006 jointly by PalTrade and the United Nations Conference on Trade and Development to represent the import/export needs of Palestinian businesses), have also done research and lobbying on this issue.

1 United Nations Office of Coordination of Humanitarian Activities.
2 United Nations Welfare Relief Organization.
3 United Nations Conference on Trade and Development.

The research I was commissioned to do arose out of these concerns and interactions. In November of 2007, Israel's Civil Administration (established by the Government of Israel in 1981 to run all non-military activities in the Occupied Territories) began a trial run of the facility that was to be opened the following month. As these new freight crossing arrangements became clearer, they met with protest and calls for improvement by several of the Palestinian stakeholders and international bodies described above concerned with facilitating Palestinian trade and economy. Complaints related to the time taken to cross were raised. While the Crossing Points Authority committed to a 45 minutes average crossing time, there were complaints that overall passage time was longer than this. Others concerns were the limited and constrained working hours (sometimes requiring a truck coming from a port to wait overnight), uncertainty regarding operations, damage to goods in the loading/unloading process, spoilage of goods, and a paletization requirement, which can cause inefficient packing or be impossible for some kinds of goods. Also, the location of a limited number of crossings was constraining, especially in the southern West Bank. Here, the single freight crossing can mean a considerable detour for shipments originating far from this westerly point. A northbound shipment from the northern edge of the Southern West Bank, close to Jerusalem, for example, or a southbound one in the southern villages, must travel unnecessarily south or north, respectively, to cross Tarqumiya, before continuing.

The discussions about improvements of freight shipping focused on the capacities of the crossings, security arrangements, and, especially, scanning technologies, crossing opening hours, and ensuring redundancy of crossings and lanes (in event of closure). The broader improvements suggested by Palestinian businesses and their advocates include a movement from specifying crossing capacities in terms of trucks per day to specification of time per truck; adapting the crossing processes to better meet commercial needs; a push for more selective risk assessment that would ensure swift passage for low risk vehicles and "known traders;" a transition from a back-to-back arrangement to a cab-swapping arrangement, or, even, drive-through by certified drivers and companies; and an attempt to develop alternative routes for international export/import through Jordan and Egypt.

Assessing the Amount and Movement of Freight from and within the West Bank

Despite the heavy investment of the Israeli authorities and the sophistication of the technology deployed at the points of passage, as well as the level of concern among many stakeholders regarding the hardening of the "border" between Israel and the West Bank, there was little knowledge about the amount and movement of freight, nor of how much might be expected once all flows were centralized to Tarqumiya, as declared. To fill this gap, and also to provide measurement of waiting times, my report was commissioned by the US State Department (USAID), through the mediation of the Economic Cooperation Foundation (ECF).[4]

4 Many of the institutions mentioned above, ranging from the mayor of Hebron and the Palestinian Shippers' Council on the one hand, and the Ministry of Defense's Passages Authority on the other, as well as international agencies such as UNOCHA (United Nations Office for the Coordination of Humanitarian Activities) were eager for the report's findings, and the results were, indeed, picked up and utilized by several of these, helping to shift Israeli policies and practices. Thus, the report itself is an example of the complex polycentric arena in which "Israeli" territorial limits are managed.

The Survey Structure

The survey (Garb 2008) had two parts. The first was a 24-hour cordon survey conducted in March 2008 of all freight vehicles into and from the southern West Bank at all six passage points marked in circles on Map 5.2 (as well as Halhul-Seir Junction). Counts were collected from all teams on a continuous half-hourly basis. In addition, time stamped photographs were taken (over 2,000 photos for a nine-hour sampling period), which allowed a fine-grained and precise time series to be reconstructed at will for any location and period. Since Palestinian trucks using the back-to-back facility did not travel past the crossing, but only transferred their loads, we also measured the time it took between arriving at the crossing, and the time they left.

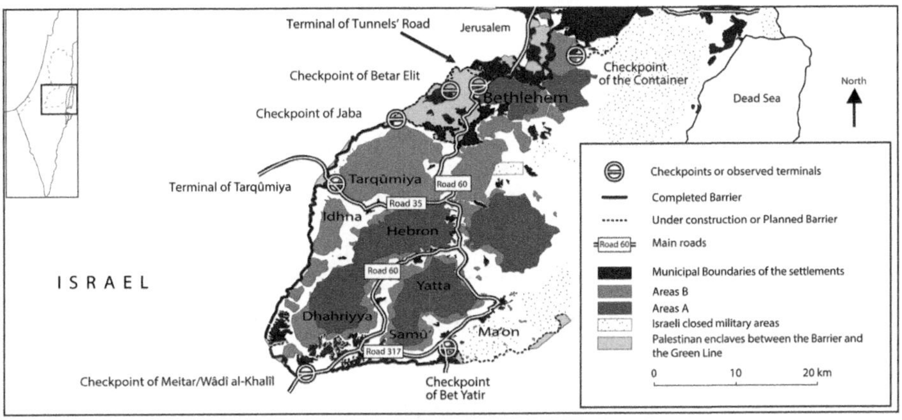

Map 5.2 Major terminals and checkpoints between the southern West Bank and Israel

Map realized by M.Barazani (CRFJ) and M. Coulon (LAMES).
Source: OCHA Information Management Unit; Map produced June 2009; Data Base and Statistics: OCHA, PA, MoP

The second part of the survey was a day-long (February 26, 2008), fine-grained survey of travel on the Route 35 highway, which was conceived as being the main approach to Tarqumiya. Surveyors were stationed at the four points (marked with numbered bubbles in Figure 5.4 along this single east-west road leading from Tarqumiya to the junction with route 60. Importantly, the Tarqumiya observation point allowed us to distinguish between the trucks that crossed Tarqumiya directly or via the back-to-back facility. Other points were chosen so as to allow us to identify the volume of the turnoffs to Tarqumiya village, Idhna, and onto Route 60.

This part of the survey aimed to quantify the turnings to and from these important distal points. The survey made clear, as discussed below, the extent to which Route 35 is a "leaky" corridor, with significant sources and sinks *between* these major interchanges. The route used provides important clues on the true (as opposed to declared) origins/destinations within the Palestinian areas of the trucks going directly through the Tarqumiya crossing

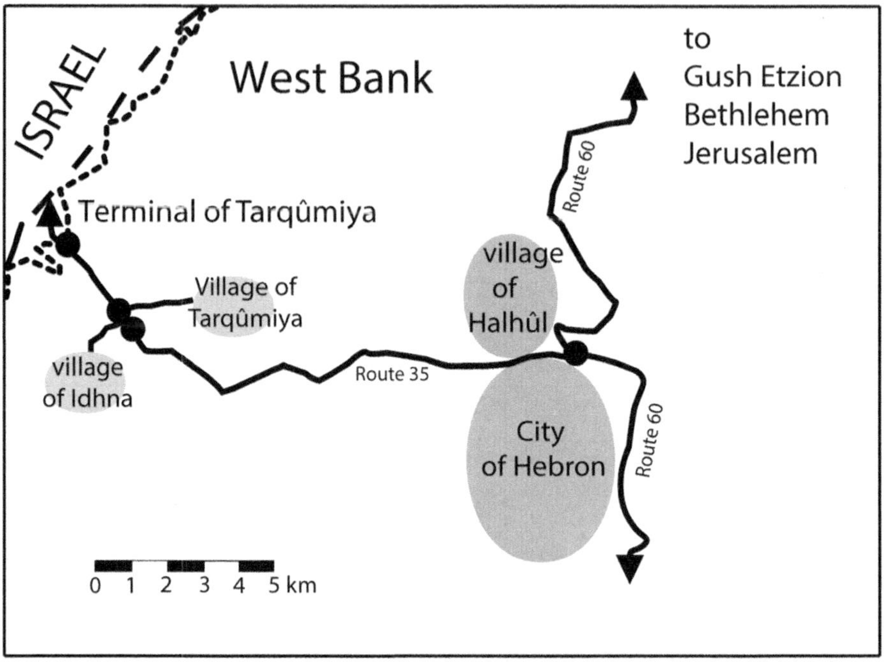

Figure 5.1 Survey points on the corridor serving Tarqûmiya

(not using the back-to-back facility). In principle, these were trucks intended for Israeli settlements or army facilities, and should have been using Route 35 (rather than side roads through Palestinian villages).[5]

Unexpected Flows, Unexpected Directions

Our report showed that, the allocation of travel between the various crossings was quite different than imagined, as were the travel patterns to and from the Tarqumiya crossing. More specifically, our study showed sustained overall levels of flow of about 120 trucks an hour (two per minute) in both directions for most of the working day (7 a.m. to 4 p.m.), with a gradual falloff of traffic after 4 p.m., and a rapid rise after 4.30 a.m.

The allocation of these flows (illustrated in Figure 5.3) was surprising to the Israeli authorities, who considered Tarqumiya as the primary crossing point in the southern West Bank. In fact, the "tunnels" crossing into southern Jerusalem carry 45 percent of the freight traffic, with Betar Elit and Tarqumiya together carrying another 35 percent. Three passages, therefore, carry 79 percent of the measured flow. The overall quantity of trucks was also much larger than expected: 3,361 trucks in the 24-hour period, whereas the Tarqumiya crossing was treating about 300 a day and ready to go up to 600 with double shifts, once all

5 Our report (Garb 2008; see www.ygarb.com) describes in more detail the methods used to ensure reliable and comprehensive counts of trucks, and the computer-facilitated visual matching procedure that allowed us to reconstruct trajectories and passage times.

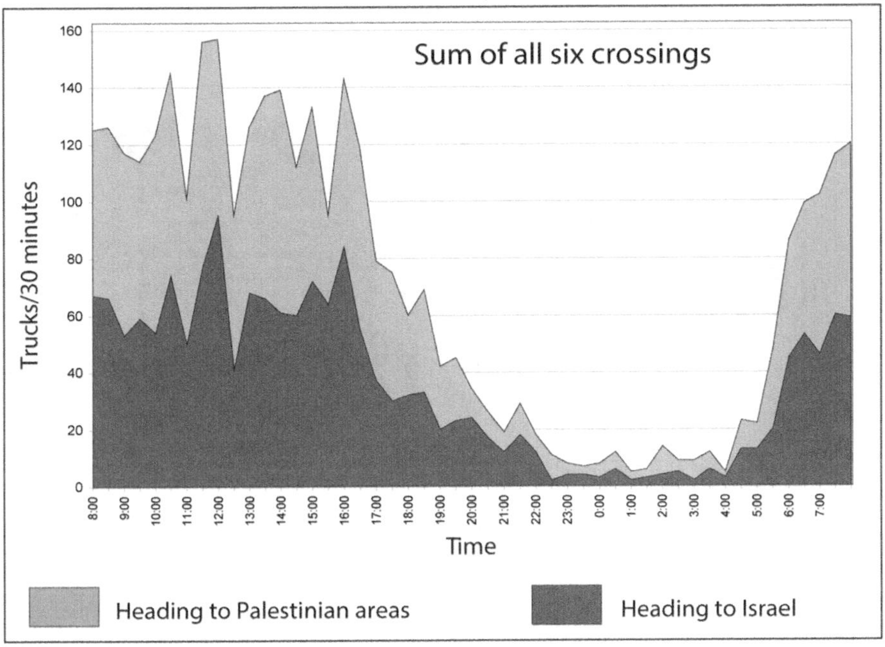

Figure 5.2 **Truck volumes in both directions between Israel and the southern West Bank through the six major crossings**

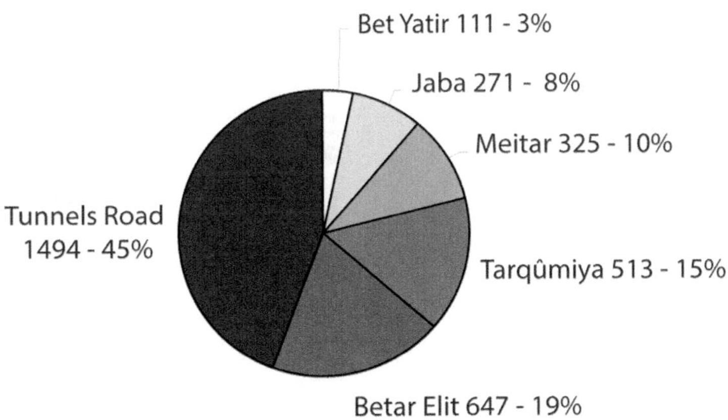

Figure 5.3 **Distribution of trucks through the six main crossings of the southern West Bank**

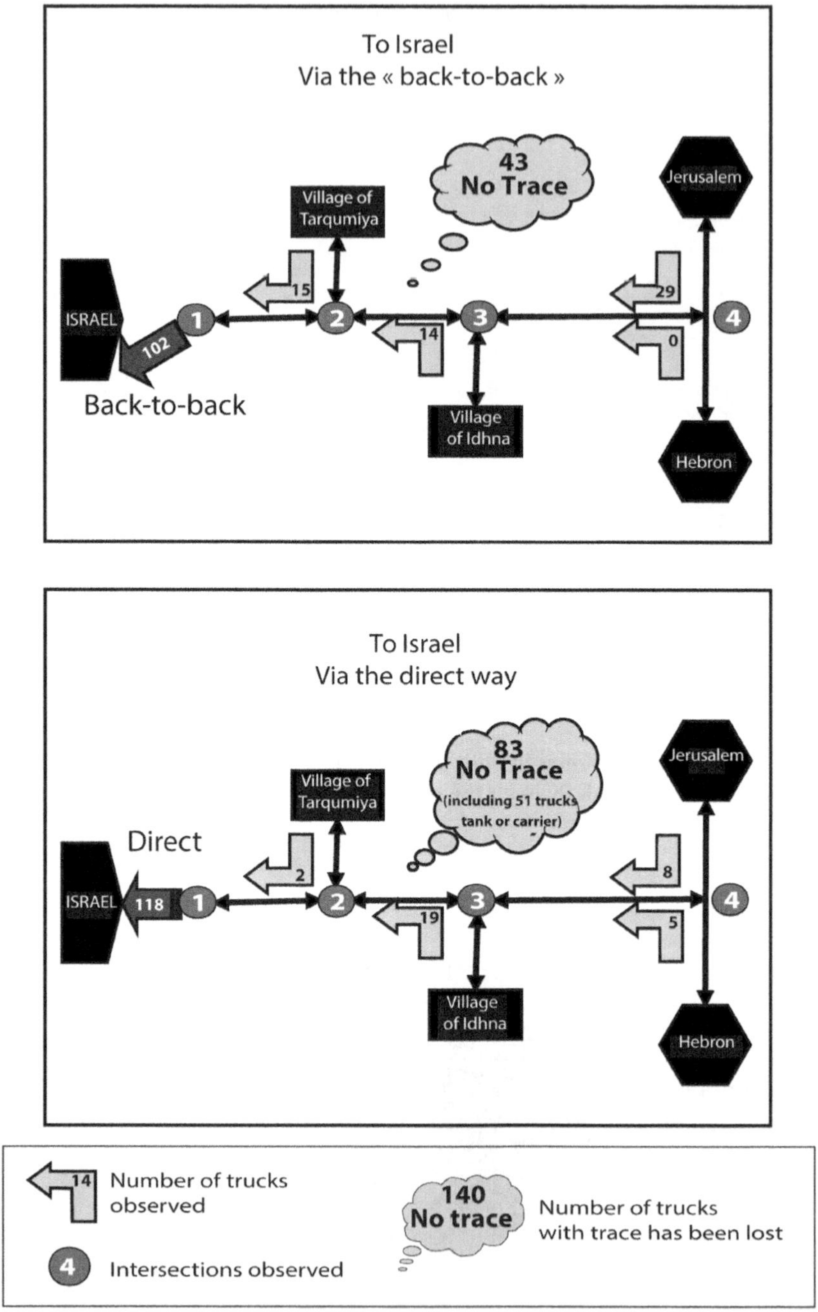

Figures 5.4a and 5.4b **Distribution of trucks' trips traveling from Palestinian enclaves to Israel along the Route 45 corridor. Numbers indicate number of trucks. The "clouds" indicate trucks that do not reach the end points of the corridor, that is they originate or have destinations between the points indicated.**

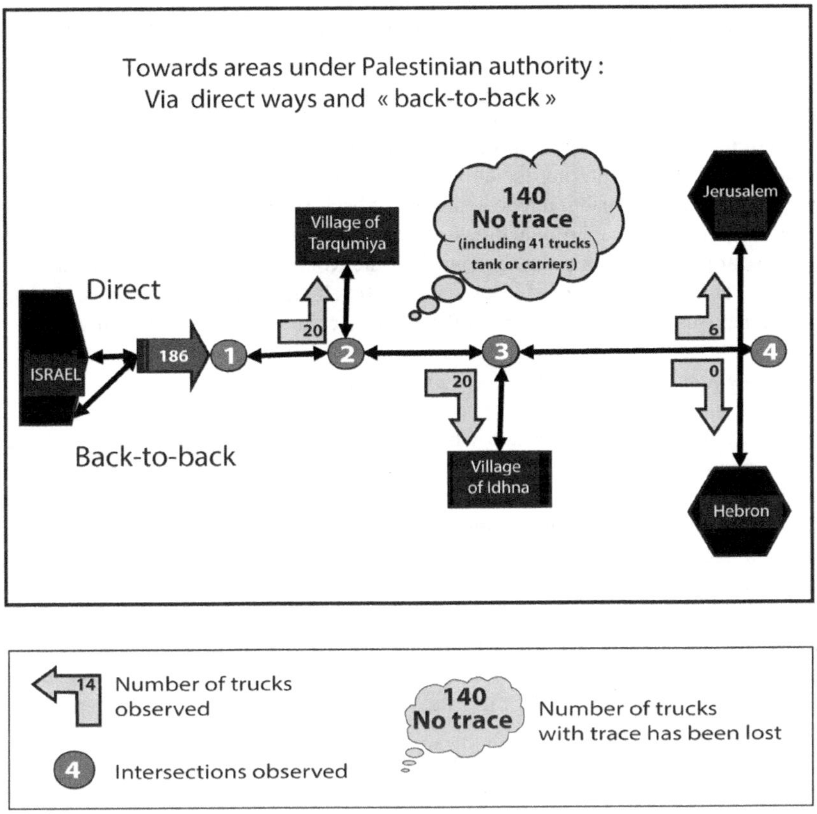

Figure 5.4c Distribution of trucks' trips traveling from Israel to Palestinian enclaves along the Route 45 corridor

of the flow was properly directed there. Even discounting for the portion of these trucks that were military or traveling from or to Israeli (i.e., settlement) locations within the territories, the volume of traffic was far more than anticipated, and once concentrated on Tarqumiya, as planned, would exceed this terminal's capacities considerably.

Additional surprises were yielded by the fine-grained study of the division of flows through the back-to-back versus direct avenues of the Tarqumiya crossing, and the approaches to this crossing along Route 35 (regarded as the main approach) versus alternative routes on small roads through the surrounding villages. The corridor was not a passage from destinations in the West Bank to the border, but had become a destination in its own right. That is, many of the observed truck trips to and from the border did not pass through this corridor, but were to the transport-intensive business activities that had sprung up along the highway corridor itself, attracted by the convenient proximity to the border. For example, various car mechanics, scrap metal stations, and other facilities benefiting from high border adjacency have sprung up at the entrance to Idhna, while a semi-formal fuel and cement transfer area flourishes adjacent to Route 35 a few hundred meters passed the Palestinian side of Tarqumiya. The Salaam fuel and frozen goods depot and the Nasser quarry along the route may also generate a fair amount of freight traffic.

The routes of trucks approaching the crossing are detailed in Figures 5.4a and 5.4b, and show that Route 35 was not the main approach corridor, as was thought. Forty-two percent of trucks entering Israel through the back-to-back and seventy percent (a very large amount) of the trucks entering Israel directly do not come from the expected entries onto Route 35. This was probably due to a number of back-roads used to shorten travel or avoid police, and to oil and quarry facilities between station 1 and 2.[6] In the other direction, from the Tarqumiya check point the "pipe" of the Route 35 corridor has "leaked" most of its content before station 3, and loses another 40 vehicles before reaching Route 60, onto which it deposits only three percent of the vehicles coming through Tarqumiya.

Porosity, Fragmentation, and Blindness of/at Border Crossings

The survey illustrated the degree to which the Crossings Authority, in charge of the new border arrangements between Israel and the Palestinian Authority had less control and knowledge of freight movement between Israeli and Palestinian areas than imagined. It indicates that most trucks with Palestinian freight were still moving through crossings they were no longer supposed to use (i.e., other than Tarqumiya), as well as using the "Israeli" lane at Tarqumiya. How does this reality accord with a fairly common conception of the emerging separation regime, one emphasizing Israel's military might, strategic foresight, panoptic surveillance capacities, and tight control? We might retain this conception, and explain away the porous situation indicated by my findings as temporary, accidental, or incidental. I think this would miss the opportunity for a deeper conceptualization of separation lines functions and flows.

Almost two decades ago, Gupta and Ferguson (1997) challenged us to think beyond the presumption that national territorial spaces are autonomous and cleanly bounded, overlapping with the peoples that inhabit them—an illusion that "enabled the power of topography to conceal successfully the topography of power." Activist and even some academic descriptions of Israeli power, which tend to bind together overly simple spatial and political topographies in this way, would benefit from Gupta and Ferguson's call. I think my survey's findings and other works in this volume provide entry points into a more complex understanding of the workings of power over people and space in the Israeli-Palestinian context, one that pays more attention to the layered and often patchy components that are often too easily lumped together into the unitary labels used for opposing sides ("Israelis," "Palestinians"). These labels are, at best, a useful shorthand, which on closer inspection seem to mask various splinterings within the categories, and cross-cutting affiliations across them. The porosity of the separation lines, as well as the seeming ignorance of key figures in the Ministry of Defense regarding these border flows derive, in part, from the disjoint fractions within each of these "sides" on the one hand, and cross-linkages between the sides on the other. Viewed in this context, a separation line is not so much a delineation between

6 In the first instance, using Route 35 connecting to Route 60 requires a detour through Halhul in order to travel south, which, added to the cost of frequent traffic fines given to truck drivers on Route 35, makes drivers shun this and travel on the Tarqûmiya village/Bet Kahil road that enters Hebron directly. Drivers enter/exit this road at a point between station 1 and 2. In the second instance, there is a quarry and fuel depot adjacent to Route 35 between station 1 and 2, so many trucks seem to be cycling to this site right at the border. This conjecture is supported by the fact that 62 percent of these "untracked" trucks are either quarry or oil vehicles.

two sides defined nationally but a site of complex transactions and filterings, restrictive and productive in different ways for different groups.

The fault lines within the "Israeli" side were often striking. For example, when I showed a graph of passage volumes over the 24-hour period at a presentation to high-level Israeli officials in the Passages Authority, there was a short flurry of genuine worried discussion between a high ranking official and his key staff member about the significant amount of traffic (20 percent) going through during hours when the passage was not staffed. Yet most low ranking soldiers serving time at the border could have told them as much. The deferential staff person promised the situation would be promptly investigated and rectified. When I presented the numbers about how much traffic was actually flowing through the various crossings, another official expressed genuine anger at being forced to "fly blind" (his phrase) in constructing the passage arrangements without valid planning information on expected traffic volumes. It was clear that blind spots, and knowledge mismatches between the "layers" of authority, such as those described in Boussard et al.'s (2003) work on organizational blindness, were very real.

Another time, the presence of a "blind spot" was made clear to me very literally. I was setting up our surveyors at one of the crossings, and stopped to explain our goals to the crossing commander, an officer who was an academic stationed there on his annual reserve duty. In a moment of scholarly camaraderie he took me aside, some 50 meters from the crossing, and pointed down into the valley below the checkpoint, which ran from the Palestinian to the Israeli side of the border: "For every one they count there," he said, pointing to the surveyors at the checkpoint, "there was one going through down there," he said, pointing to the dirt road below that ran into Israel. "That trail can carry semi-trailers." Here within a single person there was a disjunction between what he as a commanding officer was doing in supervising his soldiers' activities (checking all trucks at the crossing carefully) and what he, as an individual, could see and acknowledge (that as many trucks were passing below, unchecked). Or, perhaps, we can view this not so much as schizophrenia within an individual as a pragmatic accommodation of formal and informal systems.

These layerings create knowledge leaks on the one hand, and the inability to construct coherent knowledge on the other. The Israeli official working his shift can ignore what happens before it begins at 7 a.m., and the supervisor of all shifts at the terminal knows different things, and the soldier at the adjacent military checkpoint something different, again, and the Ministry of Defense in Tel Aviv knows less (and more) than any of them. The knowledge missing (or not transferred) in this case was not some operational detail, but actually rather foundational for the apparent functioning of the crossing regime: how many trucks would it be serving, where were they going, and why?

We were able to create this knowledge because our small team had a focused organizational goal of producing precisely this overall truck count. Our advantage was this focus and the fact that we were not large and bureaucratic. When I expressed surprise to a senior official who told me that they really did not know how many trucks were moving through the tunnels crossing into Jerusalem (my numbers, eventually, showed this to be the main crossing), he explained to me that just getting insurance coverage for a surveyor to stand at the crossing booth, which was at the unclear border of several administrative jurisdictions, would have been a major undertaking. This is an example of the ways in which a formal system blinds itself through its formality.

There are similar splinterings of actors and interests on the "Palestinian" side of the Tarqumiya area. Who do we mean when we talk about "Palestinian freight?" The business that owns the freight? The shipping company they use? Or the driver hired by this company?

For example, Palestinian businesses lobbied to reduce delays, so that more than one trip to the Ashdod Port might be done daily. But Palestinian trucking companies were disappointed when this was achieved, since the overall fleet of trucks could be reduced if each truck can move more freight in a given period. And what size of businesses are we talking about? American advisors are pushing Israeli officials to adopt the kind of "trusted trader" arrangement that the US has developed along its borders with Canada and Mexico, which would allow rapid passage for certain certified well-established businesses demonstrating large financial stakes and robust internal security measures.[7] Large Palestinian companies will be glad of such an arrangement, allowing them to benefit from shorter and more reliable passages, while smaller ones, which cannot afford the technologies and certifications involved, will be undermined by it. The Palestinian Authority might publicly decry the losses to Palestinian businesses because of the unilateral imposition of the official crossing, but has been accused of being less than adamant about this, given that trucks going through the back-to-back will pay the 17 percent VAT that informal crossers will not.

"Palestinian freight" is also entangled with Israeli trucking companies and their drivers. These benefit from the back-to-back arrangement, since they now do the portion of the trip from the crossing facility to the port, to the separation line with Gaza, or to destinations in Israel. In fact, some also benefit from travel inside the Territories, through less formal arrangements in which Israeli-licensed trucks bearing Palestinian goods travel straight through the direct passage, avoiding the wait and other annoyances of the back-to-back arrangements. Israeli Arab truck drivers are particularly well positioned to take advantage of such opportunities, since they can move legally within Israel and comfortably within the Palestinian territories.

Palestinians map and often utilize the organizational layering and epistemic patchiness around the crossings; for some, Israeli blind spots and conflicting interests are essential to their livelihood. As the most local and pragmatic level, there is the Palestinian probe truck sent to test the alertness of the supervision off the direct passage on a particular day and call back to other drivers on whether to proceed or wait, and then the more ornate and large scale informal operations that are described elsewhere in this volume. The crossing has created its own ecosystem of hybrid border life: a Palestinian scrap metal consolidator in a nearby village is supervised by a Jewish security guard under a special arrangement that allows their trucks to pass through directly; quarries and fuel depots coordinated with Israeli businesses enjoy proximity to the border and easy passage; Palestinians in Hebron who broker or forge helpful shipping documents for use at the crossing; Israeli businesses or people in Israeli settlements who will provide documents that certify Palestinian goods as originating with them, or, simply, drive the truck through themselves for 1,000 shekel. These kind of combined and uneven enterprises make it more difficult to define (and "control") the border in any simple way.

The Tarqumiya study findings and my experiences in conducting it provide a window into the complex and fragmented workings of power underlying standard constructions of Israeli power, borders, and mobility control. Neither official Israeli accounts of Tarqumiya's functioning and of the emerging freight regime, nor the protesting Palestinian counter-account of Israeli control do justice to the messy flows and forces on the ground, and the complex multi-organization (indeed, multi-national) fields that increasingly shape its operation. These simpler accounts and much of the prevailing imagery of borders play

7 I was told that the term "trusted trader" was not palatable to Israeli officials, who insisted on discussing the notion under the heading of "known trader."

on familiar ideological and emotional registers, and can sometimes be regarded as first approximations or necessary shorthand. But as Gupta and Ferguson (1997) suggest, the stylized topographies of power they contain are worryingly simplifying ones, whose construction, maintenance, and purposes deserve attention in their own right.

Here I have shown the divergence between knowledge produced from simple truck counts on the one hand, and what ostensible managers of these freight flows thought was going on, on the other. This gap pushes us toward a more nuanced, multi-layered—indeed, fragmented—account of movement and borders, and of the corresponding structural blindness they generate (or require). I would argue that such an account is important not only to better understand current arrangements, but to move more fundamentally beyond them.

Chapter 6
From Chocolate Bars to Motor Cars: Separation and Goods Trafficking between Israel and the West Bank (2007–2010)

Basel Natsheh and Cédric Parizot

Introduction

This chapter analyzes the change in the informal trafficking of standard consumer goods between Israel and the West Bank between 2007 and 2010, when the security forces of the Palestinian Authority (PA) were redeployed in the main Palestinian cities. This study aims to assess the PA's capacity to territorialize its control, and to identify forms of regulation between the Israeli and Palestinian economies, over and above formal exchanges.

Since the beginning of the 1990s the Israeli separation policy and the delimitation of the PA's zones of jurisdiction in the Gaza Strip and West Bank have raised new boundaries and limits. The Paris Protocol (1994) imposed new taxation rules on the transport of goods between Israel and the Palestinian Territories. Hence, local players had to reorganize to work around these limits or avoid the new regulations. A large number of movements and transactions that had previously been carried out in broad daylight and been tolerated by the Israeli authorities were gradually criminalized and hence went underground. These limits and rules have also caused the emergence of economic, legal and status differentials that Israeli and Palestinian economic players have exploited to develop informal activities and create more profit. The period of the Oslo Accords (1993–2000) thus saw the emergence of a "border economy" or more precisely, a "separation economy," if we acknowledge that no internationally recognized borders exist today between Israel and the Palestinian Territories.

Research into such informal trade has concentrated on the strategies Palestinian workers have used to keep on entering into Israel to work clandestinely, or on the smugglers who facilitate their entry (Bornstein 2002; Parizot 2006b, 2008b, 2014; Bontemps 2009; Amiry 2010). Some researchers have stressed the direct link between the reinforcement of Israeli security systems and the increased professionalization of the traffickers' networks (Parizot 2009b, 2014). As in other parts of the world, the increased "border" security and the differentials to which it gives rise have increased informal traffic (Chandoul et al. 1991; Andreas 2001; Bennafla and Peraldi 2008). Furthermore, such research shows that studying clandestine travel highlights the changes in the functioning of the power mechanisms deployed by Israel in the West Bank (Parizot 2014).

In contrast, little work has focused on the traffic in goods, or only on a very one-off basis concentrating on specific or localized trades such as that in cars stolen in Israel and sold in the Palestinian enclaves (Hertzog 2005) or the trade in everyday consumer goods between the West Bank and north-eastern Negev (Parizot 2006b, 2008b). They also stress how the multiplying of barriers encourages informal activities. During the 1990s, the delimitation of zones of action for the Israeli and Palestinian police forces created areas of refuge for car thieves (Abu Moaleik 2005). In the West Bank during the Second Intifada (2000–2004), the limitations placed on the movements of the Palestinian police force have encouraged the import of a large number of faulty, out-of-date goods (Laban 2005). Under siege and holed up in their barracks by the Israeli army, the Palestinian police force were unable to combat such informal practices. Lastly, the Wall and the entire security system deployed by Israel do not constitute real obstacles as they leave a number of routes open through which stolen cars can be channeled into the Palestinian enclaves (Barthe 2007). Finally, the prioritization of security has distracted the attention of the Israeli and Palestinian authorities from such traffic (Lia 2006).

This chapter continues this work by showing the close link between changing security arrangements and the development of goods trafficking. We will focus on the informal trade that channels everyday consumer goods between Israel and the Palestinian West Bank Territories. While less sensitive, everyday traffic is more sociologically relevant than gunrunning or drug pushing in that it involves more people and is thus more likely to give an accurate account of the everyday lives of Palestinians and Israelis. Such informal trade is practiced by many ordinary people who become "suitcase traders." When travelling, they take advantage of differentials in price, buying articles they can carry home to sell on their local markets. Some of them are also entrepreneurs who sell more specialist items such as building materials, electronics or medicines. They are as likely to trade in chocolate as in stolen Israeli cars that are sold for spares in the Palestinian Territories, in Israel itself and neighboring countries. Lastly, the diversity of these products' origins and destinations also demonstrates the extent to which this traffic is global.

By assessing the ability of the PA to control the flow of goods that transit through the enclaves, our first objective will be to analyze how it succeeds in asserting territorial control over its territories in the West Bank. Rather than adopting a normative approach, searching to define whether or not the PA has sufficient resources to achieve this goal, we will try to highlight the specific forms of this territorialization. A comparison with that taken by Israeli control will help us determine the nature of the territorial regime that has emerged in the Israeli and Palestinian territories in which there is not just a spatial separation between the populations, but also a ranking in their relations to space.

Ariel Handel (2009a) explains that Israeli security arrangements affects the capacity for movement of the Israeli and Palestinian populations in different ways: Israelis can move through a fluid, uninterrupted, predictable space that can be objectivized, particularly using modern cartographic science while Palestinians move in a fragmented, unstable space. The time a journey takes depends both on the constant changes of obstacles' location and the biosocial status of the traveler. Palestinian space is therefore intersubjective and maps are of no use. This research will show that the Israeli policy of separation does not simply create a structural inequality of use of space between Israeli and Palestinian citizens, but also an inequality in the implementation of control between the Israeli and Palestinian Authorities.

The question of the territorialization of control exercised by the PA has been particularly relevant since 2007 when draconian political measures were taken in order to restore its sovereignty. According to the plan promoted by General Keith Dayton of the US, and with the agreement of the Israelis, the various security services were allowed to leave their barracks and take over the management of security in the towns of Jenin, Nablus, Tulkarem, Ramallah, and Hebron. The objective of these redeployments was to strengthen the position of Mahmoud Abbas as Palestinian president and thus thwart Hamas who had just taken power in Gaza in June 2007 (Legrain 2010). The redeployments were also presented as an additional stage in the process of building a future Palestinian State. Furthermore, the boycott campaign launched at the end of 2009 by the government of Salam Fayyad on goods from the Israeli colonies marked a strong desire of the Palestinians to break their dependence on the Israeli economy and thus impose their own separation. But as we will show, the large-scale traffic of goods continued. Actually, the restrictions on movement and action imposed by Israel on the Palestinian police combined with the differentials created by the multiplication of limits since the time of the Oslo Accords encourage the continuation of such traffics, and even, their large-scale development.

The second objective of this chapter is to understand the changes in economic relations that have occurred between the Israeli and Palestinian territories over the last two decades. Between the end of the 1990s and 2000, Palestinian enclaves stopped being essential labor pools for Israeli companies. The importing of Asian, European, and African labor greatly reduced the dependence of the Israelis on Palestinian workers (Kemp and Raijman 2008). Furthermore, the Palestinian enclaves also became less attractive to Israeli companies seeking to outsource their business. They preferred to outsource their production to Jordan, Egypt, Turkey, and China who offered much cheaper labor than in the Occupied Palestinian Territories (Bouillon 2006). During this period, many workshops closed in the Gaza Strip and the West Bank. The proportion of imports from the Palestinian Territories has continued to diminish over the last 30 years to a point where it accounts for a limited share of all the goods imported into the Israeli market. In the first half of the 1980s, the proportion of imports—excluding diamonds[1]—from the West Bank and Gaza varied between 2.5 and 3.5 percent, falling to 1.5 percent in the 1990s and then to 1 percent between 2000 and 2005 (Peres Center and PalTrade 2006).

In contrast, the Palestinian market remains strategic for Israeli exporters. Although the proportion of Israeli exports to the Palestinian areas has dwindled as the conflict has worsened, these enclaves remain Israel's most important export market after the United States (Peres Center and PalTrade 2006),[2] since they remain a captive marker (Grinberg, this volume). Taking informal traffic as our starting point, we will try to identify what types of relation and regulation between the Israeli and Palestinian markets emerged over and above formal trade.

This research is based on investigations performed between 2007 and 2010 in the West Bank and Israel. Data was collected from both sides in order to follow and understand the networks on which this informal trade depends. But we did not enjoy the same freedom of movement to undertake these investigations. Furthermore, given the impossibility of carrying out surveys in the Gaza Strip, we were obliged to limit this study to the trade between the West Bank and Israel. As a resident of the West Bank, Basel Natsheh, the joint author of this article with Cédric Parizot, would not travel to Gaza. Moreover, he had to apply for a permit to visit Jerusalem and Israeli territory. He was only granted a one-month permit limited to Jerusalem in 2008. The applications he made later with the backing of the French Consulate with which he had work relations[3] were routinely rejected without Basel being given any explanations. Although Cédric Parizot's French passport enabled him to travel freely between Israel and the West Bank, in order to gain access to the Gaza Strip he had to submit to extremely complex administrative formalities the outcome of which was

1 Worth several billion dollars, the diamond trade accounts for a major share of the Israeli commercial balance.

2 The proportion of exports to the West Bank and Gaza Strip dropped from 12 percent—excluding diamonds—in 1987 to somewhere around 6.5 percent in 2005.

3 At the time, Basel Natsheh was working as head of mission at ADETEF. In 2010, he was a consultant to the Palestinian minister for the economy.

not guaranteed.[4] The mobility regime introduced by Israel since the 1990s therefore also affects the conditions under which research is done; most importantly it ranks researchers' accessibility to the field according to their status and identity.

Our approach combines anthropology and economic sociology. We carried out formal, semi-directive interviews with Palestinian customs and police officers as well as with certain representatives of the Israeli authorities. We have also taken advantage of several visits by teams of French customs officers seconded as part of a training program set up by ADETEF[5] to the Palestinian Ministry of Finance. Data was subsequently collected from Israeli and Palestinian players involved more or less actively in this informal trade. Our observation methods were therefore based on immersion and the sort of free-floating looking and listening used by anthropologists.

We will first analyze the changes in the systems of Palestinian control since the Oslo Accords in the context of the readjustment of security measures imposed by Israel. We will try to show how existing security arrangements facilitate and maintain the development of informal trade between Israel and the West Bank. We will then explain in more detail the nature of the traffic in order to understand the new forms of economic regulation they reveal between the Israeli and Palestinian markets, and beyond that, the forms of territorialization of Palestinian control compared with those of Israeli control.

"Deterritorialized" Palestinian Control

During the Second Intifada (2000–2004), the Israeli re-invasion of the Palestinian enclaves greatly reduced the PA's capacity for action. The redeployments of 2007 enabled the PA to regain partial control over the major towns on the West Bank. But in 2010 the Palestinian police and customs services were still having difficulty extending their control in both space and time. Israeli security measures leave a number of gaps in which entrepreneurs of the informal economy can hide or extend their activities. In addition, incursions by the Israeli army regularly suspend action by the Palestinian police. The Palestinian customs and police are therefore unable to directly monitor the limits of the enclaves or the crossing points for goods and people entering or leaving them. Palestinian control is thus "deterritorialized," not in the sense of breaking free of its territory to extend itself beyond it, but rather because it is incapable of deploying within the limits of that territory.

Fragmented Space and Time

During the 1990s the creation of PA's zones of jurisdiction in the West Bank resulted in a patchwork of enclaves surrounded by zones under Israeli authority. The region was divided into three types of areas named A, B, and C (see Map 6.1).[6]

4 Apart from journalists, foreign nationals wishing to enter the Gaza Strip have to go through their diplomatic representatives or the institution employing them. These representatives then coordinate with the Israeli authorities to obtain authorization to enter the coastal strip; such permission is not routinely granted.

5 ADETEF is an international cooperation body for the French ministries for the economy, the budget and sustainable development. The agency makes the professional skills of French civil servants and experts available to the governments and public authorities of certain foreign countries.

6 The A zones included more or less the centers of the six Palestinian towns of Jenin, Nablus, Tulkarem, Qalqiliya, Ramallah, and Bethlehem, while the fringes of these towns and nearly 450

This fragmentation sets the PA a major problem in trying to control the flow of people and goods from one enclave to another. In fact it creates a system of refuge in that, including as they do some 60 percent of the West Bank, the C zones offer wide, open spaces in which fugitives can flee the scrutiny of the PA or develop the activities it is trying to repress. This system also works in the opposite direction: starting in the second half of the 1990s, certain Israeli traffickers found in the autonomous Palestinian enclaves (A zones) refuges where they could evade control and pursuit by the Israeli police—particularly as Israeli citizens kept extraterritorial immunity.[7]

This system of refuge changed during the Second Intifada (2000–2005) when the PA saw its powers and fields of action greatly reduced. Coordination between the Israeli and Palestinian authorities broke down, reducing the number of Palestinian operations in the B zones or delaying their introduction. In 2002, the Israeli army started targeting the institutions of the PA, holding Yasser Arafat directly responsible for the second Palestinian uprising and the wave of suicide bombings in Israel (Cypel 2005, p. 278). During "Operation Defensive Shield" (Hebrew: *homat magen*) in which the Israeli army again invaded the A zones, security coordination with the Palestinians was frozen and the PA's barracks and security institutions besieged. Despite the partial withdrawal of the Israeli army in the following years, the Palestinian police could no longer move about or act inside the A zones without prior authorization from the Israeli army.

In summer 2007, the Palestinian police force was redeployed in the A zones of the West Bank. The police engaged in direct confrontations with groups and institutions affiliated to Hamas (Intelligence and Terrorism Information Center 2008). But regaining control over the enclaves has encountered several obstacles. The security arrangements imposed by Israel limit the PA's geographical capacity for action. Between 2009 and 2010 in the town of Hebron (Map 6.2) the Palestinian police was unable to deploy in the whole of zone H-1, which is the equivalent of an A zone. On the one hand, it could only send armed officers into an extremely limited area measuring 1.8 square kilometers lying between avenues Ein Sara, as-Salam, and Wadi At-Tufah. Outside this zone, armed Palestinian units could only operate if they had prior authorization from the coordination office (DCO) of the Israeli army, and then only for a limited period of time. In addition, they were totally prohibited from entering the industrial zone of Hebron.

The capacity for action of the PA was also limited in time because it was regularly suspended by Israeli incursions. Considering the Palestinian forces' action against Hamas and other Islamist groups inadequate (Intelligence and Terrorism Information Center 2008), the Israeli army regularly made incursions into the Palestinian enclaves. Every time an Israeli unit entered an A zone it ordered the Palestinian police force back to their barracks. In July 2009, in zone H-1 the Palestinian police were ordered to suspend their activities more than 20 times.[8] In some West Bank towns these time restrictions have been defined on a fixed timetable. In 2007, between Tulkarem and Jenin, the security arrangements agreed between

---

villages were defined as zones B. The zones C cover the remaining land in the West Bank including the Israeli settlements and the major roads. In the A zones, Israel delegated security and civil control to the PA; in the B zones, the PA was responsible for public order and the internal security of the Palestinians while Israel reserved the right to act on any questions of external security. Lastly, the C zones remained under Israeli control. In 2000, on the eve of the Second Intifada, the A zones covered 17 percent and the B zones 23 percent of the region.

7 The PA has no powers to detain or try Israeli civilians who have committed offences in the A zones (Lia 2006, p. 290); their only possibility is to hand them over to the Israeli police.

8 Interview with the coordination officer for the Hebron district, Mujahid Abu Snineh, July 2009.

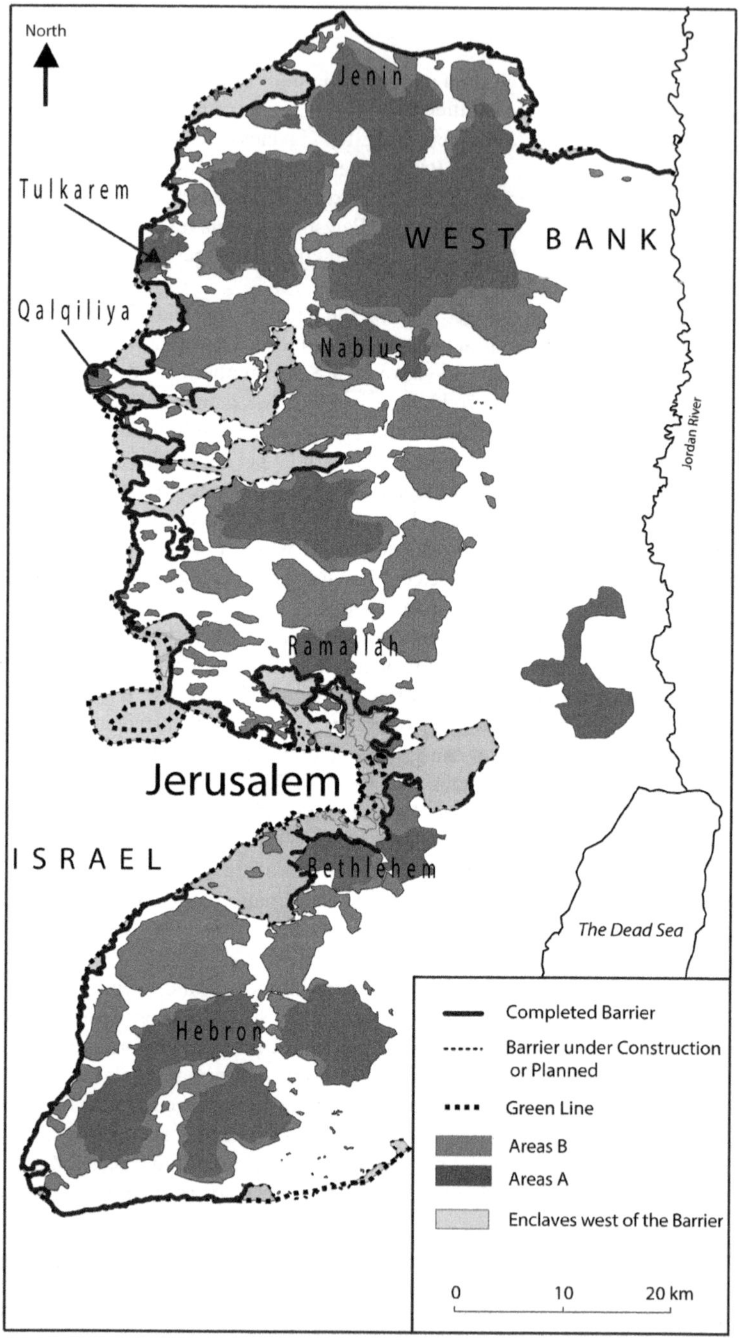

Map 6.1 A, B, and C zones in the West Bank

Map realized by M.Barazani (CRFJ) and M. Coulon (LAMES).

Source: OCHA Information Management Unit; Map produced June 2009; Data Base and Statistics: OCHA, PA, MoP.

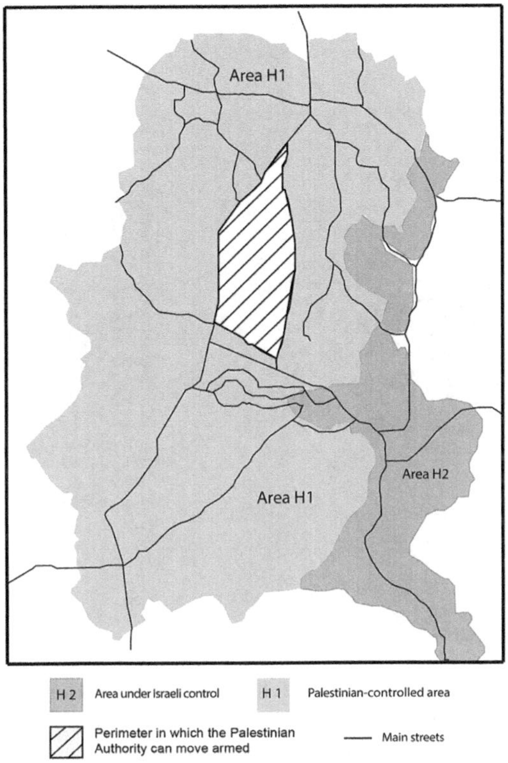

Map 6.2 Security arrangements imposed by Israel on Palestinian police in Hebron, 2007–2009

Map : Tareq Natsheh, 2009

Israel and the Palestinians divided up the time during which the Palestinian authorities could act. They gave the PA total control between 6 a.m. and midnight, but imposed coordination with the Israelis during the hours of night (International Crisis Group 2008, p. 13). Finally, the delays in implementing coordination between the Palestinian police and the Israelis constituted an additional time limit (Issacharoff and Azoulay 2008). It sometimes needed hours, even days, for the Israeli coordination office to authorize the Palestinian police to act in a B zone or in restricted areas of A zones. This impossibility of acting immediately and the uncertain nature of Israeli reaction times thus prevented the Palestinian police taking advantage of timing to deploy control strategies.

Consequently, whilst the campaign to restore Palestinian control over the large towns saw a certain return to order, the new security arrangements imposed by Israel during the post-Intifada period left the entrepreneurs of the informal economy more freedom of movement than the PA forces. At the same time, these security arrangements contributed to extending the refuge zones created during the Oslo Accords period. At this time these refuge zones lay in the C zones and certain B zones. From 2000 they grew to include more B zones and particularly the A zones. In autumn 2009, Palestinian customs were no longer

Figure 6.1 The industrial zone of Hebron. The sign in Hebrew states that Israeli citizens are not allowed to enter. In practice, this zone is also out of bounds to the Palestinian customs service

Photo: Cédric Parizot, March 2010.

allowed to operate in the regions of Yatta and Samu‘ to the south of Hebron even though they were in an A zone.[9] Similarly, even though it was in zone H-1 but out of bounds to the Palestinian police, the industrial zone to the south of Hebron has remained a favorite place to store smuggled and fake goods (Figure 6.1). The market inside zone H-2 controlled by the Israelis is also a notorious hideout for smugglers and those fleeing the Palestinian police.

Lastly, the traffickers in goods have also taken advantage of the priority given to security by the Israelis and the PA. The fight against armed Palestinian groups, whether carried out by the Israeli authorities trying to combat suicide bombings and maintain control over the region or the PA seeking to limit the action of groups who disagree with its political stance (Hamas and Islamic Jihad), thereby fail to act against breaches of ordinary law (Lia 2006). In addition, the manipulation of smuggling networks (Parizot 2008b, 2009b, 2014) offered such networks a degree of protection and encouraged the spread of their activities.

Customs without Borders

Depriving the PA of control over a large part of the West Bank, these security arrangements also prevent customs officials and customs police from controlling the entrances to and in

9 Interview with Husam Khalaileh, Hebron customs service, December 2009.

the periphery of the Palestinian enclaves. The fringes of the enclaves are often categorized as B zones. Unless the Israeli army gives its permission, the PA has no way of setting up barriers or points of control to check what goods are entering or leaving an enclave. In December 2009, of the 97 points of entry to the city of Hebron only 15 were monitored by Palestinians, but the customs police could only act freely at 8 of them: any action regarding the other 7 required coordination with the Israeli authorities.

Nor have the Palestinians any control over the points of entry for goods once they have crossed the Green Line. As part of the implementation of the policy of separation and the construction of the Wall, the Israeli army set up six crossing points in the "seam zone" to ensure the transit of goods in and out of Israel (Map 6.3): Tarqûmiya (west of Hebron), Betunia (south of Ramallah), Taybeh/Sha'ar Ephraim (south-east of Tulkarem), Al Jalameh (north of Jenin), and Bisan/Beit She'an (north of the Jordan Valley). Located in a C zone, they are managed solely by Israelis. Until 2006 they were run by the army but the privatization of the checkpoints resulted in them being transferred to private companies.[10]

The lack of immediate control over these crossing points reduces the power of the PA in applying its fiscal controls. Traders take advantage of this, frequently under-declaring the value of their goods; this is particularly easy to do as the Israeli crossing-point staff are more concerned with security checks than VAT declarations. The employees are even less motivated to check given that Israel receives nothing from these declarations and the VAT paid by Palestinian carriers.[11] Palestinian traders therefore take few risks in marking down the value of their goods in order to reduce their tax payments.

Nor do the customs police control the points of arrival and transit of goods imported from abroad. In accordance with the Paris Protocol (1994), goods imported by Palestinian traders transit via the Israeli ports of Ashdod and Haifa. A certain quantity of foodstuffs may also come from Jordan via the Al Karameh/Allenby crossing point. As these ports and crossing points are under Israeli control, customs declarations and claims for exoneration of customs duties by Palestinian traders are therefore made through Israeli customs. The Israelis then give the merchants a form that they have to hand in to the Palestinian customs personally. Given the fact that there is no interface between Israeli and Palestinian databases, the PA only has these paper declarations with which to work out how much tax is due to them each month. Badly filled-in forms do not always state the precise volume and value of the goods concerned. And given that the Israeli public and private institutions often overlook scrutinizing such customs declarations, traders tend to under-declare the volume and value of their goods. As the Israeli authorities derive no significant profit from these declarations, once again they are not motivated to be over-zealous.

At the end of the day, marking down the value of goods or concealing their real nature is facilitated by the legal provisions governing the procedure for declaring taxes and customs on the Palestinian side. Once their goods have gone through the Israeli ports and checkpoints, Palestinian traders have 45 days to make their declaration to the customs office in their region. Therefore, unless they get caught by the customs police at the entrance to

10 See Havkin, this volume.

11 Under the Paris Protocol (1994), VAT on Israeli goods sold by Israeli traders in the Palestinian Territories has to be paid to the Israeli authorities who are then responsible for transferring it each month to the PA. In contrast, Palestinian traders exporting goods to Israel have to make their VAT payments to the PA which is responsible for transferring them to the Israeli authorities. A monthly meeting is held between Israelis and Palestinians to work out the sums collected by each administration and who owes what to whom. Given that the Palestinians import more than the Israelis, the PA receives payments each month.

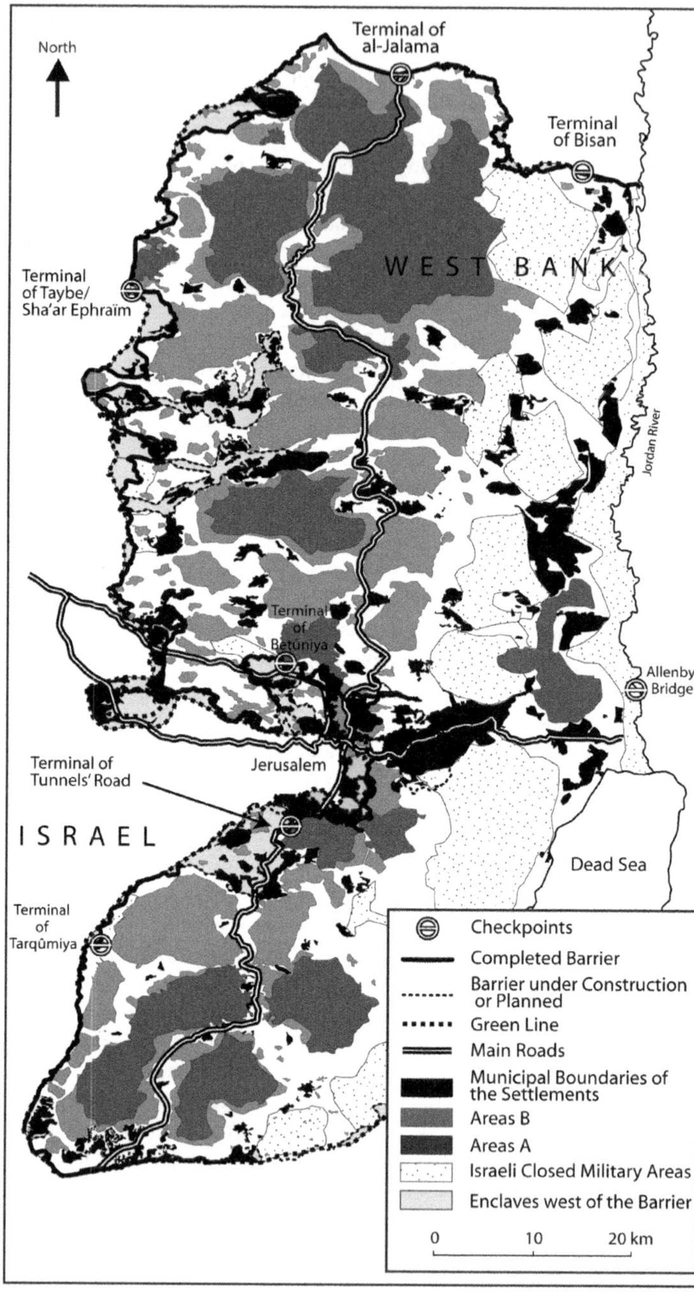

Map 6.3 Main crossings for the treatment of goods between Israel, the West Bank, and Jordan

Map realized by M.Barazani (CRFJ) and M. Coulon (LAMES).

Source: OCHA Information Management Unit; Map produced June 2009; Data Base and Statistics: OCHA, PA, MoP.

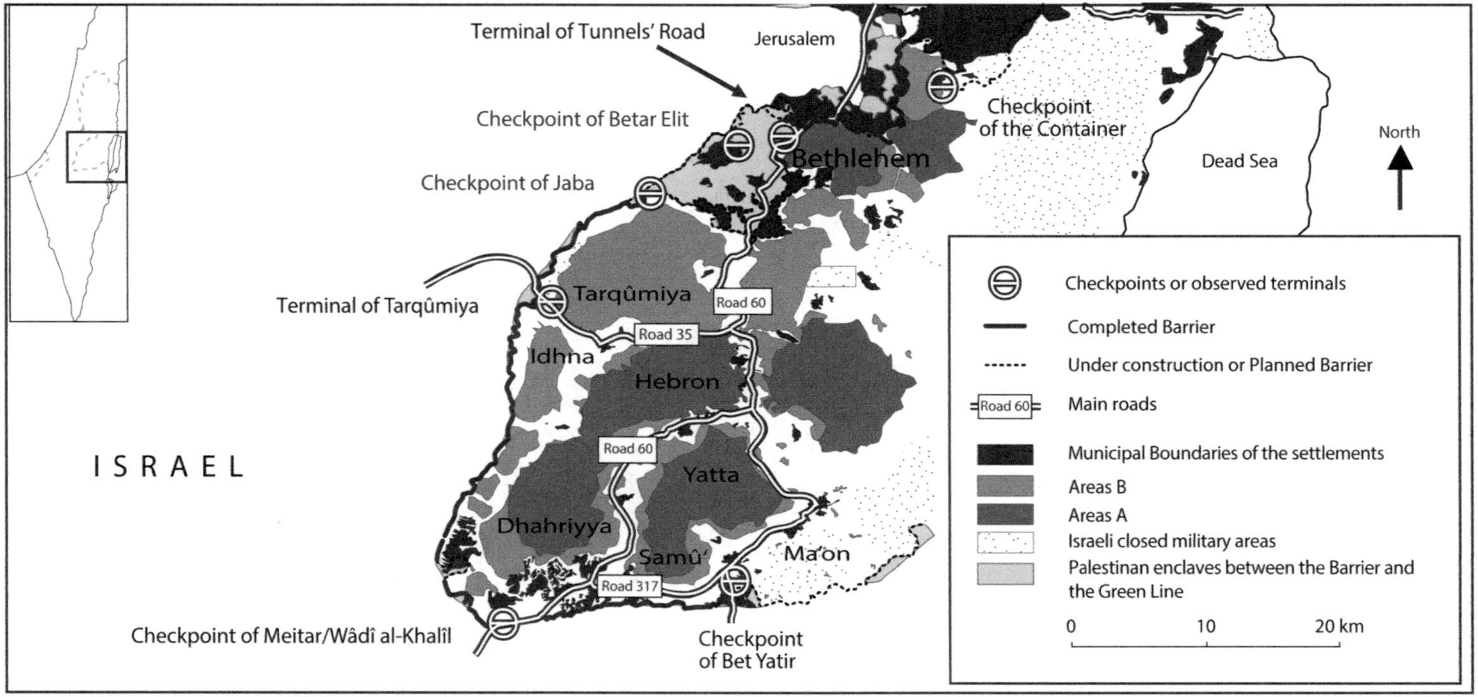

Map 6.4 **South West Bank**

Map realized by M.Barazani (CRFJ) and M. Coulon (LAMES).

Source: OCHA Information Management Unit; Map produced June 2009; Data Base and Statistics: OCHA, PA, MoP.

a Palestinian town, most traders have enough time to get rid of their goods before they have to declare them to customs. As a result, apart from the declarations presented to the Israelis by traders and those transferred to them by the Israelis, the Palestinian customs authorities have no way of checking that the effective nature and value of the goods match the declarations made by the trader or carrier.

Lastly, the considerable distance between the checkpoints and the destination of the goods—Palestinian enclaves or Israeli settlements—and the fact that traders handle both Israeli and Palestinian goods means that they can switch from the very tightly controlled system imposed on Palestinian goods to the much freer system designed for Israeli products. The example of Sâlim ash-Sharârke[12] is particularly illuminating: a resident of Yatta, he built his house in 2007. At that time the steel used in the building industry was cheaper in Israel than in the Palestinian Territories. Sâlim ash-Sharârke, therefore, bought several tons of steel from an Israeli entrepreneur in Beersheba, Moshe Azoulay, who drew up a bogus invoice made out to an Israeli living in the settlement of Ma'on which is south of the Palestinian town of Yatta (see Map 6.4).

Officially intended for an Israeli settlement, the consignment had to go through the control system imposed on goods destined for the settlements: it was therefore able to evade the restrictions and checks to which goods going to the Palestinian enclaves are subject; i.e., submit to the back-to-back procedure at Tarqûmiya.[13] Instead it was able to take a much faster road, passing much further south through the checkpoint of Meitar/Wadi al-Khalîl. It took the Routes 60 and 317 heading to the Israeli settlement of Ma'on. But instead of continuing to Ma'on he entered the Palestinian enclave of Yatta via an entry that is not controlled by the PA. As the periphery of this enclave is defined as a B zone, the customs police cannot set up checkpoints on it. To summarize, by using this informal way of importing his steel, Sâlim benefited in three ways: he acquired the goods more cheaply than those sold locally in the Palestinian enclaves, he slashed his transport costs, and also avoided paying VAT because the goods were officially destined for an entrepreneur and not a private individual.

The Israeli security mechanisms deployed from 2000 are paradoxical in their effects. On the one hand, they constituted an obstacle to the development of trade between the West Bank and the outside world. The back-to-back system of transferring products has slowed down flow, damaged foodstuffs, and pushed prices up (PalTrade 2009). On the other hand, because they prioritize security and keep Palestinian customs at bay, they open the way to various forms of traffic. Firstly, the location of crossing points on the seam zone rather than at the entrance to the Palestinian enclaves reduces Palestinian control over the validity of the declared value of goods, and, secondly, the Israeli capacity to check the final destination of vehicles carrying goods on the Palestinian side. On leaving the checkpoint, a lorry may either head towards an Israeli settlement or a Palestinian enclave (see Garb, this volume). These configurations therefore encourage fraud in the control systems: firstly in terms of the value and type of goods and, secondly, the identification and destination of the products (Israeli or Palestinian).

The fight against these informal practices has proved particularly difficult since there is very little coordination between the Israelis and Palestinians. Combined with the

12 All proper names cited in this article have been changed to protect the individuals concerned.

13 According to that procedure, goods going or coming from to PA areas are unloaded on one side of the checkpoint in order to go through detection and security checks, then they are reloaded on the other side of the checkpoint onto a second lorry.

Palestinians' limited capacity to collect information, the lack of coordination constitutes a major obstacle for the Palestinians in terms of projecting control beyond the enclaves into the heart of the informal networks. This is particularly true given that the customs authorities do not have an extended or structured system of informers. Since they lack financial means they cannot recruit informers on a regular basis. Moreover, due to the fact they have no control over the means of communication or the borders, it is difficult for them to manipulate the smugglers to obtain information on their rivals. Even though they are considered an authority regulating crossings between the enclaves and zones controlled by Israel, they do not have the same powers as the Israelis who are able to negotiate the closure or opening of channels with one set of players in exchange for information on the activities of others.

Informal Trade and New Israeli-Palestinian Economic Regulation

The study of informal trafficking enables us to assess the unilateral character of the separation between Israelis and Palestinians. In the absence of borders and controls on the periphery of the enclaves, the PA does not have any real means of regulating the Israeli products coming onto its markets—which explains why the Palestinian economy remains largely enslaved to the Israeli economy. However, unlike the Oslo Accords period (1993–2000), new forms of economic regulation have appeared: the crisis in the affected Palestinian enclaves and the new security arrangements, together with the legal and tariff differentials they create, open the way to new traffic which builds upon the new economic complementarities between Israeli and Palestinian territories.

Palestinian Enclaves: A Continuing "Captive" Market

Products smuggled from Israeli zones to the Palestinian enclaves include all types of everyday consumer goods: foodstuffs (mineral water, confectionery, meat, flour, eggs, etc.), toys, tobacco (cigarettes, hookah tobacco), clothes, cosmetics, pharmaceutical products. They can be found on most Palestinian stalls and markets. Because they are part of the informal economy it is difficult to evaluate volumes accurately. Apart from the loss in taxes they represent, such products pose a real problem to health and public safety: certain traffickers take advantage of the customs officials' inability to control health and technical standards to introduce perished or defective products into the enclaves.

This type of traffic existed well before the creation of the PA in 1994. It was reduced in the second half of the 1990s following police and customs pressure before developing once again during the Second Intifada. Between 2000 and 2005, the Israeli army was focused on fighting the Palestinian resistance and did not intervene against the smugglers. On the contrary, certain groups were mobilized in order to reinforce the networks of informers for the Israeli intelligence services. During this period people started noticing the widespread arrival of perished foodstuffs such as meat, milk, and vegetables (Laban 2005). From 2007 onwards the partial takeover of A zones in the West Bank has enabled the Palestinian authorities to resume the fight against smuggling. However this has proved particularly difficult because punishment for these crimes is still based on Jordanian laws prior to 1967, which do not condemn practices such as "re-labelling"[14] of goods. Lastly, due to the precarious economic situation of the Palestinian population, there is still considerable demand for cheap smuggled goods.

The trade in cars intended for the breaker's yard in Israel but which are sold in the Palestinian enclaves is another example of how Israeli products are "recycled." The traffic is all the more difficult to combat because it also plays on the duplication of control systems and that of Israeli and Palestinian areas of jurisdiction. Palestinian mechanics buy the cars from Israeli mechanics for spare parts. The cars are then shipped legally into the Palestinian enclaves. The real traffic only begins when a Palestinian trader sells a car to a customer who uses it to drive around in. Such cars or lorries are still being sold in B and C zones to Palestinians who do not have the money to buy a vehicle registered with the Ministry of Transport. The owners of such vehicles can drive without much fear of being stopped: other than in the large Palestinian towns where the Palestinian police make frequent confiscations (Ma'an News Agency 2009), it is impossible or difficult to check such vehicles. And this trade is encouraged when the Israeli police show a certain degree of tolerance towards drivers provided they do not drive on the main highways used by the Israeli settlers. In certain zones, such as south of the Hebron Hills or in certain villages to the west of Ramallah, we observed that these vehicles account for over half of the cars on the road.

The fight against such informal trade is particularly difficult because Israeli settlements in the West Bank often act as support or transit bases. Israeli settlers are sometimes the main players in the traffic, trading directly in products manufactured in the settlements. This trade began when the first Israeli settlements were created in the 1970s. In the south of Hebron Hills the settlements often sell their livestock (cattle, sheep, poultry) and milk to the Palestinian populations in the region, those in the Jordan Valley supply the Palestinian towns and villages with their fruit and vegetable products while those in the Ariel region to the south of Nablus trade in manufactured goods.

The PA began fighting this type of trade in the 1990s due to its informal nature and the financial losses incurred. In 2009, the fight took on a political dimension. On December 8, 2009, Salam Fayyad's government officially announced a boycott of these products, promising to remove them from Palestinian shops. In January 2010, the Palestinian prime minister set up a special fund to support this cause and in May he launched an awareness campaign while on the ground Palestinian customs officials increased their efforts.

The challenge remains, however, difficult to meet. Even though the networks of police informers often enable goods to be tracked, the police are unable to intervene on the outskirts of the enclaves and their entry points and can only act once the goods arrive in the Palestinian shops. The customs police cannot launch an operation to intercept trade in the Israeli towns or on the main highways in the West Bank. Palestinian traders may therefore load and ship products without any fear of getting caught. The interconnectedness between Israeli and Palestinian territories also increases the points of contact between the settlements and enclaves. In Hebron the situation is particularly significant: settlements are located both around and inside the city itself and provide both support bases and many points of entry to the markets of the Palestinian town.

The boycott announced by the Fayyad government has therefore had a more political than economic impact. In a situation in which Palestinian customs police have only limited

14 In order to evade controls by the PA certain entrepreneurs specialize in manufacturing packaging and labels to repackage goods that have been tampered with.

control on the enclave boundaries and points of entry, the only possibility of effectively implementing such a boycott is to ensure an awareness campaign is organized successfully. The value of Palestinian customs police takings remains limited compared to trade values: in May 2010 the takings totaled 5 million dollars (Zacharia 2010), whereas as annual sales of this type of product total 200 million dollars in the West Bank. The simple fact of having to appeal to the people to fight the import of products from the settlements is another illustration of the PA's weak capacity to protect its market.

The interceptions mainly affect small companies and farms in the settlements for which trade with the Palestinian enclaves accounts for a significant proportion of their business. On the other hand they have less impact on companies with a larger turnover for whom trade with the enclaves is secondary. In economic terms, the effects of the boycott are consequently sporadic. At a political level protests by settlers running small farms have triggered angry reactions within the Netanyahu administration. But this mobilization is linked more to the capacity for mobilization and pressure from Israeli settlers than the impact of the boycott policy. To summarize, unless there is massive mobilization by the Palestinian population as part of the boycott, it is unlikely to achieve its goal.

Globalization of the Border Economy

Far from restricting itself to the Israeli and Palestinian territories, goods traffic is becoming part of international informal trade thanks to the changing legal and control systems since the beginning of the 1990s. We will focus on two types of trade: stolen cars and fake goods. In both cases Palestinian enclaves act as transit zones for the development of informal trade in Israel or between Israel and other countries. Trafficking, therefore, shows not only that new forms of complementarity are emerging between the Israeli and Palestinian markets, but that the complementarities between these two spaces are also used by international smuggling networks.

When the PA's zones of jurisdiction were defined, the enclaves became strategic zones for the development of car theft and the informal market in spare parts. Lorries and cars stolen in Israel were dispatched to the enclaves either to have a makeover and be sold to private individuals or to be broken up into spare parts. Vehicles were mainly sold in the heart of the Palestinian market while spare parts were intended for both the Palestinian and Israeli markets. Because the Israeli police refused to intervene in the enclaves, in the initial years of the peace process the enclaves could act as hiding places. In the Gaza Strip, Palestinians trafficking vehicles felt safe once they were over the "border" (Abu Moaleik 2004). In the West Bank spare parts dealers began running massive open-air breakers yards. Between 1993 and 1997 the trafficking grew considerably: the number of vehicles stolen in Israel went from just under 25,000 to over 45,000 a year.[15]

During the Second Intifada (2000–2005) the lockdowns and restrictions on movement imposed by the army between the Occupied Territories and Israel reduced the trafficking. From 2002, the invasion of the Palestinian enclaves enabled the Israeli police to confiscate directly in A and B zones of the West Bank over 1,600 vehicles stolen in Israel (Shahar 2003). The number of cars stolen each year fell to around 25,000 in 2004 (Katz et al. 2006). The Israeli police have since continued to intervene regularly in the Palestinian enclaves. The operations are led by the Etkar unit. This unit, which specializes in the fight against

15 Hatzofe, Israel Insurance Association. Available at: http://www.iris.org.il/cartheft.htm, accessed February 28, 2010.

car theft, was created in 1998 and dismantled in 2004 due to budget cuts. It was finally reinstated in 2005 when 40 percent of its budget was funded by Israeli insurance companies. It has since operated unilaterally without any coordination with the PA.[16] Since 2007, car thieves have also had to deal with the Palestinian police who have also adopted a more uncompromising attitude (Ma'an News Agency 2009).

Traffickers have continued operating despite increasingly difficult conditions. In terms of value, that is, in proportion to the number of vehicles on the Israeli roads, action by the Israeli and Palestinian police has reduced the number of thefts to a level lower than that of 1994. According to Israeli police statistics, whereas in 1994, when 19.4 vehicles in every 1,000 were stolen, in 2009, the figure was only 9.8. On the other hand, in terms of volume the number of stolen vehicles remains higher than in 1994: less than 25,000, compared with approximately 30,000 in 2007.[17]

It is particularly difficult to stop such trafficking, firstly due to the territorial and security situation. The overlapping zones of jurisdiction and the many crossings between them make it difficult to control vehicles, which also increases the number of crossing and circumvention points (Barthe 2007). There is still considerable demand while the traffickers create complex networks involving Israeli and Palestinian collaboration. Certain networks are currently run by large Israeli families who subcontract to petty criminals from various countries.[18] Most trafficking of stolen cars is nowadays structured by the demands of the Israeli market, which is the traffickers' main outlet: it supplies the Israeli mechanics who provide their customers with spare parts. The cars are stolen in Israel and taken by an Israeli or Palestinian driver to a breakers yard in the West Bank outside the range of the Palestinian police or protected by corrupt police officers. Then, the parts are shipped back in small quantities to Israeli garages that specialize in selling spares. The system is so efficient that an Israeli mechanic in Tel Aviv can order a part in the morning and have it delivered via the West Bank the same evening.

The traffic is becoming international. Since about 2005, spare parts that were previously intended for the Israeli and Palestinian markets alone have been exported by some networks to neighboring Arabic countries. In 2006, the Etkar unit intercepted a network of traffickers exporting spare parts to Jordan and then on to Iraq (Katz et al. 2006).

The West Bank has also become a transit zone for importing fake goods to Israel. These are mainly Chinese clothes imported into A zones and then forwarded to Israel. In 2009, Israeli traders often took advantage of weaknesses in the control system to import copies of various American and European brands from China. They used Palestinian intermediaries to contact the Chinese factories that subcontract for the brands and order batches of sports shoes manufactured over and above the quotas determined by the brand. Because they came straight from the factory, the shoes could be purchased at a lower cost from the Israeli franchises. Once they left the factory, the batches were loaded into a container in the name of a Palestinian trader and forwarded to the Port of Ashdod and finally shipped to the West Bank. After the container had gone through customs at Ashdod it was sent to Hebron so that the fake shoes could then be sent to Tel Aviv, Jerusalem, Haifa, or another Israeli city. The fake shoes were intended for shops that were well established and were therefore sold alongside shoes that were legally imported.

---

16 Interview with a police officer from the Etkar unit, Tel Aviv, March 2009.

17 Statistics published on the official site of the Israeli police: http://www.police.gov.il/mehozot/agafAHM/yahidotArtziot/Documents/etekar.pdf, accessed February 28, 2010.

18 Ibid.

Working with Palestinians minimized the risks taken by the Israeli sponsors. Palestinian goods could not be penalized because they were not subject to Israeli commercial legislation, and there were no Palestinian regulations governing fakes. If Israeli customs found any illegal products, the only risk was that the goods would be destroyed. Since late 2009 and early 2010, however, Israeli customs have become more vigilant and stricter in resisting this type of trade: containers of clothes are opened much more frequently and inspected more carefully. In order to cope with the increased surveillance traders have set up new strategies: they are importing increasing numbers of unlabeled products and labeling them in the Palestinian enclaves or Israel.

Figure 6.2 Production of fake clothing labels, West Bank
Credit: Cédric Parizot, March 2010.

The production of fake labels in the Palestinian enclaves has developed since the 1990s in a context that has seen a considerable drop in the demand for textiles in the local workshops. The fact that Asian markets are now open to Israel following the Oslo Accords (1994–2000)[19] has enabled Israeli entrepreneurs to relocate their production, which previously took place

19 Following the launch of the Oslo Accords, the State of Israel has managed to develop economic and commercial relations with countries that previously refused to trade with it.

in factories and workshops in the West Bank, to countries such as Jordan, Turkey, and China (Bouillon 2006). Israeli entrepreneurs were attracted by both the cheap labor and the fact that the goods could be shipped very easily between these markets and Israel. In the second half of the 1990s the implementation of the Israeli closure policy in the West Bank and Gaza Strip affected the movement of people and goods. To avoid closing down their workshops and factories, certain Palestinian entrepreneurs began producing labels in the A zones. The absence of Palestinian legislation on fake goods and the difficulty of access for the Israeli police offered total impunity. The production was mainly used by Palestinian traders who wanted to add value to the goods they sold in a market where the demand for international brands is constantly increasing. The rest was intended for the Israeli and foreign markets.

The development in the traffic of fake products and labels between Israeli and Palestinian territories and countries like China shows how certain economic players have managed to grasp the opportunities presented by the system of separation to ensure their continued economic survival or to create more profits. In the West Bank they enable many of them to continue business in a context where Palestinian entrepreneurs are needed less to perform subcontracting work for the Israeli market. In Israel trafficking enables certain Israeli traders to increase their margins either to bear the cost of competition or to make more profit. In a way, trafficking is part of a process proper to globalization described as the "informalisation" of the formal sector. To cope with competition entrepreneurs include more and more informal trading in their formal business (Mercier 2009).

Conclusion

In 2010, PA policies to control the flow of goods and people remain very deterritorialized. This deterritorialization is not synonymous with power: it does not refer to the capacity to break geographical boundaries, but to the PA's inability to project control over its territory or stabilize it over time. This perspective makes it possible to better assess the limits of the redeployment of Palestinian forces since summer 2007, which have created the illusion that Mahmoud Abbas's power and sovereignty have been restored within the enclaves of the West Bank. In actual fact the security measures maintained by Israel have not enabled such a process to come to fruition.

The deterritorialization of Palestinian control is in sharp contrast to that of Israeli control. Israel maintains firm control over 60 percent of the West Bank, i.e., in the C zones that include the main highways and zones administered by the settlements. Furthermore, the Israeli army has displaced its long-term control beyond the boundaries defined during the Oslo Accords period because not only does it frequently intervene in the A and B zones of the Palestinian enclaves, but also because it regulates the movements of the PA in the A zones.

The imbalance between Israelis and Palestinians in the territorialization of control shows that the Israeli separation policy has produced a territorial regime that cannot be understood using the categories of analysis proper to the political imagination of the modern State. The separation does not result in the emergence of two separate territories in which only the control of the State that claims prevails. On the contrary, it maintains a single territory in which several players with unequal degrees of power, i.e., Israelis, Palestinians, international players, smugglers, etc., act.

Depending on their identity and the zone in question, the movement of people or goods is likely to be monitored and defined by the Israelis and/or the PA. The way in which the regime actually operates appears even more complex if the intervention of the other formal and informal players involved is taken into consideration. In the previous chapter, Yaacov Garb highlighted the various players involved in the formal management of goods shipped between Israel and the West Bank. In this chapter we have emphasized the role of the "informal entrepreneurs" of the separation. They should also be taken into consideration, given that, by making it easier to work around the physical and legal obstacles, they also participate in regulating the crossings.

The power to regulate of these informal entrepreneurs in the informal market is particularly significant as it directly affects the restructuring of relations between the Israeli and Palestinian economies. These entrepreneurs play on the legal, commercial, and fiscal differentials between the territories and control systems and help create new forms of complementarity between Israeli and Palestinian markets. As we have shown, they help maintain the Palestinian market not only as a captive market but as a market in which perished and defective goods are recycled. They also contribute to maintaining trade between the Israeli settlements and Palestinian enclaves, which the PA finds very difficult to combat. Lastly, they give the enclaves the status of transit zones. This trafficking concerns more players as it does not only affect those involved in the informal economy. In an increasingly globalized context, such types of trade are becoming more and more strategic, both for large and small entrepreneurs, in Israel and the Palestinian enclaves alike.

Acknowledgments

We would like to thank the following colleagues and friends for the interest and patience they have shown in the many re-reads, comments, and advice that helped us to write this chapter: Arnaud Garcette, Aurélia Dusserre, Benoit Laborde, Delphine Mercier, Malika Hattab, Stéphanie Latte Abdallah, and Sylvie Mazella. We would also like to thank Mathieu Coulon and Marjolaine Barazani (Maps 6.1, 6.3 and 6.4) and Tareq Natsheh (Map 6.2) for their help in drawing the maps.

Chapter 7
The Rise and Fall of Gaza's Tunnel Economy (2007–2014)

Nicolas Pelham

Introduction: Palestine's Struggle for Trade Routes

This chapter analyzes the rise of Gaza's tunnel economy in response to siege and its even more searing collapse with Egypt's military takeover in July 2013. It traces the three-stage transformation of Gaza's economy, from a formal to an informal economy when regional and global powers boycotted the rule of the Palestinian Islamist movement, Hamas, and then to drip-fed dependency, when Egypt's armed forces destroyed the tunnel complex in the summer of 2013. It argues that contrary to the intentions of its architects, the siege led to the reconfiguration of Gaza's socioeconomic hierarchy and fostered alternative trade conduits. However, ultimately the tunnel complex failed to deliver the economic independence Gaza's leaders had sought.

The chapter is divided into three sections. The first traces the surge in tunnel-burrowing that followed the internationally-assisted blockade Israel imposed on Gaza in the wake of Hamas's 2007 takeover. It examines how Hamas cadres led the efforts to dig itself out of the siege, and used the proceeds from the trade to finance its rule. And it explores how the trading route away from Israel and towards Egypt marked a fundamental shift in the allocation of resources in Gaza, diverting the economy from the traditional and cosmopolitan mercantile class plugged into the international markets to Gaza's Bedouin underclass plugged into clan networks spanning the Sinai peninsula and beyond into northeast Africa. As a result of the siege, Gaza lost contact with the formal contact but established a parallel informal economy whose tentacles could reach at least as far, and with far less red tape.

The second section explores the consolidation of the tunnel economy following Israel's bombing campaign against the tunnel infrastructure in its Cast Lead offensive of the winter of 2008–2009. It details how the resulting repairs and rehabilitation resulted in an overhaul of the supply lines. So substantial was the upgrade that by mid-2010 more goods were flowing into Gaza through the tunnels more cheaply than flowed from Israel prior to its imposition of the siege. The tunnels became the conduits for over two-thirds of Gaza's imports, including 90 percent of the enclave's construction materials and such basics as fuel.

As supply outpaced demand, however, tunnel operators became a victim of their own success. Israel's easing of the closures amid international protests at its lethal enforcement of the blockade against the Mavi Marmara flotilla in May 2010 hastened the tunnel economy's contraction. The slowdown continued with the electoral successes of the Muslim Brotherhood in Egypt in 2012, and President Mohammed Morsi's sympathetic consideration of the establishment of an above-ground free-trade zone on the Egyptian–Gazan border. By mid-2013, less than a third of the 1,200 tunnels functioning at the height of the tunnel boom remained in operation.

The final section explores the potentially fatal blow to the tunnel economy dealt by General Abdelfatah al-Sisi's July 2013 overthrow of the Morsi regime and offensive against his Bedouin opponents in Sinai. Determined to cut off their source of financing and perceiving continued Islamist rule on Egypt's northeast border a national security threat, he ordered his forces to incapacitate the tunnel economy. Hundreds of tunnels were destroyed and thousands of tons of tunnel-bound traffic impounded in the resulting assault. Gaza's economy plummeted to a point resembling the worst days of its siege.

Though an ingenuous and resourceful experiment which briefly offered the hope of regeneration, the tunnels have failed to be the panacea that Hamas's rulers had promised. Only Gaza's reintegration into the global economy and the lifting of its multi-layered siege can secure the development from which Gaza's 1.8 million have been for so long deprived.

I am a journalist by profession and not a social scientist, and this chapter is based on the sources that I collected as a journalist: I have indeed made dozens of field-trips to Gaza and hundreds of interviews with people associated with the tunnel economy. In particular, it draws on findings of an unpublished study of Gaza's political economy for the UN Office for the Coordination of Humanitarian Affairs (OCHA) on which I was privileged to work during 2010. This comprised the most comprehensive survey to date of Gaza's tunnel economy—including a sample of over 500 Gazans involved in tunnel-related activity, including officials from Gaza's regulatory authorities and local and national government departments.

The Rise of Gaza's Tunnel Economy

For millennia Rafah had been a stopover for merchants plying the cross-continental trade route between Asia to Africa. Israel's establishment in 1948 did not entirely sever the tie, for Gaza was attached to Egypt until 1967, and Israel's occupation thereafter maintained the connection to Sinai. Bedouin families and clans traversing the borders continued to mingle and marry. Only in 1982 when Egypt and Israel carved their border along Gaza's southern edge cutting Rafah in two as part of the Camp David Accords did separation really set in.

No sooner had the Camp David agreement bifurcated Rafah town, than the Bedouin families straddling the border began burrowing underneath, particularly at the mid-point of the 14-kilometer border, where the earth is softest. Israel's first recorded discovery of a tunnel occurred in 1983, within a year of the implementation of the accord. Using their residential houses for cover, Gazans dug underneath the ground floors to a depth of about 15 meters, headed south for a few dozen meters, and surfaced on the Egyptian side of the border often in a relative's house, grove, or chicken coop to avoid detection. By the late 1980s, tunnel operators were importing such basics as processed cheese, which was subsidized in Egypt and taxed in Israel, and probably some contraband too, including drugs, gold, and weapons.

Israel's incremental disengagement from Gaza quickened the expansion. Following the signing of the Oslo Accords in 1993, Israel built a barrier around Gaza, restricting access through Israel's terminals which periodically closed. Israel's response to the Al-Aqsa (Second) Intifada which erupted in September 2000 offered a foretaste of the siege on Gaza to come. Israel upgraded the barrier along its borders, destroyed Gaza's seaport and airport in 2001, and in 2004 ploughed through 1,500 homes along the 14-kilometer border with Egypt to carve out a 100-meter-wide Philadelphia corridor, cordoned from Egypt with a 7-meter high wall. Lockdowns lasted months. For the first time the tunnels served as safety valves for wholesalers to address scarce supplies. Political factions built their own for smuggling weapons and cash, as well as military operations, such as detonation of a mine under a tank. A senior PA security official and a prominent Fatah leader in Gaza, Sami Abu Samhadana, from a Bedouin clan straddling the Rafah frontier, oversaw much of the expansion, highlighting the fusion of security and business interests which was to become a hallmark of future development.

The closures intensified following, first, Israel's September 2005 disengagement, Hamas's electoral triumph in January 2006 and assumption of government, and the takeover of Gaza by Hamas's military wing, the Ezz al-Din al-Qassam Brigades, in June 2007. Israel closed Erez terminal to Gaza's laborers in March 2006, shut Karni, Gaza's prime crossing for goods, for half of the six months prior to Gilad Shalit's capture in June 2006, and closed it altogether immediately after.[1] Following a salvo of rocket fire from Gaza in November 2007, Israel cut food supplies by half, and severed fuel imports. In January 2008, after rocket attacks on Israel's town of Sderot, Israel announced a total blockade, prohibiting all but seven types of humanitarian supplies. Most basics, from toilet paper to pasta, were banned.[2]

Hamas's initially looked above ground to Egypt. Hopes of formal arrangements proved slight. The Mubarak regime was nervous both of Israeli intentions to shunt responsibility for Gaza onto Egypt and of national security risks posed by the rise of Islamists on their northeastern front, and joined Israel's goods boycott. Besieged on all sides and without fuel, Gazans abandoned their cars on the roadside and travelled by donkey. Failing to win Egypt's assent, Hamas forced the issue. In January 2008, the Ezz al-Din al-Qassam Brigades, the armed wing of the ruling Palestinian Islamist movement, Hamas, bulldozed the Philadelphia corridor barricade, opening a passage for hundreds of thousands to spill into Sinai on a shopping spree pent-up after months of denial. But the respite was brief. Within 11 days Egyptian forces had herded them back. A reinforced contingent of Egyptian army forces locked Gaza's gates, and built a fortified border wall.

The event marked a turning point in the decision to move conduits underground. The rudimentary network already in operation provided a base for expansion, and in the months that followed the Qassam Brigades oversaw a program of industrial-scale burrowing. Each tunnel cost between $80,000 and $200,000 to dig, and in search of investment, Hamas turned its preachers into fundraisers, using mosques and charitable networks to promote funds. Preachers attracted investors with offers of unrealistically high rates of returns.[3] They extolled tunnel operations as a "resistance" activity, and hailed workers killed on the job as "martyrs."[4] Gaza's already limited electricity was diverted to power hoists.

1 The UN's humanitarian agency OCHA reported that Karni was closed for 54 percent of scheduled days during 2006, compared to only 18 percent and 19 percent of days in 2004 and 2005 respectively. Exports collapsed to only 4 percent of the 2005/6 harvest (UNOCHA 2006, p. 3).

2 These were wheat flour, rice, sugar, vegetable oil, dairy products and milk, and legumes. The number of truckloads entering Gaza fell from 12,000 per month on the eve of disengagement, to 7,000 per month following Shalit's capture a year later, and a mere 2,000 by November 2007.

3 In the summer of 2008, preachers in Bureij refugee camp launched a pyramid scheme attracting Gazans offering returns of 30–50 percent per month. Gazans reportedly sold their wives' jewelry and family cars to take cash in on the dividends. However, Israel's bombardment of the tunnels in January 2009 prompted a rash of withdrawals. Unable to pay, brokers fled and the scheme collapsed. Belatedly, the Hamas government offered compensation of 16 percent on investments of some $400 million. There were few payouts.

4 Gaza's authorities required tunnel operators to provide rudimentary life insurance policies for tunnel laborers, recommending compensation of $9,000 to $11,000 to families of married workers

Construction workers barred from working in Israel after Israel closed Erez terminal to laborers in March 2006 found new employment in the tunnels at rates not so far below those in Israel. Teams of six laborers working round the clock in two 12-hour shifts could dig an average of 10 to 15 meters each day.

Within weeks, peddlers appeared on Gaza's roadsides selling petrol in plastic bottles, laced in sand from the tunnels. But once opened, the tunnels were constantly upgraded. By mid-2008, the larger ones sported internal lighting, intercoms for communication, and generators to maintain operations in the event of frequent power cuts. The tunnels' rough hewn edges were smoothed to reduce damage to imports. So congested were parts of the border that in-place diggers burrowed tunnels one on top of the other, using Google Earth to map routes and ensure that they stayed on course. Fuel which had arrived in jerry cans was pumped through three-quarter-inch pipes at a rate of 20,000 liters per hour.

As noted above, the tunnels still had military uses. The Qassam Brigades reportedly included a tunnel warfare commando unit. But their primary purpose changed. From weapons smuggling enterprises, the tunnels rapidly burgeoned into what one trader described as "the lungs through which Gaza breathes." According to Ziad Jarghoun, by the eve of Israel's December 2008 Gaza war, their number had grown from 20 in mid-2005 to around 500,[5] increasing underground trade from an average of $30 million per year, to $36 million per month.[6] The rewards were significant. Fully operational, a tunnel could generate the cost of its construction in a month, with earnings split equally between Gazan and Egyptian operators. With demand far in excess of supply, tunnel operators stood to earn $50 for ferrying a 50 kg sack through the tunnels. The PA's continued salary payments to 75,000 employees, including all security personnel suspended on full pay, sustained Gaza's liquidity and purchasing power.[7] Free of interference from Israeli controls and red tape, the trade cycle was often faster than Israel's pre-blockade route. Goods could arrive within three to five days of placing an order. When Israel reduced gas supplies, smuggled canisters quickly surfaced on the market. Vaccines quickly arrived following reports of disease sweeping chicken farms. And ahead of Muslim holidays, traders imported toys, live sheep, and fresh beef.

The tunnels not only spared Gaza's economy from de-development, they redeveloped it. As established trading routes via the Israeli Port of Ashdod waned as they shifted to Sinai, Gaza's commercial ties with Egypt revived after a 40-year lapse. Merchants switched suppliers from Israeli and European to Egyptian, Chinese, and Turkish outlets.

The mercantile class changed as rapidly as the merchandise. The shifting trade routes distributed wealth and economic power away from the traditional business elite, hitherto linked to Israel, towards an underground economy. Gaza's established merchants excelled in foreign languages, exposure to the wider world and higher education. By contrast the new masters of Gaza's supply lines were skilled in smuggling, often came from an underclass of semi-nomadic Bedouin stock with family connections straddling the border,[8] and enjoyed backing from Gaza's new rulers, Hamas. As such, the tunnels became a key

---

killed in the tunnels.

5 Interview, tunnel operator, Rafah (Egyptian side), June 2011.

6 Ziad Jarghoun is a PFLP Central Committee member.

7 Including payments for health and education, economists estimate the PA injects some $1.5 billion annually into Gaza. Interview, Ali Abu Shahla, then secretary general of the Palestinian Businessmen Association, Gaza City, May 2012.

8 Some of the key families were refugees removed from the Negev in 1948, able to tap into Bedouin clan networks across the Sinai and beyond into Libya and the Horn of Africa.

driver of upward mobility and social change, empowering groups that were previously marginalized and spawning a new class of nouveau riche middlemen. PA officials claimed the tunnel economy created 1,800 new millionaires (Al-Ahram 2013).

To finance their rapid expansion, tunnel operators established small joint ventures of between four to fifteen stakeholders drawn from a broad cross-section of Gazan society to build and operate one or two commercial tunnels. Lawyers drew up contacts stipulating the price per share, the number of partners, and system for distributing shareholder profits. Partners interviewed for this study include porters at the Rafah land crossing, security officers working in the former PA administration, university graduates, NGO workers, and diggers who had made good. One interviewee, Abu Ahmed, who had previously earned 30–70 shekels a day as a taxi driver, invested his wife's jewelry worth $20,000 to become an equal partner with nine others in building a tunnel. Many Gazans who had previously relied on work with Israel also cut their losses. Former agricultural workers who had lost their jobs worked Israel's land ploughed earnings into the tunnels, and construction contractors whose businesses had collapsed due to Israel's suspension of construction materials hired labor for building tunnels. Stung by Israel's blockade, some merchants invested too, as did other moneyed Gazans drawn by the prospect of quick profits.

As the Hamas government legalized the tunnel economy, so too they sought to regulate it.[9] It adopted a two-tier approach to the tunnels. It sidestepped control of factional tunnels, which after the Qassam Brigade chased away Fatah's forces in June 2006 were largely run by the Qassam Brigades. At the same time and not without clashes, the government gradually weaned control of the commercial tunnels from the Qassam Brigades. The Interior Ministry established a Tunnel Affairs Commission to act as the regulatory authority. It enforced a blacklist of imports from Egypt, including weapons, alcohol and tramadol, a sedative much used in Gaza. In response to public concern at a rising toll of casualties in the tunnels, particularly of child workers, it issued guidelines intended to ensure safe working conditions. Over time it quarantined the site with fencing, and assigned some 300 internal security personnel in black fatigues to restrict access, check documentation at entry points, and patrol tunnel mouths on motorbike. It introduced a system of tunnel licensing to prevent unauthorized digging. It monitored the market for evidence of hoarding and price fixing, particularly of fuel, and enforced price stabilization. Violations were punished with tunnel closure and fines. "We used to earn thousands smuggling shipments of hand guns, grenades, bullets, and TNT," said a tunnel operator in business since the Second Intifada. "But it's not worth the risk to be prosecuted by Hamas."[10]

The Commission also taxed imports, to the point where they became the Hamas government's primary revenue source. The Commission weighed trucks on an electronic weigh station buried in the sand, and required haulers to present chits for their cargoes before exiting the tunnel zone. Rafah municipality charged operators a one-off license fee of 10,000 NIS per tunnel ($2850), and 1,000–3,000 NIS for connection to the electricity grid. Customs duties were charged on Egyptian-subsidized fuel (initially of about 0.5 NIS per liter) gas (of 30 NIS per canister), tobacco (3 NIS per pack) and generators. In addition, Gaza's authorities levied a 14.5 percent value-added sales tax on all goods.

9 Hamas officials claim they controlled the entire operation except the section, east of Rafah terminal, which falls inside Israel's self-declared buffer zone, and thereby benefits from Israel's protection. When Hamas's monitors approach, smugglers reportedly open fire on Israel. Israel shoots back, helpfully chasing Hamas's monitors away.

10 Interview, tunnel operator, Rafah, April 2010.

Hamas's oversight and taxing of hitherto tax-free tunnels did not go unchallenged. In late November 2007, armed clashes erupted between Hamas security forces and members of the al-Sha'er clan in Rafah, after Hamas destroyed two of its tunnels.[11] But, for the most part, the rapidly-expanding business opportunities Hamas's rule brought largely trumped lingering resentment at the intrusion. Unemployment in Rafah, on the Egyptian border, fell from about 50 percent on the eve of the takeover to 20 percent by December 2008 (Al-Surani 2008). Trade routes pointing north switched south. A decade earlier, all but one percent of Gaza's total imports came from or via Israel (US Department of State 1997). By the eve of the Gaza war the ratio had nearly reversed.

Egypt and Israel both adopted contradictory policies towards the tunnel traffic. The reorientation of Gaza's trading routes to Egypt enabled Israel to tighten its closures around the enclave and thus deepen the divide between Palestine's cloven halves: Gaza and the West Bank. The tunnel economy also helped Israel highlight Hamas's status as outlaws, beyond the realm of legitimate commerce. Hamas's build-up of business interests and an asset base also gave Israel leverage to pressure it to accept its dictates and halt fire.[12] Egypt had no more desire than Israel to shoulder Gaza, and treating Gaza as a political football punted responsibility back towards Israel.[13] But in the waning months of the Mubarak regime its ability to impose discipline was weaker. The tunnels offered copious opportunities for bribes from a hitherto unprofitable region from traders wishing to avoid state controls.[14] Government sponsored in Gaza, the tunnels remained clandestine on Egypt's side of the border. Egypt established checkpoints along the northern Sinai coastal road leading to Gaza and repeatedly reported its forces had uncovered weapons stockpiles stashed in caches in northeast Sinai close to the Rafah border (Salim 2010). At the same time much slipped through, for a price. While the tunnels mouths moved from basements into the open terrain in Gaza's Philadelphia corridor—protected from the elements by white canvas coverings, in Egypt the tunnel mouths remained covered or else snaked up to a kilometer inside Egypt.

Of greater concern was the tunnel economy's impact on Sinai. Excluded from the formal sector, such as tourism and government service, Sinai's marginalized Bedouin found new sources of economic growth in first sustaining and then rebuilding Gaza. Israel's heightened curbs on Bedouin smuggling from Sinai, including the erection of new fortified positions, the stationing of two additional battalions and beginning in December 2010 construction of a 240-kilometer iron wall, plugging Sinai's smuggling routes to the Negev, further increased their dependence on Gaza. Economic ties revived long-standing solidarity between the tribes straddling the border, and gave them common cause. "We're Palestinians working for the sake of Palestine," said a tunnel laborer on Egypt's side of Rafah.[15] Families in Rafah,

11 Interview, Palestinian observer, Rafah, November 2007.

12 A third of the 48 targets Israel's armed forces struck inside Gaza in the first half of 2010 were tunnels. IDF intensifies "terror tunnel" attacks in Gaza. *Yediot Achronot*, July 1, 2010. From 2006–2009, at least seven people were killed in Israeli airstrikes on the tunnels, and more than 250 injured (Al-Mezan Centre for Human Rights 2009).

13 According to the World Health Organization, 63 people, including 22 children, died waiting to access medical care outside of Gaza between February 2008 and June 2010.

14 According to operators, Egyptian security officers charged $2000 to $3000 not to detonate exposed tunnels. "My cousin bought $150,000 worth of carbonated drinks, and half an hour before its dispatch to Gaza, the police seized his truck. Not only did he lose the consignment, he had to pay the police the costs of transport to the dumpsite and the wages of the laborers who destroyed it." Interview, Gaza wholesaler, Rafah, April 2010. "If Egypt really wanted to shut the tunnels down, it could easily do it." Interview, fuel distributor, Rafah, April 2010.

15 Interview, tunnel operator, Rafah (Egyptian side), July 2011.

a town divided by Israel's withdrawal to the international border in 1982, reunited, going back and forth via the tunnels for supper. The Sawarka, Ramailat, and Tarabeen tribes, whose land straddled the Gaza-Egypt border, were the prime beneficiaries, creaming off a substantial slice of the $1 billion in goods (and weapons) that flowed into Gaza annually. The plush new hamlets sporting Chinese-style pagoda roofs near the border bore testimony of the profits. "A decade ago, my whole clan had three cars. Nowadays each household does," says a Bedouin trader in Rafah.[16] Such were the ties, that Palestinian businessmen laundered profits from smuggling buying property not only in Gaza but northern Sinai.

The increase in economic and military clout deeply worried Egypt's authorities. So too did fears of seepage of militants and weaponry from Gaza into the Sinai, endangering its lucrative southern resorts (Ali 2009). On occasion, Bedouin operators tapped into armed clan defense committees versed in Sinai's topography through centuries of roaming the terrain and well-armed to subvert Egyptian security efforts. There were sporadic reports of clashes between Bedouin irregulars and Egyptian forces seizing contraband.[17]

From Operation Cast Lead to the Arab Spring

Israel's Cast Lead offensive during winter 2008–2009 provided the tunnel economy with its greatest test to date. Although the tunnels reportedly served as an escape hatch for some Hamas officials, aerial bombardment severely damaged the network, halted commercial traffic, and precipitated a run on the tunnel construction investment funds, and, as noted in note 4, precipitated their collapse. As part of the internationally-brokered ceasefire, Israel secured US agreement to curb smuggling to Gaza and international policing of the Red Sea to prevent Iranian shipments via Sudan. Egypt also committed to construct a 25-meter deep underground steel barrier supervised by US military engineers along its Gaza border, aimed at plugging the tunnels within a year. By the end of 2010, the Egyptian authorities claimed to have sabotaged some 600 tunnels, plugging entrances with solid waste, sand, or explosives, and flooding passages with sewage (Ma'an News Agency 2010c). Use of tear gas and other crowd control techniques inside the tunnels resulted in several publicized deaths. "Last week four boys were gassed in the tunnels," said a head teacher in Rafah. "The Egyptian police just closed the escape hatch, and left them to suffocate."[18] Tunnel operators, too, reported increased surveillance. "In the past, they would look the other way when a lorry stopped to unload at a tunnel mouth," said one. "The war marked a turning point in how Egypt's security deals with us."[19]

But although the tunnel economy suffered a temporary setback, they did not torpedo it. Egypt's declared counter-measures never quite materialized. Egypt blamed stony ground for hampering the hammering of steel plates more than four meters underground

16 Interview, Rafah (Egyptian side), July 2011.

17 For instance, Bedouin gunmen opened fire on Egyptian forces after they seized a consignment of tunnel-bound cars (BBC 2010b).

18 Interview, Mazen Abu Amarneh, Hamas official and schoolmaster, Rafah, May 2010. "The Egyptian police suddenly raided one of the tunnels I was in and poured a noxious liquid inside. I was badly burnt and couldn't work for three months." Interview, tunnel worker, aged 20, Rafah, April 2010. Associated Press reported dozens of deaths by Egyptian security forces. "Hamas: Egypt gasses 4 smugglers," Associated Press, April 29, 2010. Egypt's authorities denied the claims.

19 Interview, tunnel operator, Rafah, April 2010.

(Associated Press 2010).[20] When they did, tunnel operators quickly cut holes with blow torches, nullifying the multi-million dollar project at a cost of a few thousand dollars (BBC 2010a). Tellingly, construction slowed in the area where tunnel activity was most concentrated. Sympathetic media coverage of the siege and condemnation of the Mubarak regime further dampened Egypt's political will. Frustrated, US Congress abandoned the underground wall project in mid-2011.[21]

Moreover, the measures only spurred Palestinians to dig longer and deeper tunnels, less prone to detection or sabotage. Some extended 1.5 kilometers and 25 meters underground. The ceasefire that followed Israel's 2008–2009 winter offensive gave a breathing space for tunnel repairs unfettered by fears of Israeli attack. While the international community put its promise of reconstruction on hold, Hamas reduced levies to kickstart their own.[22] Operators reinforced tunnels first with wooden planks, then breeze blocks, then metal to support a widening of tunnels to accommodate the import of construction materials. They replaced rope ladders flung down the shafts with electric lifts, and the four-meter long sledges, or *shahata*, pulled by winches with carts running on tracks, much as in coalmines. Incidence of death, damage, and loss declined, too, since Egypt's security forces found longer tunnels harder to find, and tunnel conditions had improved. Economies of scale further lowered the costs. By the summer of 2011, 60 percent of traders reported prices had fallen to equal to or below the pre-siege cost of goods coming from Israel.[23]

Within two years of Israel's 2008–2009 winter offensive, tunnel capacity had increased tenfold, leading to a surge in consumerism. The number of tunnels transporting livestock rose from 3 in 2008 to at least 30 by mid-2010. Fuel which had cost four times the price in Israel in 2007 sold at a quarter Israel's price by 2010. Prices for Turkish cement plummeted from $1,500 per ton at the height of the closures in mid-2008 to the pre-siege price of $100. The cost of transporting a 50 kg sack of goods through the tunnels fell from $50 to $5. As capacity improved, and prices fell to within a range Gazans deemed affordable, so demand grew. Between 2008 and 2010, traders of household goods reported a 60 percent increase in imports via the tunnels. Cars—hitherto cut into three and welded together in Gaza—arrived whole, dragged through the tunnels by bulldozers.

By mid-2010, Gaza's retailers reported they had alleviated "to a reasonable extent or more" shortages resulting from Israeli restrictions. Wholesalers surveyed in April 2010 reported that the tunnels accounted for 68 percent of all goods available in Gaza's markets, including 90 percent of all construction goods, fuel, and household appliances, 70 percent of its clothes and office supplies, 60 percent of its food and 17 percent of medicines. One in four merchants stocked goods solely transported via the tunnels. "There are at least 1,500 underground tunnels now," says an owner. "Most are bigger and better than ever before, and all of them are open for business. The result is more competition, more price wars."[24]

20 By the scheduled date of completion a year later, only half the length had been dug.

21 US army engineers pulled in mid-2010 with only four kilometers complete (Ma'an News Agency, 2010a).

22 The incentives led to a brisk trade in damaged tunnels. A tunnel of 950m in length and 22m in depth damaged during the Cast Lead bombardment fetched about $40,000.

23 Some goods imported via the tunnels to Gaza—particularly basics and fuel heavily subsidized in Egypt—sold for much less than imports from Israel. For instance, in mid-2011, a 25 kg bag of rice sold for 50 NIS imported via the tunnels and 85 NIS imported from Israel. Wheat flour coming from the tunnels cost 40 percent less than flour imported from Israel; and some chemical fertilizers sold for less than half the price (World Food Programme 2011).

24 Interview, tunnel owner, Gaza City, April 2010.

As of 2012, goods traffic via the tunnels was four times that through Israel's sold crossing of Kerem Shalom.[25]

Increasingly able to transport heavy raw materials in bulk, the upgraded tunnels also facilitated Gaza's reconstruction. While world leaders promised billions at showcase conferences in Paris and Sharm el-Sheikh's luxury hotels but failed to nudge Israel into lifting its ban on construction materials, large tunnels in 2010 shifted 170 tons of raw materials each per day. Gazans began rebuilding their war-torn enclave themselves. Instead of importing gravel, Gazans made their own by pummeling war rubble, turning the shell-shattered Erez industrial park and bombarded EU-funded airport into new buildings. When those were exhausted, tunnel operators began shifting gravel too. By mid-2011, 3,000 tons of gravel, 500 tons of steel rods, and 3,000 tons of cement were arriving per day. Gaza morphed into a building site, meeting five years of pent-up demand and war damage.[26] Finance Minister Ala Refati aimed to raise construction from 10,000 units in 2011 to 20,000 by 2013.[27] Other sectors too saw benefits. Farmers circumvented Israel's ban on seeds, pesticides, irrigation pipes, and basic agricultural tools such as hoes and buckets. Factories imported spare parts: by October 2011, half of the 1,400 factories Israel destroyed in its January 2009 offensive were back in production, Hamas officials claimed. The owner of a plastics factory even increased his work force above its pre-blockade size. Contractors complained of a shortage of constructions workers; the International Labour Office reported a deficit of 5,000 (International Labour Office 2012).

All told, tunnel refurbishment precipitated a recovery as rapid as Gaza's earlier decline. Prior to Israel's Cast Lead offensive, the tunnels had at best provided limited relief from a severe contraction of 39 percent in real GDP per capita between 2005 and 2009.[28] After it, the tunnels facilitated what a September 2011 World Bank report described as "exceptionally high growth" of 28 percent in the first half of 2011 (World Bank 2011, p.7). Unemployment in Gaza dropped from 45 percent before the Cast Lead offensive to 32 percent by mid-2011 (UNRWA 2011, p. 3). Rafah's markets bristled with shoppers and café-goers late into the night; backstreet ATM's distributed hundred dollar bills.

By 2010, the markets were saturated. Falling labor costs resulting from the growing recourse to cheaper Egyptian labor to dig and shunt goods through the tunnels, economies of scale and the attraction of fresh investors sharply bolstered supply. In 2010, $100,000 could buy construction of a 1.2 km tunnel able to transport 150 tons, four times the length and four times the capacity the same money could buy in 2007. As noted, competition pushed down prices even faster. At the same time demand was falling. Israel's decision to end its ban on commercial goods in June 2010 following Turkish-led outrage at its killing of nine nationals aboard the Mavi Marmara flotilla bound for Gaza triggered a market glut. As retailers shifted back to imports from Israel,[29] tunnel operators suspended work: by the end of 2010 work had halted at over half of Gaza's 1,100 tunnels (Barzak 2010). Many that survived operated part-time. Tunnels increasingly focused on goods that Israel taxed heavily, such as fuel,

25 The situation of workers of the occupied Arab territories, International Labour Office, Geneva, May 2012.

26 Israel destroyed 6,000 housing units in Operation Cast Lead. As of 2011, UN Habitat estimated that Gaza needed 60,000 housing units, requiring 158,450 truckloads of cement.

27 Interview, Gaza City, February 2012.

28 Interview, UN official, Gaza, May 2012.

29 A major food wholesaler in Gaza reported that whereas he imported 95 percent of his stock through the tunnels at the height of the blockade, by the end of 2010 85 percent of his imports came from Israel.

or banned altogether such as construction materials, exports, and items that Israel defined as "dual-use" such as spare parts. "Israel's blacklist is the smugglers' green list," says a prominent Gaza businessman, who imports Egyptian cacti through the tunnels.[30]

Egypt's February 2011 uprising against Mubarak threatened to be the coup de grâce. As Egypt's security apparatus in Sinai took flight, border controls collapsed. Tunnel operators enjoyed a respite from demands for bribes and attacks on tunnel mouths. Egyptian operators sentenced in absentia received amnesties.[31] With overland restrictions still in place, increasingly, commercial tunnels offered fast-track access for people as well as goods.[32] While passage overland required months of prior coordination, the tunnels took days. Passenger costs of hundreds of dollars fell to 100 NIS ($30) for the 200-yard crossing (Pelham 2011). The Tunnel Affairs Commission regulated passage, sending crossing times via SMS messages to mobile phones. While the Rafah crossing closed at 5 p.m., the tunnels operated around the clock, and was of particular help for the 35 percent of male applicants aged 15–40 that the Egyptian authorities often barred on security grounds. After five years of closure, Gazans travelled south for business and holiday. Travel agents offered package tours to the Sinai Riviera. Visiting academics from Pakistan arrived for lecturers in Gaza's universities.

The relaxation had its limitations. It did little to revive Gaza's production for export markets and nothing to address the ban on trade with Gaza's historic markets in the West Bank.[33] There was limited export to Egypt of scrap metal (smelted in Sinai and reimported as steel rods for construction and possibly also military use),[34] dapple racing horses (which all but disappeared from Gaza due to high Egyptian demand), and surplus produce from Gaza's drive for food self-sufficiency: watermelons, apples, and eggs. However, agricultural produce aside, Egypt's lower labor costs and purchasing power rendered most Gazan goods uncompetitive.[35] Much of the Egypt-bound traffic consisted of re-exports from Israel sought after in Egypt, such as shoes and hair gel (Yaghi 2010).

Islamist electoral victories in Egypt in 2011 and 2012 fuelled Gaza's hopes of a final breakout from their seven-year jail. Repeatedly, Hamas officials proposed closing the tunnels if Egypt would open its overland borders to normal traffic, reintegrating Gaza into an international formal economy. Egyptian and Gaza officials held talks on establishing a free-trade trade zone straddling the border,[36] and an open highway from Gaza extending

30 Interview, Ali Abu Shahla, secretary general of the Palestinian Businessmen Association, Gaza City, May 2012.

31 "If you didn't pay, they'd imprison you." Interview, tunnel operator, Rafah (Egyptian side), July 2011. He claimed that he was strung from his arms during 10 days of detention, and paid a $45,000 bribe to avoid his 5-year prison sentence.

32 In May 2011, Egypt allowed 600 Palestinians into Egypt through the Rafah terminal daily, up from 250.

33 The 2005 Agreement on Movement and Access (AMA) provided for 400 trucks of exports to leave Gaza per day, but in five years of siege, only 1,156 trucks of exports left Gaza via Israel's crossings. Data collated from UNOCHA website: http://www.ochaopt.org/dbs/Crossings/CommodityReports.aspx?id=1010003, accessed May 15 2014.

34 Site inspection and interview, tunnel operator, Rafah, December 2011.

35 Prior to June 2007, Gaza's textile, furniture, and agricultural exporters relied on Israeli middlemen to sell in Israel, the West Bank, or abroad.

36 Interview, Gaza Chamber of Commerce official Maher al-Tabbaa, Gaza City, February 2012. Businessmen from the Chambers of Commerce of Gaza and Ismailia established the Egypt-Palestine Company, aimed at establishing a 1000-dunam free-trade zone. Egyptian officials estimated bilateral trade could double to $2 billion annually, surpassing the value of US military aid.

along the southern Mediterranean coast to Morocco. In an attempt to show willing, Hamas committed to halt the unlicensed import of stolen cars.[37]

But as with its 2006 electoral triumph, Hamas's euphoria was quickly punctured as the military retained its grip. Increasingly Egypt's military and old-regime cadres blamed Gaza for Egypt's ills—for fomenting Sinai's Bedouin revolt, freeing Islamists including Mohammed Morsi, its brief president, from jail, and consuming Egypt's subsidized fuel, leading to grueling Nile Valley shortages.[38] In February 2012, Egypt's intelligence forces overturned Hamas's agreement with Morsi's government to link Gaza to the trans-Sinai natural gas pipeline, with the assistance of a $70m loan from the Islamic Development Bank. Far from relaxing their border controls, Egypt's army revived the worst days of the siege. In the first quarter of 2013, the tunnels provided 65 percent of Gaza's flour, 98 percent of its sugar and 100 percent of its steel and cement (Breuer 2013). By the second quarter, Egypt's generals had recouped sufficient power vis-à-vis President Morsi to curb the bulk of the flows. Soldiers blocked fuel-tankers bound for Gaza from crossing the Suez Canal, and what fuel trickled failed to reverse blackouts of 16 hours per day. With Gaza's fate increasingly intertwined with Egypt's, Hamas increasingly seemed fated to be that of the Brotherhood's.

For weeks after Morsi's overthrow in July 2013, hundreds of awnings covering the mouths of a vast tunnel complex beneath the Gaza-Egyptian border continued to flutter in the wind. But the dust cloud hanging over the site had gone. The thousands of workers who shoveled raw materials for Gaza's reconstruction had abandoned their posts, and the bulldozers that bucketed supplies into juggernauts lay parked in the sand. The door on the hut where customs officials processed the chits haulers required to leave the cordoned site was locked. After five years of dramatic growth, Egypt's soldiers bulldozed homes housing the tunnel mouths and detonated explosives at the mouths of 90 percent of Gaza's 300 functioning tunnels. Egypt's armed forces also targeted tunnels no longer in use. By October 2013, Egypt's border guard chief, Major General Ahmed Ibrahim, claimed to have demolished 794 tunnels (Al-Ahram 2013).

Overnight, Gaza's 1.8 million people lost the conduits that supplied them with half their needs, and almost all their fuel and construction materials. Drivers left their cars outside petrol stations, for the day they might reopen. Police and official cars queued too, as Hamas's own reserves ran dry. Unemployment climbed again, as some 20,000–30,000 construction workers were laid off (Euromid Observer 2013). "Egypt's rope around Gaza's neck," read the headline of *Al-Risala*, a newspaper run by Hamas, above a front-page photo of two Egyptian battleships apparently heading Gaza's way. After two decades of assaults on the tunnel economy, General Abdelfatah al-Sisi had succeeded where all previous efforts had failed.

Winners and Losers

In an economy blighted by systemic unemployment arising from Israel's ban on Gaza's workers, bombardment of its manufacturing base, closure of export markets and a marked

37 Interview, Ali Refati, national economy minister, Hamas government, Gaza City, February 2012.

38 Gaza also imported substantial quantities of Egyptian-subsidized wheat-flour. See "From Eased or Uneased – One Year On," World Food Programme report, June 2011. However, Gazan economists noted that Egypt's subsidy bill was offset by some $250 million per year Gazans paid in customs and sales taxes on goods from Egypt. Interview, Palestinian businessman, Gaza City, May 2012.

slowdown in donor-funded development projects, the tunnels briefly emerged as Gaza's largest non-governmental employer. For a time, tunnel workers were Gaza's best paid laborers. In 2008, the average daily wage for a tunnel worker was $75, five times Gaza's median wage according to official Palestinian figures, and more than West Bank Palestinians earned building Israel's Jewish settlements.[39] School drop-outs scrimping 20 NIS as street peddlers earned 10 times as much laboring in the tunnels. Market saturation and recourse to Egyptian tunnel haulers noted above subsequently depressed daily wages to closer to 80 NIS, though this was still quadruple a farmhand's wage. With each fully-functioning tunnel employing 20–30 people,[40] the tunnel industry at its height employed an estimated 5,000 tunnel owners and 25,000 workers, supporting some 150,000 dependents, or 10 percent of the population (Ma'an News Agency 2010b).[41]

Such was the turnaround, that Gaza's economy briefly boomed. New hotels, restaurants, and beach cafes sprouted in Gaza City, attracting custom not only from Gaza's new tunnel elite, but exiles returning to Gaza (sometimes via the tunnels), and even North Sinai tourists.[42] The Gaza zoo replenished its stock, not least with a lion. A new luxury hotel, al-Mashtal, optimistically bought cocktail glasses. Visiting West Bank businessmen complained that the latest model sports cars, occasionally with Egyptian plates, cruised Gaza's streets. Real estate prices tripled.[43]

Nevertheless, the macroeconomic figures disguised both the wide disparities in wealth distribution and Gaza's economic transformation. While Beit Hanoun, Gaza's closed gateway to Israel sunk into depression, Rafah, hitherto the enclave's most depressed city, boomed. As noted, Gaza's traditional mercantile elite who had long curried ties with Israeli and West Bank suppliers found their financial and political clout increasingly threatened by a new generation of smugglers with ties down the Darb al-Arbaeen into Sudan.[44] Traders proficient in foreign languages acquired through travel and a higher education flagged, challenged by a new bourgeoisie of smugglers whose clans straddled the border. Some entered the retail trade, using their ability to fix prices through the tunnels

39 According to the Palestinian Central Bureau of Statistics (2010), the average median daily wage in Gaza was 56.9 NIS ($15) in the first quarter of 2010.

40 In a standard tunnel in mid-2010, a team of 15 to 20 porters packed goods arriving from Egypt in 50 kg sacks and stacked them for haulage through the tunnels. A second team of 10 haulers (*saheb*) inside the tunnel moved the goods through the tunnel. Porters worked two- to four-hour shifts and are paid less than haulers who work ten- to fifteen-hour shifts. Haulers loaded goods onto sledges (*shayata*) usually 4 to 10 m long and 80 to 120 cm wide. These were attached to steel ropes connected to an electric motor. In short tunnels this was placed at the far end of the tunnel, but as tunnels have grown in length and changed direction, tunnel operators installed additional engines in small chambers (*dishmas*) interspersed at at least 100 meter intervals. The chambers, normally 1.5 square meters, also contained a second smaller engine for pulling empty sledges back after unloading. In each chamber, a worker operated the engine and shifted the sledges from one rope to the next. At the far end of the tunnel, two or three workers loaded goods into metal or wooden containers, or cages, for hoisting by electric winch to the surface. Once above ground, two to four workers unloaded the hoisted goods. An average truckload of 30 tons could take eight hours to shift through a standard tunnel.

41 By 2012, the number had fallen to 9,000 (International Labour Office 2012).

42 "It's nearer and cheaper than Alexandria, and Gaza's girls are more liberal." Interview, resident aged 23, Rafah (Egyptian side), July 2011.

43 Bassim Khoury (2009), quoted "Le Hamas ne sait plus quoi faire de son argent."

44 The Darb al-Arbaeen, the 40 days road, is the traditional caravan route through the desert connecting Sudan to Egypt. See Dan Ephron, "Easing the Import Blockade Won't Help Gaza," *Newsweek*, June 23, 2010. "If we close the borders with Israel we will loose our connection with our West Bank brothers." Interview, Ali Abu Shahla, Gaza City, May 2012.

to undercut merchants importing from the formal economy. (Some even distributed their own catalogues direct to consumers to attract custom.) "No matter what we do, we cannot compete with the tunnel owners. They have decreased our income by 70 percent at least," complained Ala' Abu Halima, a long-standing Gaza merchant specializing in agricultural inputs.[45] In desperation, some old-time merchants sought to partner in tunnel cooperatives, though most feared any association with tunnels would jeopardize relations with Israeli and western trading partners. In short, the tunnels became a key driver of upward mobility and social change, empowering marginalized groups and spawning a class of nouveau riche.

By contrast, the UN and the international community shed influence. Banned by US restrictions from buying tunnel produce, they were forced to rely on the trickle of goods Israel let through. UN officials noted the paradox that US-led measures were hitting hardest those whose clout they claimed to enhance. In the words of the UN's Middle East peace process coordinator, Robert Sherry (2010): "Smugglers and militants control commerce. While international agencies and local contractors who wish to procure goods through legitimate crossings too often stand idle due to the Israeli closure."

Even after Israel relaxed its ban on construction materials for aid projects in Gaza, Israeli delays continued to hamper UN outreach, ensuring that in the struggle for hearts and minds, Hamas often was first on the scene.

As international influence waned, Hamas's leaders increasingly took center-stage as Gaza's source of patronage. They reconfigured Gaza's trade routes.[46] Steadily rising tariffs helped boost government revenues. As Gaza's economy rebounded, revenues grew from \$150–\$200 million in 2009 to \$340 million in 2011. From a guerrilla force armed with a social and charitable network to garner popular support, Hamas morphed into a governing authority with a functioning army, bureaucracy and economy. Flush with liquidity, its leaders repaired the infrastructure Israel bombed in operation Cast Lead, including parliament and many mosques. It upgraded Salah al-Din Road, the Rafah to Gaza highway, and even began laying grass in city centers and installing traffic lights. The government also acquired a degree of fiscal independence from the international community, its own Damascus-based leadership, and even Gaza's military wing.[47]

By orientating trade ties towards Egypt, Hamas claimed to have fulfilled a key undertaking in its 2006 election manifesto to revive Gaza's ties with the Islamic world and cut Gaza loose from Israel and the Paris protocols, which bound Palestine's economy to Israel's. In the words of a Hamas official, "The siege is a blessing in disguise. It is weaning us off of Israel and sixty years of aid, and helping us to help ourselves."[48] Thanks to the growing dependence of Sinai's Bedouin on trade with Gaza, Hamas could particularly project its influence into Sinai. Factional tunnels may also have helped the Qassam Brigades and other armed groups build up military clout inside the peninsula as well. As the Mubarak regime collapsed, Gazans increasingly spoke of Hamas's "strategic depth" inside Egypt's periphery.

---

45 Interview, Gaza City, April 2010.

46 For instance, they circumvented global restrictions on financial transfers by importing laptops, and selling them at cost.

47 In 2011, the Hamas government cajoled the Interior Ministry to deposit all receipts into a single Treasury account, rather than its own coffers in return for Interior Ministry prioritization in disbursements. Interview, Huda Naim, deputy head of the PLC commission, October 2011. National Economy Minister Ali Refati assigned 200 customs and excise officers to monitor the collective of levies. Interview, Gaza City, October 11, 2011.

48 Interview, Mazen Abu Amarneh, middle-ranking Hamas official, Rafah, June 2010.

That said, the tunnel economy also stained Hamas's reputation for financial propriety. It came under fire for prioritizing its financial and governance interests over the Islamic resistance of its namesake. It used its control over tunnel licenses to ensure its members received posts on the boards of tunnel cooperatives, earning it unfavorable comparisons to the PA whose corruption it had promised to sweep away. "This is not the old style radical movement," noted a Gaza economist. "Hamas has acquired a business venture."[49] Calculations of its tunnel earnings suggested Hamas raised far more money than it declared.[50] Some accused the Qassam Brigades of the profiteering associated with their predecessors, the Fatah security chiefs who held sway in Gaza until the Hamas military takeover in June 2007. On occasion, individual Qassam Brigade commanders and their Interior Ministry counterparts fought for control of the tunnel holdings. In the words of a Salafi Jihadi militant from Gaza's Middle Area:

> Before entering government, Hamas's cadres focussed on religious sermons and memorising Quran. Now they are most interested in money, tunnel business and fraud. Hamas used to talk about paradise, but now they think about buying land, cars and apartments. Before they prayed in the mosque, now they pray at home.[51]

Calls for accountability have mounted as the Haniyeh government has increased the tax burden. When fuel shortages intensified in Spring 2012, allegations abounded that Hamas leaders received uninterrupted supplies of electricity and petrol stations continued to operate for the exclusive use of Hamas members. Recrimination ran high that Hamas profited from the tunnels more than its population. A cavalier approach to child labor, which the movement did little to curb, and some 200 tunnel fatalities further harmed the movement's standing with human rights groups (International Labour Office 2012).[52] Safety controls on imports appear similarly lax, although the Tunnel Commission insists a 16-man contingent carries out sporadic spot checks.[53] Yet perhaps the greatest criticism is that the tunnel economy was never more than a temporary fix. While freeing Gaza from its neighbors' stranglehold, it failed to provide a long-term solution to Gaza's isolation or the tools for sustained, export-driven growth. The manufacturing base failed to revive,[54] and without access to former markets it was hard to see how it can. Ultimately borne by international donors and passed on to their taxpayers, the high costs of food aid and other support could only be saved by opening the border to normal trade. Despite improved access, most of Gaza's 240,000 refugee youth have never even left the enclave, and 51 percent of them remain unemployed.[55]

---

49 Interview, Omar Shaaban, October 10, 2011.

50 In mid-2011, Economy Minister Refati claimed Hamas raised $17 million of its $25 million monthly current expenditure locally, and receiving the shortfall from Iran and other donors. Local business and some NGOs made further indirect contributions. Interview, Refati, February 12, 2012. However, businessmen estimate total domestic revenues at over $50 million per month, including vehicle license fees ($10 million), land registration ($6 million), municipal and sales taxes, and duties on cigarettes ($5 million). Interview, Palestinian construction magnate, Gaza City, February 12, 2012.

51 Interview, Gaza City, May 2010.

52 On a site inspection in December 2011 the use of children was much in evidence. As in Victorian mines, tunnel operators preferred children on account of their nimbler bodies.

53 Interview, Tunnel Affairs Commission official, Rafah, July 2010.

54 In 2011, manufacturing outputs in Gaza remained at only 68 percent of their 2006 level. UN data, provided May 2012.

55 "Youth are denied the chance to meet others outside their own community, to broaden their horizons, experience new cultures, or see things from a different perspective." Interview, Christopher

While Israel's GNP soared from $13,800 in 1993 to over $32,000 today, Gaza's decreased from $1,230 to $1,074. Had productivity increased at pre-Oslo levels, its real GNP per capita would have been an estimated 88 percent higher (Oxfam 2013).

The End of the Tunnel Era?

The peaks and troughs of Gaza's tunnel economy came to an abrupt halt with Egypt's overthrow of Morsi and launch of its Sinai operation starting in July 2013. Three years of exponential growth and even tentative development shifted into reverse. Construction ground to a halt; Hamas lost its revenue base, and Gaza its strategic safety valve from Israeli pressure.

Having geared its economy to the tunnels, Hamas struggled to finance its rule. Bereft of much of the $1million per day it had earned in tunnel dues, in August 2013 the Haniyeh government put its 46,000-strong army and bureaucracy on half pay, and in early 2014 delayed paying even that, sparking rare public sector protests. Initially it sought to increase taxes on the trickle of goods that still managed to cross. Cigarette taxes tripled in a week; cement prices quadrupled. It also feared that the increased hardship could provoke rising discontent. Instead of the promised free-trade zone with Egypt, Gaza faced a buffer zone, or *cordon sanitaire*. Without fuel, Gaza's power plant shut down, increasing blackouts to some 16 hours per day. In places, the sewage system collapsed, spilling into the street. In parallel with their disruption of passenger flows underground, Egypt's security forces closed the Rafah terminal. Claustrophobic Gaza was an open-air prison again.

At a time of such radical oscillations in the region, predicting scenarios is a hazardous exercise. But unlike previous shocks to the tunnel economy, which Hamas always managed to subvert, this latest assault felt terminal. Fearing potential unrest, Hamas's siege mentality revived. Only months after their triumphal tours feted on the shoulders of the faithful of the region's leading mosques, Hamas's leaders prepared for lock down again. Despairing of their politicians to find an exit and determined to buck the region-wide Islamist downfall, the military wing flexed its muscles. The first Islamist movement to take power on the Mediterranean now speaks of making a last stand. Its forces erected night-time checkpoints in the center of Gaza City, closed news agencies and detained a widening circle of suspected opponents. The head of a newly-opened Egyptian community association in Gaza City was hauled in for questioning. The Qassam Brigades staged military parades, firing guns into the air and giving the Brotherhood's four-finger salute.

Whether or not Hamas can survive without the tunnels will largely depend on its ability to adapt. In a sign of its readiness to adapt to greater dependency on Israel, its finance minister committed to introduce a tax on imports from Israel, in effect promoting double taxation since Israel already collected taxes on goods crossing into Gaza to fund President Abbas's Palestinian Authority. Israel showed a tentative interest in obliging, compensating for Egypt's contraction of supply lines, by easing its restrictions. Construction materials began to sporadically flow again from Israel into Gaza. For the first time ever, 400 truckloads passed over its Kerem Shalom crossing in day. "If demand grows, we're ready to step in," said an Israeli army officer.[56]

Gunness, UNRWA spokesman, May 2012.

56 Interview, Erez, September 3, 2013.

Such professed altruism had its limits. Following Morsi's downfall, Israel reneged on upholding the terms of the ceasefire agreement the Egyptian president had helped broker between Israel and Hamas in November 2012. This had provided for the phased opening of Gaza's crossings with Israel. But though trade rose, it remained severely restricted. Israel continued to prevent the passage of raw materials for commercial use, and halted supplies to donor projects as well, after announcing its discovery of a tunnel apparently for military use heading from Gaza into Israel. With tunnel traffic all but terminated from Egypt, Gaza's development, bar a Qatari-financed road project, largely ground to a halt.

Nevertheless, the common interests in preventing a Hamas's collapse and Gaza's slide into its pre-2007 security chaos seem clear. Israel appears to share Hamas's interest in perpetuating its rule. As noted above, Israel sought to maintain Palestine's divided rule and exclude Gaza outside any arrangements it made with President Abbas in the West Bank. And by acting as the aid conduit for Gaza, it sees a potential spin-off in improving ties with Qatar and particularly Turkey, the prime sponsors of Gaza's reconstruction.

For its part, neither the Palestinian Authority nor the international community sees an interest in reviving the tunnel economy. President Abbas commended General Sisi's efforts on a visit to Cairo in November 2013, hopeful perhaps that Hamas's woes would be his gain. And while attempting to ease the resulting hardship, the United Nations welcomed the shift in Gaza's supply lines back to Israel, as a first step to its reintegration within the framework of the Israel-Palestine paradigm. In their statements, the UN joined Egypt and the Palestinian Authority in highlighting the illegality of the tunnels, and calling on Israel to reopen the crossings. "The tunnels were an exception," says a UN official, "Israel has to assume its responsibility for Gaza's needs.[57]

Ultimately Gaza's rehabilitation will depend on its re-entry to the formal economy. While the tunnels prevented Gaza's collapse, they fuelled centrifugal forces in the region, fostering the Bedouin uprising in the Sinai that threatens to destabilize Egypt and regional Jihadi militancy as well as the erosion of central authority through bribery and corruption. To this end, all parties—Egypt, Israel, Gaza, the Palestinian Authority, and the UN—should work to formalize Gaza's economic relations, and end the curbs and policies of exclusion which foster the informal economies. As its economy normalizes, so might its politics. Security for all, Israel included, can better be achieved by monitoring formal access and movement of people and goods above ground than hidden below it. In short, while the tunnels served as a homemade driver of Gaza's reintegration into the region and a makeshift dynamo for regeneration, Gaza's economic and commercial rehabilitation can best be served by their demise.

57 Interview, UN official, Gaza City, September 4, 2013.

Chapter 8
Economic Discourses and the Construction of Borders in the Israeli Palestinian Space since the 1967 Occupation

Lev Luis Grinberg

Introduction

The military control of Palestinian borders aiming to supervise the movement of people, commodities, capital and explosives has been the major question in the construction of the Israeli domination regime in the West Bank (WB) and Gaza Strip (GS) since June 1967. Despite the dominant security discourse and legitimization, I will argue here that the economic interests have been of dominant importance. The pattern of domination established in 1967—a captive market incorrectly called a "customs union"—has suffered significant changes since its formal Palestinian legitimization by the "Paris protocol" on May 1994, and since then it has deteriorated into a structurally unstable regime that I have proposed to call a "strangling envelope" (Grinberg 2007a).

Following my previous research on the Israeli political economy and sociology of the Israeli–Palestinian conflict (Grinberg 1991, 1993b, 2008, 2010) I will analyze here the institutionalization of the military-economic domination regime of the Palestinians after the occupation of the West Bank and Gaza in June 1967, and its transformation following the "peace process" (1993–2000). I will show how the poor interpretation and implementation of the accords established a new regime of "strangled envelopment," leading to its most salient form in Gaza since the withdrawal of Israeli military and settler forces in 2005. The patterns of the military-economic domination regime and its changes were shaped by the interests, power relations and compromises between three powerful actors: the military, the dominant economic groups, and the ruling party. I will analyze here the institutionalization of the domination pattern and its changes following Palestinian resistance. I will analyze the incorporation of the WB and GS to the Israeli economy immediately after the 1967 war, the changes of the domination pattern following the First Intifada (1987) that ended in the Paris protocol (1994), and the reactions to the Second Intifada (2000) that ended in the Gaza withdrawal (2005), the economic closure of Gaza, and its isolation from the WB.

The political-economic analysis suggests criticism of the models proposed by economists that developed "peace plans" for Palestinian economic independence (Fischer et al. 1993; Arnon and Bamya 2007). My criticism of these approaches is that they neglect the economic interests of Israeli dominant groups, and their decisive power in the Jewish state. This neglect is very common in the neoliberal economic models that neglect ethnic and class conflicts, and tend to ignore the domination of capital and the power of institutions in shaping state policies. The neoliberal economic approach blames the politicians for the bad implementation of their plans, and they present the potential success of their plans conditioned by the political will and capacity of implementation of politicians. I will argue that these economic plans are based on wrong theoretical assumptions about the reasons for the present situation, and by doing so their suggestions are constantly manipulated to legitimize and improve the military-economic regime of domination instead of changing it.

Many teams of economists at the World Bank and academic institutes in the United States and Israel struggled to find solutions for the Palestinian economy in the period between the First Intifada of late 1987 and the 1993 Oslo Accords (Fischer et al. 1993, 1994; World Bank 1993). They meticulously analyzed the Palestinian economy's problems and potential, but ignored the main obstacle: the determination of Israel's economic and military elite to continue ruling the Palestinians. The elegant term "customs envelope" is an economic misnomer, I argue, for what may be more aptly dubbed a "strangling envelope" preventing free competition between Israelis and Palestinians. The most salient example of the economists' negligence is the Ben-Shahar Committee report (1993) that established the guidelines for the Israeli negotiators in the Paris meetings and is the basis of the economic agreement called the Paris Protocols. The participants in the committee were academics economists, military and civil officers of the state, and representatives of different business interests. The committee made a big effort to show that there are two options for economic relations, a customs union (like the EU) and a "free-trade area" (the US-Mexican model), that they have exactly the same impact on the Palestinians, and in both cases they will benefit from "small economy" advantages.[1] They forgot to say that without a recognized border it is impossible to have a "free-trade area," despite the fact that the report emphasized that the only political guideline they had is to prevent the demarcation of a border between the Israeli and Palestinian economies. In other words, they consciously did "ideological" work, to justify the "customs union" model. They did so, however, as Professor Zussman properly criticized, at the same time when it had no chance to be implemented, because the military imposed a long-range closure of the borders, Palestinians could not enter Israel to work and the entrance of Palestinian products was limited (Zussman 1994).

Despite being well aware of the injustices inherent in these power relations, Israeli economists continue to rely on normative models designed to "solve the problem," rather than analyze the forces which have created and which perpetuate the problem and prevent a solution. I argue that these economists' search for a solution misses the point, which is that many Israelis profit from the Palestinians' ongoing dependence on Israel. While normative neoliberal economics is obsessed with designing abstract models in which politics is supposed to stop interfering with the proper workings of the so-called free market, political economics analyzes power relations and struggle among social groups and organizations which depend on the state for shaping the economic field characterized by conflict, coercion and domination.[2] We therefore face a fundamental contradiction: in political economics, power relations are used to explain differences and conflicts among social groups, while normative neoliberal economics presumes that politics and power relations among groups are obstacles that prevent free markets from functioning properly, and they are therefore

---

1 The analytical concept of the "advantage of small economies" is related to classical economic theory. It assumes that countries that can produce cheap products will benefit from opening their economies to free commerce with economies that can produce more elaborated products. This theory aims to demonstrate that opening the markets to the import of British products will develop small economies.

2 This is obviously a rather general definition, and there are many others. Political economics is a broad-based discipline which has yet to be formally recognized by universities. Beginning in the 1980s, it is being studied mainly outside economics departments, in departments of political science, international relations, history, sociology, geography, labor studies, management, and social work. Its founders were Karl Marx and Max Weber. See Grinberg 1996 for an introductory discussion.

neutralized or removed from the analytic framework. This criticism is relevant in an era when both the Israeli and Palestinian prime ministers, Benyamin Netanyahu and Salam Fayyad, hold neoliberal economic views and developed strategies for "economic peace." I will argue here that "economic peace" is an oxymoron, because the crucial source of the conflict is economic. In this chapter I rely on the political-economic framework to analyze past relations between Israelis and Palestinians as well as to point to present and future difficulties.

The Institutionalization of the Military-Economic Regime 1967–1987

Let me start from the last day of the so called Six-Day War, June 11, 1967.[3] The first economic decision to incorporate the occupied Palestinian population into the Israeli economy was to impose the use of the Israeli currency. The next move, after the closure of the West Bank and Gaza to the import of external commodities was its opening to Israeli products at the end of June 1967.[4] These moves created the "customs union" or more precisely the "captive market." Later on, after the destruction of the wall that separated East and West Jerusalem, the municipality employed the Palestinian workers previously employed in the Jordanian municipality, and the military administration set up in the West Bank employed the civil public workers of the Kingdom of Jordan. The destruction of the wall was the materialization of the opening of the borders, and the most important penetration of workers took place at the beginning through the open movement in Jerusalem, especially through the construction industry.

Since then the IDF became the articulator of the Israeli-Palestinian economy due to its legal status as the sovereign authority of the occupied territories. The military has had to articulate its own institutional interests to provide security to the Israeli citizens, with the economic interests of the dominant Israeli classes and the economic needs of the Palestinian occupied population. Here is the linkage between security and the economy, which shaped the military-economic domination pattern.

The military administration was interested in guaranteeing the economic wellbeing of the Palestinian population in order to prevent their recruitment to resistance organizations. In addition, the military had to take into consideration the economic interests of different Israeli classes, and the political interests of the ruling party. The first solution unanimously approved at the end of 1967 was the opening of the bridges over the Jordan River to export Palestinian agricultural products and commerce with Jordan (Gazit 1985. This move was made to prevent the import of Palestinian products to the Israeli markets, which might damage Israeli producers, mainly in agriculture, due to the advantage of the "small economy." The need to prevent the entrance of cheap Palestinian products to the Israeli markets became one of the most important principles of the military-economic pattern of domination that was imposed to control Palestinian movement.

The most serious question was related to the employment of Palestinian workers; the military was interested in their employment, however it was indifferent where they were employed. In other words the question was if they would be employed within the sovereign borders of the Israeli State, or in the West Bank (WB) and Gaza Strip (GS). The Israeli producers wanted to employ them within Israel, and opposed their employment in WB and GS by Palestinians, because they were afraid that the products might enter the Israeli

3 This part is based on my book in Hebrew *The Histadrut Above All* (Grinberg 1993a).

4 Histadrut Archives (file IV-204-5 64, 26.6.1967).

markets due to the advantages of the "small economy." The entry of Palestinian workers was legalized under pressure from the security establishment and Israeli employers, albeit not without prior bargaining with Israeli stakeholders threatened by this sudden influx of cheap labor. The move was completed with relative ease, however, as both private and unionized Israeli employers were very keen to employ Palestinians, and the Ministry of Finance sought to reduce wages in "Israel," which had risen sharply in 1960–1965. The only stumbling block was the Histadrut labor union federation, which at the time was the second most powerful institution in Israel after the IDF. Its leaders supported the government's wish to employ Palestinians inside "Israel" since they had a business interest in employing cheap, non-unionized workers, mainly in the construction, industry and farming sectors, where it was a central employer. They were faced by relatively weaker unions, mainly of farming and construction workers, opposed to employing non-Israelis since this was liable to affect their pension funds. The solution was to create a joint mechanism involving the Histadrut, the Ministry of Security and the Ministry of Labor to jointly supervise the entry of Palestinian workers and share the tax deductions from their wages (Grinberg 1993a).

In this way a relatively effective military economic regime of domination was established within one year. The Israeli military played a crucial political role in articulating the economic interests of Israelis and Palestinians, and controlling all the borders of Israel, of the WB and GS, and those between them. The regime was not at all a customs union, but it was presented as such. It was not a customs union because it was imposed on the Palestinians according to Israeli interests to neutralize the advantage of the "small economy." The Palestinian economy was maintained in a dependent relation with the Israeli economy: investments in the WB and GS were prevented, and Palestinian products were prevented from entering Israeli markets. Palestinians could not import cheap products from the international markets and were constructed as a captive market for the expensive Israeli products. Employment within sovereign Israel and technical support for agricultural production were the two main sources of economic growth without development. The military control of Palestinian external borders was crucial for the functioning of this regime. Between 1967 and 1985, the Palestinian economy grew considerably although it did not develop, that is, there was no industrialization, no new technologies were introduced and the employment of skilled workers remained at low levels.[5] A "black" economy flourished alongside the formal economy. About half of the Palestinian workers managed to enter "Israel" unsupervised to be employed by tax-evading employers. Moreover, despite military controls, commodities managed to penetrate the Israeli market thanks to cooperation between Palestinian producers and Israeli merchants. Conversely, cooperation between Israeli producers and Palestinian merchants enabled the former to export goods disguised as Palestinian to Arab countries.

To conclude, the formal envelopment economy was forced on the Palestinians by the IDF, while the black economy was the fruit of Israeli-Palestinian cooperation. Until 1987, movement between "Israel" and the "territories" was relatively unrestricted. Military-economic domination was counterbalanced by labor exports to "Israel" and Persian Gulf countries, and the black economy. By the mid-1980s, however, Israel and the Persian Gulf countries suffered an economic downturn, which, together with the First Intifada (1987) (largely precipitated by this crisis), led to partial reorganization of this domination pattern by the Paris Protocol.

---

5 The distinction between growth and development is one of the key concepts of the dependency theory. See Cardoso and Faletto 1979.

The First Intifada and the Oslo Peace Accords 1987–2000

The relative economic stability and growth that prevailed before the First Intifada were important factors in the failure of Palestinian resistance to recruit activists and to encourage popular mobilization against the occupation. The Palestinian economy was also supported by the "export" of some of the labor force to the Gulf Arab states and money transfers from there. The relative stability was shocked by an economic crisis in the Gulf and within Israel that shrunk the export of labor forces and the flow of money and lead to unemployment and recession. The economic crisis was combined with a political crisis in the PLO, due to their expulsion to Tunis after the Israeli siege of Beirut in 1982.

These circumstances created a political vacuum which was filled by Palestinian civil society organizations prepared to take action against Israel's military-economic domination. The struggle was led by youth organizations in the refugee camps, women's groups and labor unions headed by political parties, which initiated a popular, unarmed uprising with mass demonstrations and general strikes (Nassar and Heacock 1990). Among other things, their aims were to break free of the economic dependency on Israel, and disrupt the regular movement of workers into "Israel" and the workings of the captive market. They also acted to prevent Israel from receiving Palestinian tax money.

The political meaning of the First Intifada was the attempt to liberate the occupied population from the military-economic regime of domination, as a strategy that could lead to national sovereignty over the occupied territories. The goal was to prevent the entrance of Israeli soldiers into Palestinian cities, towns and villages, to prevent the consumption of Israeli products and encourage Palestinian production, to prevent regular supply of Palestinians labor to the Israel economy and to prevent public services in the Palestinian cities provided by Palestinians being hired by the Israeli administration. The Israeli military response was also mainly economic, by the restriction of permits to enter and work inside the sovereign state. In this way the military punished militants and young people directly, and indirectly the whole population by shrinking jobs and money transfers. In order to do so the control of movement of workers in the borders was increased. The Intifada clearly demarcated the areas of the Palestinian resistance, and by default the borders of the sovereign Israeli State. Following the demarcation of the Intifada borders the PLO recognized Israel de facto in 1988, when it declared an independent state in the West Bank and Gaza Strip.

The Intifada leaders emphasized the need to develop local industries, even if this should mean temporary reduction in the population's standard of living. This is why Palestinian villages and towns were closed to Israelis and, conversely, why Palestinian workers were prevented from working in "Israel." Despite the successful mobilization of the Palestinians the power of the Israeli military succeeded partially in preventing the entrance of Palestinian workers to Israel and to cause significant damage to the Palestinian economy. The number of Gazan Palestinian workers in Israel was reduced from 45 percent to 37 percent due to the relative success of control of the entrance of illegal workers (Arnon and Weinblatt 2001). However, the Israeli economy was also damaged by the Intifada, which prevented the expected investments after the liberalization of the economy in 1985, production fell and a deep recession started, when levels of unemployment grew from 6.1 percent in 1987 to 8.9 percent in 1989.

Within half a year, the mass demonstrations gradually subsided, while the more violent aspect of the uprising gained momentum. The strikes did not stop. The Intifada dealt a severe blow to the Israeli economy, as it reduced local sales of Israeli goods as well as black exports to Arab countries, disrupted the entry of Palestinian workers into "Israel" and even affected tourism. All these exacerbated the recession which began in 1985 (Grinberg 1991). In this sense, it may be argued that one of the Intifada's key objectives was achieved. The economic crisis made many Israelis, particularly businessmen and economists rethink Israel's economic relations with the Palestinians. Israeli, Palestinian and North American economists who realized that the envelopment policy had become obsolete, began searching for models for independent development of the Palestinian economy. Several studies examined the relative advantages of the Palestinian economy as the smaller of the two and suggested new ways of managing it in peacetime (Fischer et al. 1993, 1994; World Bank 1993; Ben-Shahar Committee 1993). The most comprehensive was conducted by the World Bank (1993), demonstrating how Israeli policy had deliberately prevented the Palestinian economy from developing and suggesting ways to promote development, such as foreign investments, industrial and agricultural development, and, above all, reformed relations with Israel.

The two alternatives suggested for the economic envelopment were a customs union and a free-trade zone. The first meant a uniform import and export policy, as well as unrestricted movement of commodities and people between "Israel" and the "territories." This policy would have enabled the Palestinians to enjoy their advantages as a smaller economy—cheaper labor and lower commodity prices—both by free access to employment in "Israel" and by selling farming and industrial goods therein. The second alternative—free-trade zone—would have enabled the Palestinians to import and export freely. Under this alternative, control of border crossings would have ensured that only goods actually produced in the "territories" would have entered "Israel," rather than cheaper imports from across the PA's borders with Jordan and Egypt. Compared to the first alternative, the second would have been preferable, in that it would have liberated the Palestinians of the need to buy Israeli produce and allowed them to import cheaper commodities. However, its implementation would have required the demarcation of a clear borderline, mutual control of which would have restricted the movement of commodities not produced in Israel or Palestine.

In preparation to negotiating future economic relations with the Palestinians, the Israeli government appointed a committee headed by Tel Aviv University economist Professor Haim Ben-Shahar. The committee weighed the pros and cons of the two alternatives, but only superficially, since already in the introduction to its concluding report, its members made it clear that political constraints prevented them from demarcating the border between "Israel" and the "territories," and it was obvious that the free-trade model could not be implemented without joint border controls (Ben-Shahar Committee 1993). In practice, the committee's detailed work was designed to prove that the Palestinians would still be able to reap the advantages of being the smaller economy. However, the report glossed over the fact that without free movement of people and goods, all such advantages are lost and the economy becomes suffocated. There was also no mention of the obvious fact that any advantage offered to the Palestinians would come at the expense of Israeli producers selling to the captive market, who would now have to compete against cheaper Palestinian products. The main innovation in the agreement eventually signed in 1994 was financing the PA's services using customs levies charged by Israel for goods imported into the area governed by the PA. However, the Paris Protocol's main shortcoming lay in the fact that it ignored the reality of life in the "territories," the closure policy (Roy 2001), and Israel's continued control of all resources.

The Paris Protocols were based mainly on a mutual understanding of the Israeli economic interest to reactivate the economy and to enter globalization, despite the disagreement on fundamental issues like future borders, independence and economic relations. From its inception the economic motivation of the peace agreements was its most crucial feature (Shafir and Peled 2000, 2002; Ram 2008; Grinberg 2010). The most popular slogan of the peace process was that it will succeed if the Palestinians significantly improve their quality of life, and will have a lot to lose if they reject Israeli demands.

Despite this common wisdom the Israeli negotiators were in a powerful position to impose their own point of view of the desired agreement on the Palestinians, which reflected the interests of the dominant classes. The economic agreement maintained the "captured market" that forced the Palestinians to buy expensive Israeli products, but promised them free movement of workers and products to Israel, which was supposed to create the "small economy" advantage effect. The problem was that given the constant closure of borders the Israelis could benefit from the "captive market," but the Palestinians could not benefit from Israeli employment and open markets.

During the 1990s, the closure of WB and GS borders became a standard collective punishment of the military against the Palestinians. The Paris economic protocol (1994) was signed two months after the Hebron massacre,[6] and after two months of almost hermetic closure, meaning that when the agreement was signed it was already irrelevant. In reaction to the suicide bombings (that started in April 1994) the military continued to use the closure of borders as economic pressure on the population and the Palestinian Authorities.

At that time, Israeli economist Zvi Zussman (1994) published a remarkable article, in which he wrote that the customs envelope is bound to fail given the existing closure conditions, and that it would be better to implement the free-trade zone alternative. This meant allowing Palestinian border police to control the entry of Israeli citizens into the "territories," something the Israeli government was naturally opposed to since this would mean recognizing the Green Line as an international border and the PA as sovereign beyond it. Ever since Yasser Arafat entered Gaza as PA chairman in July 1994, Israel's actual military-economic control policy worked in direct contravention of the Paris Protocol: the IDF tightened the closures hoping that the civilians' suffering would cause the PA to confront Hamas.

In the absence of the regular supply of cheap labor the Israeli employers demanded the import of workers from other non-developed countries, like the Philippines, Thailand, Eastern Europe, Latin America, Africa, India, and China (Rozenhek 1999). The imported workers (called foreign workers in Israel) who replaced Palestinian workers within a few years were cheaper and more docile. One of the most important incentives for Palestinians to agree to the customs union concept, the Israeli demand for labor, was neutralized. The replacement of Palestinian workers was the result of power relations between the military, the employers and the ruling party: the military wanted to use Palestinian economic dependency to fight terror, the employers wanted cheap workers, and the ruling party wanted to maintain the image of a peace process.

The customs union model was never implemented due to the closure policy and because Palestinian workers were replaced by immigrant workers from other countries. Nevertheless,

6 The massacre in Hebron took place at the end of February 1994 in the tomb of the patriarchs, when a Jewish settler dressed as a soldier started shooting Muslims at prayers. He succeeded in killing 29 worshipers before he was killed. The massacre sparked riots immediately afterward all over the Occupied Territories and the IDF killed another 11 Palestinians. Rioting continued for two weeks with another 15 demonstrators killed by the IDF, all this while imposing a closure and intermittent curfew. The total closure of the territories continued until May.

the rhetoric arguing for the importance of a free market economy to the success of the peace process was hardly ever challenged. As could be expected, the Palestinian delegates to the economic talks in Paris demanded that a free-trade zone be established, but their Israeli counterparts rejected that demand pending an agreement on the borders issue. Thus, in practice, the Paris Protocol legitimized the relations of dependency established by Israel since 1967, as well as the economic envelopment, only this time the situation on the ground was much more severe than before 1987: since 1994, free entry into "Israel" for Palestinian workers never resumed. Moreover, the entry of Palestinian goods was restricted even further due to the closure policy, yet, smuggling became more developed as its merchants encountered more barriers.

A free-trade zone would require bilateral border control. In reality, the borders continued to be unilaterally controlled by Israel, allowing Israeli goods and people to cross unrestricted, and preventing the same for Palestinians. The cessation of employment in "Israel" was temporarily compensated for by investments in the PA-controlled areas, mainly in Palestinian-financed construction projects and donations by European countries, mostly spent on financing an oversized bureaucracy. Within its unilaterally controlled borders, the Palestinian economy survived with great difficulty thanks to external subsidies of 2.3 billion dollars per year sent by donor states (Fischer, Alonso-Gamo, and Erickson von Allmen 2001).

The Oslo agreements did not improve the economic situation of the Palestinians, moreover, the dependency on Israeli products and on the job markets increased and the PA budget became dependent on the Israeli collection of customs, because it controls the external borders of the WB and GS. The control of the movement of Palestinian workers and commodities was improved by the segmentation of Palestinian lands into A, B, and C areas[7] that allowed the IDF to build check-posts between the areas, not only on the Green Line, in the passage from the occupied territories to sovereign Israel. European, US, and Arab State donors to the Palestinian economy supported the strangled economy, unable to grow and develop, and by doing so they indirectly subsidized the benefits for Israel from the military-economic domination regime.

The Second Intifada and the Strangling of Gaza 2000–2008

The economic situation in the "territories" took a turn for the worse after the Second Intifada broke out in 2000: economic activity shrunk, unemployment was way up, foreign investments ceased, as did employment in and trade with "Israel," as well as tourism. GDP growth felt from 8 percent to −1 percent, and unemployment in the WB rose from 9 percent to 28 percent (between 1999 and 2002) and from 16 percent to 38 percent in the Gaza Strip (Arnon 2007). In order to further tighten its stranglehold over the Palestinian population, the IDF developed a more refined form of the closure policy: the so-called encirclement (in Hebrew *keter*) policy,[8] which meant roadblocks disconnected Palestinians towns and villages, and cut the West Bank off from the Gaza Strip.

In the absence of a recognized border between Israel and Palestine and a mutual control of movements between sovereign states the peace agreements became an effective way

7 Area A was in control of the PA, Area C of the Israeli Military, and area B was under civil control of the PA and security control of the IDF.

8 A Hebrew neologism coined for that purpose.

to legitimize the uneven military-economic regime of Israeli domination. The absence of any positive political vision of future independence, and increasing economic dependency, were two important factors in the beginning of the Second Intifada. However, instead of a political opening it provoked increasingly violent repression. The Second Intifada did not have a clear economic goal, like the first, and it also did not mark a clear border between Israel and Palestine, due to the October 2000 riots of the Palestinians inside the borders of the sovereign state.

Despite the absence of Palestinian economic goals, and of intentional economic repression by the military, the economic effects of the Second Intifada for the Palestinians and Israelis were even stronger than in the first. The reason is precisely the mutual penetrated borders. Since the Oslo agreements the Palestinian territories were fragmented in many enclaves following the division into A, B, and C areas and the IDF was able to surround every city, town, and village and disconnect it from the others by means of road blocks and checkpoints. But also the Palestinians were able to penetrate the Israeli borders, because during the Oslo years they learned to bypass the closure obstacles looking for work in Israel. Together with the workers came bombs that endangered the lives of Israeli citizens. The mutual penetration of borders was the main reason for the economic recession provoked by the Second Intifada: the fragmentation of the Palestinian areas prevented movement of workers and commodities not only to Israel but also between Palestinians. The security threat of Palestinian bombs inside sovereign Israel convinced the Israelis to stay at home, and to reduce their economic activities creating a recession.

The only economic relief for the Israelis existed when the Palestinians declared a ceasefire in the summer of 2003 and, after Arafat's death, since 2005. During periods of explicit or implicit ceasefire and complete absence of terrorist threats, Israelis restarted their normal economic activities, mainly consumption in malls and restaurants, buying cars and houses, and investments. All this reactivated the economy.

However, the Israeli strategy to exit the Gaza Strip in the summer of 2005 emphasized both the Palestinian dependency on Israel, international cooperation with the Israeli strangling regime, and the supremacy of the economic interests vis-à-vis security considerations. According to the published data, Sharon's plan of withdrawal assumed some form of economic sovereignty for the Palestinian Authorities in Gaza (Benziman 2004; Shelah 2005). It included the opening of an airport and seaport. However, the Chamber of Commerce, the industry organization, and the treasury state officers opposed the idea of economic independence for the Gaza Strip. They strongly argued and defended the Paris Protocol of "Customs Union" as the only viable form of economic relations with the Palestinians (Grinberg 2010).

In order to understand the economic motivations of the dominant social groups in Israel we must understand two factors: the economic interests of these groups and the military capacities and incapacities to control borders. The Israeli producers are interested in maintaining the Palestinian "captive market" because they can force them to buy expensive products by closing the external borders and preventing the import of cheaper products from neighboring countries. Israel is able to close the external borders due to the existence of a sovereign state on the other side which is also interested in sealing the border. The negotiations on the passage of commodities from Egypt to Gaza started after the unilateral withdrawal, and it demonstrated that Israel is much more concerned about controlling the entry of commodities than people, money and even weapons. The agreement on the passages to and from Gaza was mediated by the UN and signed in

November 2005, but never implemented (Guttman and Keinon 2006; Smooha 2007).[9] In January 2008 the wall was destroyed and tens of thousands of Palestinians crossed the border to buy products in the Egyptian city of Rafah, but the border was closed again in a few days (Al-Mughrabi 2008). The symbolic act of blowing up the wall that separates Gaza from Egypt was an act of protest against the Israeli-Egyptian cooperation in strangling the Gazan economy and the flow of citizens buying everything they could in Rafah showed the desperate situation of the population.

The strangling economy of Gaza originated, however, in a more complex problem that was expected to appear if the withdrawal was not done unilaterally but as a result of previous negotiations. The Palestinians demanded a "free-trade area" to be built in Gaza, because the borders are agreed and recognized and the movement of commodities from Egypt and to Israel can be controlled by the Palestinian Authority. They also had an agreement of open passage from Gaza to the West Bank, guaranteed by the Oslo Accords and constant promises from the Israelis. The problem is that Israel can control the external borders of the West Bank and the Gaza Strip with Jordan and Egypt,[10] it can control the borders between Gaza and Israel, and it can control the passage between Gaza and the West Bank. The uncontrolled border is between the West Bank and sovereign Israel. If the Palestinians were allowed to have freedom to import cheap commodities to Gaza, and were allowed to sell them in the West Bank, it was expected that these commodities would be easily smuggled into Israel. Here is the crucial reason for Israel's imposition of the strangling envelops: the danger is the potential damage caused by the illegal entrance of cheap products to Israeli markets after they were legally imported by the Palestinians. The creation of a captive market for expensive Israeli products is not only due to the interest of profit from sales, but mainly to prevent the smuggling of cheap products from Egypt and Jordan into Israel. The expected profits losses are not only to Israeli producers, but also to importers, and government revenues from customs.

Here is my suggested explanation of the economic interest of powerful Israeli groups in maintaining the occupation and strangling the Palestinian economy. If the Palestinian were allowed to develop their economy and sell their products to Israel, they would benefit from the advantages of a small economy and some Israeli producers would lose in the market competition. If a free-trade area is established the danger is of illegal penetration of products from the West Bank to Israel, due to the difficulties of controlling the borders. The economic problem is how to seal the WB borders with Israel, and this is one of the additional motives behind the construction of the separation wall by Israel. But the wall is only a partial solution to this problem, because it cannot completely seal the border and prevent trafficking. Apparently Israel will continue to demand the agreed control of external borders of Palestine and prevent the free import of cheap products to the West Bank for the foreseeable future.

Conclusion

Israel was able to evacuate Gaza thanks to the presence of a clearly marked and recognized border between the two, and Israel's lack of territorial claims on Gaza, as well as its unwillingness to continue running the daily lives of the local Palestinians, risking its

9 See also "Agreed Documents on Movement and Access from and to Gaza" (November 15, 2005).

10 The tunnels that facilitate smuggling products from Egypt to Gaza have been an exception, but the important point here is that these products cannot enter Israel.

soldiers' lives in the process. All that would never suffice to secure Palestinian sovereignty, if only economic, as long as Israel continues to rule the West Bank, crave (part of) its territory and exploit its captive market. The Palestinians depend on Israel for work, but Israeli employers are no longer interested in Palestinian workers as they are now permitted to import cheaper and more docile workers from developing countries further afield. Under these circumstances, the situation in Gaza—with its easily controlled border—is much more severe than in the West Bank, where workers and commodities still manage to cross into "Israel." If Israel manages to complete the "Separation Wall" now under construction and seal the West Bank as well, total envelopment will extend here as well, completing Israel's stranglehold over the territories.

The major obstacle to achieve a peace agreement is precisely the advantage of the "small economy" if markets were free. In order to prevent this advantage, powerful economic groups have influenced the state and military policies aiming to prevent Palestinian development. The strangling economy is aimed at preventing Palestinian investment and industrial production, in order to block potential penetration of Palestinian products into Israel or of cheap imported products to the occupied territories and through them to Israel. The present pattern of military economic domination, the unilateral withdrawal, and the distinction between Gaza and the West Bank, reflect the incapacity of the military to control the borders of Jerusalem and the West Bank. In an era of free market ideologies it is very difficult to propose an economic agreement with strong state institutions controlling the economy. Maybe the new Keynesian ideas of state intervention brought about by the world economic crisis are good news for the prospects of Israeli-Palestinian peace.

PART III
Stories at the Margins

Chapter 9
Operationalizing Nationalism: The Security Practice and the Imagined Figure of the "Arab" Enemy among Israeli "Security Amateurs"[1]

Dganit Manor

Introduction

March 30, 2005, 6.30 p.m. A black Renault Clio rides at full speed across muddy fields not far from "Ir Darom," located approximately 12 kilometers from the Gaza border.[2] Yaakov, an armed policeman without a uniform, who was called in a rush from his home due to a "terror" alert, is sitting in the back. With a profound glance he looks for suspicious figures in the dark. Next to the driver is sitting Chief Inspector Livni, a volunteer's officer and policeman in blue uniform. Due to information regarding suspicious footprints, Livni activates volunteers with his cellphone and police radio network to block the area with police barriers. His expression is serious even though he does not think that it is a "real" event. Instead of a "terrorist," it might just be footprints of "*shabahim*"—short for "*shoim bilti hukiim*" in Hebrew, or "illegal over-stayers"—in Israeli territory. Usually those *shabahim* are Palestinian youngsters who come to work in Israel. Bumping into those illegal over-stayers is a common experience for the security people and not surprising considering their numbers.[3] With one eye, Chief Inspector Livni makes sure that I, the driver—an amateur security woman and volunteer in the Civil Guard—am not getting too deep into the mud in my black car, newly mobilized for the police force.[4] Following the popular image of the anthropologist who experiences adventures overseas, I always welcome "action events" with a smile. But this time my smile was not wide because I was afraid of the risk of bumping into a "real" terrorist. This experience demonstrates the significance of security,

1 This chapter is based on a presentation for the annual meeting of the project "Mobility, Borders and Conflicts in the Israeli Palestinian spaces and their periphery" Aix-en-Provence, November 13–15, 2008. I would like to thank the participants for their challenging questions and useful suggestions, especially Shira Havkin and Cédric Parizot. Also I would like to thank Eyal Ben-Ari for his enlightening remarks, and the reader of this collective book.

2 Ir Darom, which means "South Town" is the pseudonym I use for the city in which I conducted my field work. Following ethnographic codes of ethics, I use a pseudonym in order to protect the privacy of the people in the field. Also, using a pseudonym helps to emphasize the social and cultural image of it, in Israel.

3 The number of Palestinian workers in Israel in 2000 was 145,100. During the Second Intifada (2000–2005), the tightening of movement restrictions brought a drop in the number of Palestinians crossing the Green Line. In 2003, their number fell to approximately 43,000 people. It increased to 68,100 in 2007. Yet, since 2007, only West Bank workers are allowed to enter Israel (Parizot 2010).

4 The Civil Guard is a volunteer organization of Israeli citizens which assists in daily police work. In 2004 it had 70,000 members. It is a subdivision of the Israel Police. Between 1974 and 2004, over half a million citizens volunteered for the Civil Guard (Israel 2005).

Palestinians, terrorists, and borders, as important categories which people use to shape everyday life in Israel.

In this chapter I would like to show that security practices, such as chases, barriers, and patrols are based on the national construction of the enemy and its imaginary appearance. The way it operates confirms its logic that the Arab enemy is dangerous; whether a peaceful Palestinian worker is being captured, or a man trying to commit an attack against an Israeli target is being neutralized. Even though those practices are identified as those of security professionals, they are known to most of Israeli citizens from their own experiences as soldiers or from media reports.

This story occurred during the fieldwork I conducted for my PhD between 2002 and 2005. The research focused on volunteerism in the Civil Guard in the southern region of Israel, close to the Gaza Strip. Following the logic of studying the hegemonic discourse from its margins and borderlands where categories of gender, race, and nationality become questionable (Rosaldo 1993; Lavie and Swedenburg 1996), I chose to focus on volunteers in the Civil Guard in the south of Israel. Volunteers at "South Town" are marginalized in public discourse due to ethnic and militaristic definitions of center-periphery in Israeli society. First, they are the inhabitants of a city known as a "development town" and identified with the socioeconomic periphery of Israeli society. "Development town" is a term used to refer to the new settlements that were built in Israel during the 1950s in order to expand the population of the country's peripheral areas and to ease development pressure on the country's crowded center. These settlements played an important role in the construction of the Israeli hegemonic national identity, and the marginality of their inhabitants. Thus, even today, the people of "South Town" are conceived as peripheral Mizrahim (Oriental Jews) people dependent on the outside and public institutions (Ben-Ari and Bilu 1997a, 1997b; Hasson 1998). Second, as citizens who join the security forces of the State of Israel, they are located on the border between what scholars consider military and civil sectors of society. While combatant soldiers achieve symbolic capital and a position of power in Israeli society, other people who deal with security such as women soldiers and soldiers in blue-collar jobs are considered as marginal (Sasson-Levy 2002). Thus, despite their training, the Civil Guard's volunteers are thought of as amateurs and marginal in the security field.

Margins and borderlands have become sites of academic interest since the late 1980s. They are considered to be spaces of contradictions, displacement, deterritorialization, and symbols of power, and play a central role in the construction of national, ethnic and state identities (Gupta and Ferguson 1997; Yiftachel and Meir 1998; Donnan and Wilson 1999). Following those scholars leads to understanding margins, borders, and borderlands as places, both geographical and cognitive, that people imagine and use in the conceptualization of locales and identities.

In this chapter, I will show that in Israel geographic, social, and cognitive borders are central to the security practice and the construction of national and local identity. Considering their marginality as "Security Amateurs" and inhabitants of the southern periphery of Israel, the volunteers use the security practice to redefine their city as a dangerous site and themselves as security professionals who protect it. Those efforts of re-definitions are full of contradictions and the irony of marginal people using hegemonic codes. From an analytic point of view, they are important because they enable researchers to reveal the national logic embedded at the center of security practice. By presenting a few ethnographic examples, I show that the security practice is based on the "operationalizing of nationalism." This concept articulates the process in which local security experts use, re-appropriate, and internalize, national stereotypes to outline the figure of the dangerous enemy as a calculated risk that can be controlled and neutralized.

The chase after illegal Palestinians workers and other suspects or even their footprints is an example of this "operationalizing." The chase itself embodies the presence of the absent enemy and transforms the security theory into a solid "truth." Thus, even though the volunteers never capture Arab "terrorists," the chase after them leads to a relocation of the volunteers and their city at the heart of the security map. The first part of this chapter is dedicated to the marginal characteristics of the volunteers in the Civil Guard and their exclusion both as peripheral Mizrahim (Oriental Jews) and security "amateurs" who work with the police. Thus, even though they perform a prestigious task, securing Israeli citizens and land, they are considered to be marginal. Studying them is important for researchers in order to understand the hegemonic discourse about militarism in Israel in general and the construction of the Arab enemy in particular. The second part of this chapter deals with the construction of the professional security discourse, based on the combination of fears, national and Oriental stereotypes, and rational calculations of risks and probabilities. The use of rational logic in calculations of probabilities transforms stereotypes and fears into a dangerous reality. The security amateurs use it in order to relocate their city and themselves to the center of dangerous Israeli land and society. The last part of the chapter focuses on the important role played by peaceful civilians who are identified as "the enemy." Even though they do not blow themselves up, they are considered as a probable flesh and blood enemy. Thus operationalizing nationalism confirms its logic by the transformation of its stereotypic foundations into a solid and unquestionable reality.

Marginal Characteristics of "Security Amateurs"

According to many scholars, war, security and the military are central to the construction of Israeli society and its national identity (Ben-Ari et al. 2001). Military service is an emblem of pure patriotism and is one of the major symbols of the collective (Kimmerling 1993). In this culture of security, the combat soldier has achieved hegemonic status and is identified with good citizenship (Sasson-Levy 2008).

However, in spite of this hegemonic construction, there are people who deal with security but do not enjoy the same power and prestige such as female soldiers and male soldiers in blue-collar jobs (Sasson-Levy 2002; Levy 2007a). While currently those soldiers get academic attention, there is hardly any research dedicated to other categories of people who deal with security like the policemen and volunteers in the Civil Guard. This academic neglect also indicates that the Civil Guard's volunteers and their policemen operators do not enjoy a hegemonic status or social prestige even though policemen have been in charge of "homeland security" in Israel since 1974.

The marginal image of the police is connected to the civilians' interpretation of law enforcement. In Israel, as in other countries, civilians, and especially those who get traffic tickets, protest and despise policemen for their lack of professionalism, for handling insignificant matters and for their stupidity. As a volunteer in a blue police uniform, I often heard responses from unsatisfied civilians claiming that policemen, instead of handling security (the "truly" and only important matter in Israel), deal with petty enforcement of traffic laws. A recent survey indicates that only 40 percent of the respondents thought that the police fulfill their role successfully, half indicated that they function in their area of residence successfully and only 40 percent indicated that their confidence in the police is high (Smith and Arian 2007).

As I learnt from my research, many civilians use the sacred status of security in Israel in order to resist the law and order which policemen claim to represent. Sometimes they even use their military past to demonstrate their loyalty to the state, unlike the petty policemen. Once I heard an angry civilian who got a ticket for speeding telling the policeman: "who are you? I served in the Golani Brigade!"[5] In fact, many civilians who receive a ticket from policemen criticize traffic law enforcement. It is common to hear people telling them to: "Go and buy medicine" with the money from the fine, wishing the law representative sickness instead of good health.

The unprofessional image of the regular policeperson is also expressed in the public imagination. For example, in cinematography, the image of the policeman is represented by the character of Azulai in the classical movie *Kishon*. Azulai is an anti-hero Mizrahi (Oriental) policeman, uneducated, lacking professional abilities and authority, and as such he is unable to get promoted. Recently, the public image of the police has been further tarnished since the violent repression of the 2002 October riots, in which 13 Israeli-Arabs and a Palestinian were killed by policemen in Galilee.

While police professionalism is considered doubtful, the volunteers are definitively considered to be amateurs. After only 24 hours of training they can wear a police uniform and can even detain civilians for two hours at the police station. After additional training they obtain the authority to give traffic tickets. Their public conception as amateurs is intensified with their image as old men. A survey carried out by the police indicated that while only one-fifth of the traffic enforcement volunteers are pensioners, their activity is very prominent due to their old age and the multiple hours they spend volunteering (Sauer and Goren 2003). This survey also found that policemen think that the volunteer's old age and their prominence in their activities are problematic to their functioning and harm the image of all the volunteers in the Civil Guard. The image of the old man symbolizes the volunteers' marginality for two main reasons: First, it contradicts the heroic and masculine figure of the combatant soldier (Sasson-Levy 2008). Second, old people are thought of as being sick and in need of medical care (Hazan 1994). Therefore, old people are perceived as inefficient especially dealing with security, an area where physical ability is required. This public image of the volunteers as ineffective is intensified with public and officialdom criticism of the efficiency of their operation (Yanay 1993).

Finally, the marginality of policemen and volunteers of "South Town" is intensified by their marginal geographic space from a security perspective. There was no suicide attack in this city despite its proximity to the Palestinian border. This marginality from a security point of view is added to the social-geographic marginality as Mizrahim (Oriental Jews) and inhabitants of a "development town," as discussed earlier. The inhabitants of "South Town" and its surroundings belong to the low- and middleclass in Israeli society.[6] Most of the old volunteers emigrated from North Africa. There is also a group of volunteers who emigrated from the former Soviet Union in the 1990s. Another prominent group includes volunteers who work in the education field such as teachers and headmasters of the local schools. Most of the others have work at the local factories or as bus drivers. Some are unemployed.

5 The Golani Brigade is prestigious infantry unit within Israeli representations, and the contribution of its soldiers to Israeli national security is considered to be very high.

6 According to the Israeli Central Bureau of Statistics report from 2006, "South Town" is located with Arab villages at the low cluster, out of 10 clusters distributed according to its standard of living.

All those marginal characteristics of the volunteers in the Civil Guard at the southern periphery of Israel, leads to their exclusion both as peripheral Mizrahim and security "amateurs" who work with the police. Thus, even though they perform a prestigious task, securing Israeli citizens and land, they are considered to be marginal. In order to relocate themselves and their city at the heart of the security map the volunteers and their operators, the policemen, adopt the orientalist-professional discourse of security. A close observation of their daily security activities enables us to understand how the logic of the security is based on the operationalizing of nationalism.

Security Professionalism, Enemy, and Orientalism

The description of the chase which opened this chapter is a good example of the daily routine of the security men. The central figure recreated during this professional security activity is the "dangerous terrorist enemy." In this section I show that this figure is based on national and orientalist stereotypes. Its danger is confirmed as "real" mainly by its operationalization, as the chase after the "footprints of the enemy" demonstrates. In other words, the performance of the security practice confirms the security logic based on national and orientalist stereotypes as "real."

"Warning Scripts" and Security Professionalism

The transformation of stereotypes to real flesh and blood figures in the security discourse is carried out using what I suggest be called "warning scripts" and the ways of operating that follows them. "Warning scripts" are the central expressions which specify security professionalism, based on the rationalization of fears and probability (Ericson 1994). The volunteers learn the "warning scripts" during their training and refresh it at the beginning of every security mission in the framework of briefings.

These briefings symbolically transfer them from their status as civilians to security professionals. They occur before every mission or shift and take place at the local police station or in the case of securing a local event—on site. The person giving the briefing at large events is the patrol police officer or a substitute. The briefings have a routine structure: they start with a report on the latest important events from a security point of view such as terrorist attacks and rocket falls. Afterwards, the person giving the briefing explains the details and significance of the mission. He also gives details about the current alerts and refreshes some of the security procedures. As I will show later, during the briefing the figure of the Arab enemy is recreated in their imagination by the effort to identify and neutralize his probable dangerous abilities.

There are many local events at "South Town" celebrated according to the local, national and religious Israeli-Jewish calendar respectively with the participation of dozens, hundreds and even thousands of residents, Israeli citizens and Jews, sometimes from all over the world. Also there are celebrations and parades while bringing a new Torah scroll to the many local synagogues. In the summer there are local events organized by the municipality, with the participation of dozens of residents, mostly youths.

One of these events was a rock concert in October 2004. Despite the police estimation, only dozens of local teenagers arrived at the "Sons garden," the local site for memorials, celebrations and performances. About an hour before the arrival of the band, Chief Inspector Zohar started to brief the Civil Guard's volunteers, policemen, and hired security guards who were invited to secure the event. This briefing was the same as the dozens of briefings I participated in, and ended with "open-fire regulations." Chief Inspector Zohar explained to all of the security men that they should look for a man with Oriental traits that many Israelis are familiar with, meaning that he has black hair and eyes and brown skin color. Once a suspect is identified, they should shout: "stop, stop or I'll shoot." Zohar also explained that: "If the suspect runs away, then you must shoot in the air. If he keeps running, shoot him in his legs. If a human life is endangered then shoot in order to kill."

These "open-fire regulations" are aimed at recognizing the enemy and neutralizing them. They are known to most of the participants from their obligatory military service. The effort to recognize the enemy in order to neutralize it has an important symbolic meaning. Scholars like Ben-Ari and others (Harle 1994; Bar and Ben-Ari 2005) use the concept of "Enemy recognition" to note a performative procedure which brings to light and constructs the enemy's "true face." It gives the enemy a face with the concrete features which will make it into an appropriate target of hate and struggle on the one hand; on the other hand, "enemy recognition" enables the security men to turn the vague category of the enemy into a category that can be handled, punished and even destroyed. In all, it turns the treatment of a terrorist attack from a warning about the probability of dangerous event into a certain one.

"Warning Scripts" and "Orientalism"

The "enemy recognition" taking place during opening briefings of security tasks is based on orientalist stereotypes. The "oriental traits," such as black hair and brown skin color, used as visual signs for "enemy recognition," are taken from the imagination of the "national other" in Israeli-Zionist discourse. As in every national discourse this figure is defined as fundamentally different from the national "us" (Harle 1994). The figure of the Israeli is imagined to have a western look of blond hair and blue eyes (Almog 2000). This western image of Israeli identity has been analyzed by Smooha (1984), in his discussion of ethnicity in the Israeli Army. Smooha revealed that since the establishment of the Israeli military, Israelis consider it to be a western and advanced military, opposed to mentally and technologically inferior Arab armies. This cultural construction recreates the Israeli-Zionist national identity as composed of those traits. In this discourse, the figure of the "other" is used for the creation of the national self not only through its opposition, but also by subjugating the other in a moral and cultural way in a similar manner to the style of thought known as "orientalism." In other words, the attribution of mental and technological inferiority to Arab armies is based on orientalism, an ontological and epistemological distinction between, on the one hand, the national other, the "Orient," imagined as "inferior," "cruel," and "sensual"; and on the other hand, the "western national self," the "Occident" imagined as an "enlightened" and "progressive" self (Said 1978).

Zionist discourse is based on the construction of Israeliness as "western" and "modern" by distinguishing between it and an "Oriental" and "primitive" identity (Rabinowitz 2003). According to Almog (2000), in the years before the establishment of the Israeli state, the Arab was perceived not only as a member of an inferior culture but also as a bloodthirsty dangerous enemy and a new version of the biblical cruel enemy of Amalek, while the Zionists saw themselves as helpless victims. The consolidation of this perception was part of delineating their identity.

In this framework, the category of "enemy" indicates not only a fundamentally national other, but also a struggle between the "good" identified with it and the "bad" represented by the stranger (Harle 1994; Shamir 2005). Oriental traits according to Zionist ideology indicate this strangeness and are used as the foundation of security practice. Most of the participants at the briefing of Chief Inspector Zohar accepted this "enemy recognition" as natural. I thought that it was amusing because it matched the Ashkenazi elite's imagination of "South Town" inhabitants as inferior Mizrahim. One of the guards also found it amusing and asked in a cynical tone "what happens if a redhead Arab comes?" His ironic question about the color of the enemy's hair revealed that "enemy recognition" is based on the imagination of the national other according to its ideological stereotype. Stereotypes, according to Herzfeld (1992), are the means through which national ideology can present itself as familiar and jeopardized by equally familiar enemies. Recognizing the enemy according to its Oriental traits is the conversion of national stereotypes, of characteristics imagined as objective descriptions of an unchanging reality (Herzfeld 1992). This stereotyping is not unique to the Zionist imagination. According to Herzfeld people adopt rhetorical strategies on the basis of presumed national character everywhere. The efficiency of these stereotypes lies in their appeal to the conventions of collective representations which oppose the Arabs to an imagined western Jewish figure. This way security functions as a discursive practice creating a national identity and as an integral part of the processes of constituting subjectivity and its boundaries (Buzan and Waever 1997; Balzacq 2005; Stern 2006).

In recent years, the allegedly concrete traits of pressure and wearing large clothes were added to the "recognition of the enemy." Beyond the alleged Oriental appearance, Chief Inspector Zohar asked everybody to pay attention to confused, nervous, and sweaty people wearing large coats. These signs were believed to be of people hiding an explosive belt underneath their clothes and planning a suicide attack. Since there was a chance of rainfall that day according to the weather broadcast, he corrected himself and asked us to suspect only people with strange clothes.

The "warning scripts" are a central part of the operationalizing of nationalism, since they help to turn national stereotypes into a routine practice. The "warning scripts" repeat themselves not only in "open-fire regulations" but also in the form of special alerts. They blur the distinction between events that already happened and those that probably may happen in the future. Before everybody took their position, Chief Inspector Zohar said that "there is a general warning regarding a terror attack in the centers of a crowd. Thus, even though there are not any specific terror alerts for this event, we can never know what will happen and be ready for every possibility." This saying intensified the significance of the mission to its fulfillers. For those security amateurs, the distinction between a concrete alert specifying that a "terror attack" will occur, and a general warning concerning its probability, was not relevant. It seemed that the enemy could reach them everywhere, without any early warning. Thus, practices such as securing events, chasing suspicious figures, patrolling, and setting checkpoints begins with briefings. In those briefings, the imagined figure of the enemy and their visual traits is being repeated and confirmed as a "natural phenomenon."

"Warning Scripts" and the Relocation of Oneself within the National Security Map

The adoption of "professional discourse" enables the policemen and volunteers in the Civil Guard at "South Town" to redefine their home town as located in the middle of the Israeli terror attacks map at the time when there were no attacks or missiles. Thus, they connect alerts regarding other places to the security activity in "South Town" even though there was no specific alarm. A typical example occurred during a routine briefing before the evening shift at the police station of "South Town" in September 2004. It took place at the police investigators office with the participation of the person giving the briefing, Sergeant Major

Ya'acov, one of the regular duty shift commanders at the station, three volunteers in the Civil Guard, a new policewoman, and the "officer in charge."[7] In the same way as senior officers brief at the beginning of public event missions, Sergeant Major Ya'acov opened his briefing with a summary of the security events which occurred in the previous week: The extermination mission of people considered as "terrorists" by the Israeli military in Gaza and the dropping of three Qassam rockets in the area nearby. The participants of those briefings, as inhabitants of this area and as Israeli citizens are familiar with this security everyday reality from their personal experience and media reports. Presenting those events at the beginning of the briefing specified the local space as an integral part of dangerous national space together with the definition of the time during the local routine as part of the national situation of emergency. The definition of the time as "exceptional" is connected to the construction of an apocalyptic reality which has been at the center of the western public consciousness since September 11, 2001 (Aretxaga 2001). In Israel it is also connected to the legitimacy of militarism, and to the normalization of the experience of war (Lomski-Feder 2004).

In his brief, Ya'acov also summarized events that had not happened and had not occurred. In a determined tone he said that during his shift the evening before, "There was almost a "Parash Mongoly" (Mongolian rider in Hebrew)," referring to a well-known alert of penetration of a "terrorist" (which actually means a Palestinian) into Israeli territory.[8] The expression of the participants became serious: there was a "security" event in their area. In other words, the report of what is considered as an alert to the possibility of "terrorist penetration" transformed into a "real event" in Ya'acov's briefing. According to Hier (2003), contemporary risks come to consciousness in scientific thought. This briefing is a good example of this scientific-professional construction. The possibility of "terror attack" by the enemy is transformed into a calculated risk, blurring the difference between what might happen and what already occurred. According to Beck (2003) the concept of risk presupposes decisions that attempt to make the unforeseeable consequences of human decisions foreseeable and controllable. Also, it creates a sense of persistent danger, everywhere. Thus, the proximity of real events, alerts, and calculated risks, confirmed the dangerous nature of the Arab enemy. It also enables a re-definition of the local space as dangerous and the local time as that of an emergency. Above all it helped to relocate the activity of the police men and volunteers in the Civil Guard to the core of the security forces.

Notes on the Embodiment of the Absent Enemy

The chase after the specter of the footprints occurred on the March 30, 2005, not far from the Gaza border and ended without anything heroic or dramatic—no terrorist was exposed. Also, for the other examples given, the "warning scripts" were not fulfilled. During the three-year period of field work (2002–2005), no armed "terrorist" was caught, but despite the absence of the enemy, volunteers remained alert. This chase after the specter of the footprints ended at night after the military representatives instructed the police commanders to go back to their "routine activities." Even though the probability of the "terror attack"

7 During the evening and night shifts there is a duty roster between the officers and the senior policemen who replace the head of the station when he is absent, as the senior authority in the field.

8 The name "Mongolian rider" is a fabricated name to a real alert.

did not transform into a certainty and nobody recognized a true flesh and blood "enemy terrorist," policemen and volunteers in the Civil Guard accepted these Orientalist and national constructions.

As I argued before, the imagined figure of the enemy is approved and becomes "real" not by catching flesh and blood terrorists, but by identifying people with a profile that supports the construction of Palestinian civilians as enemies, and therefore with the probability of becoming suicide bombers. Thus, the security amateurs pay special attention to Israeli citizens with Oriental appearance such as the Bedouins living in Israel and the Palestinians from the occupied territories.[9] A common mistake to come out of the Oriental and national "enemy recognition" discourse among security people is to treat even Jewish immigrants from Ethiopia as "suspects." Even though these people are checked and released by security people, their interpellation has a symbolic significance. The fact that they are checked underscores the location of Israeli society's boundaries as Jewish, Western and democratic (Rabinowitz 2003). According to the worst case scenario logic which constructs the security and risks point of view, it is also important to neutralize Israeli Bedouins and Israeli Palestinians, the national "dangerous" others who are not Jewish but identified with the enemy and symbolically, are equal the detention of a terrorist. Therefore, catching them enables the volunteers to relocate themselves as security professionals.

In order to emphasize this point let me conclude with my best friend Ira's adventure: Ira, with other volunteers from "Tlamim" regional council and "South Town" policemen, proudly shared with her colleagues the success of one of the night shifts in which she participated in capturing *shabahim*, or illegal over-stayers, which was organized by the local security officer with the participation of local policemen and volunteers in the Civil Guard. The photo (see Figure 9.1) she sent by mail to her friends can teach us about this embodiment of the enemy. This photo is reminiscent of the genre of a school year photo: in the first line are seated in half a circle tired and suntanned boys, most of them wear caps and their eyes are downcast. They are young workers who entered without a permit into Israel to work in the region. In the second line stand the "security amateurs" suntanned and fresh, most of them wear the uniform of the Border Guard and their eyes are shining. Three of the "security amateurs" illuminate, with a flash light, their nocturnal catch, emphasizing the power relations between them. This "catch" symbolizes the enemy, even though no explosives or weapons were found on these boys. The pride of catching them enable volunteers and policemen to locate themselves, at least from a local point of view, as an integral and significant part of the professional security forces. It also confirms the national and Oriental logic embedded in security practice.

9 Bedouins are Palestinians who remained in Israel after 1949 and obtained citizenship in the 1950s. Yet, as they are Arabs, many Jews in Israel look at them as a "fifth column." Some even regard Bedouin hanging around or getting engaged to Jewish women as endangering the purity of the blood in Israel.

Figure 9.1 Volunteers from "Tlamim" regional council and "South Town" policemen, showing the Palestinian workers they caught without a permit

Credit: Rafi Babian, 2005.[10]

Conclusion

Catching the Palestinian boys who tried to cross geographic and cognitive borders plays the role of fighting a flesh and blood enemy recognized as the dangerous national other.

Security practices, such as night shifts, chases, barriers, and patrols are based on the national construction of the enemy and its imaginary appearance. On daily occasions, such as this late night-time capture, this construction is being confirmed: The Arab enemy is dangerous, whether a young peaceful Palestinian worker is being captured, or a terrorist trying to commit an attack against an Israeli target is being neutralized.

The memory of the catch itself is the trophy for long nights of chasing the specter of the abstract enemy. It enables "security amateurs" to redefine the time as an emergency, to relocate their city as a dangerous place and themselves as professionals who guard their homeland. Also, it helps to bridge the gap in the security discourse between immediate danger and the heroic actions needed to fight them opposed to the boring daily routine trying to identify the enemy when nothing happens.

This social construction of professionalism, based on partial truth and experiences, helps to justify the state of war as an exterior phenomenon which is forced on Israeli society. Also, it blurs the power relations between relatively economically-established Israeli Jews and unemployed Palestinians who are involved in this situation and turns the conflict between them into a complicated loop which is almost impossible to solve. This is an important dimension of security practice based on the combination of fear, Orientalist and national stereotypes, together with the discourse of rational calculation of probabilities and risk.

The use of professional security discourse on the margins enables a re-definition of an insignificant space as a central scene of security activity. It reinforces the borders of the national self and the imagined community, not only by the sense of belonging by practicing security, but also by imagining its enemies to be represented by civilians with the imagined traits of the dangerous enemy.

10 I would like to thank Rafi Babian for giving his permission to use this photo.

Chapter 10
Identity, Solidarity, and Socioeconomic Networks across the Separation Lines: A Study of Relations between Palestinians in Israel and in the Occupied Territories

Elisabeth Marteu

Introduction

Over the past 60 years, relations between Israeli Palestinians and those in the Occupied Territories (OPT) have undergone phases of varying degrees of proximity around issues of territory, identity, and politics. Despite the progressive public affirmation of a common Palestinian identity, these populations have followed different social and political trajectories. At the end of the first Arab–Israeli War in 1949, practically all ties were cut between those Palestinians who remained under Israeli administration and those living under Jordanian or Egyptian authority. Inter-Palestinian relations only resumed from 1967 onwards when the Israeli occupation of the Palestinian Territories made it easier to move between Israel, the West Bank, and Gaza. In the early 1990s, Israel developed a new separation policy aimed at protecting itself from the demographic threat posed by Palestinians in the OPT (Crousaz 2005). The Israeli authorities took initial measures designed to detach itself from these territories and restrict the entries of Palestinians into Israel. The Oslo Accords in 1993 contributed to this transition by breaking up the Palestinian territories and further restricting freedom of movement for Palestinians from the West Bank and Gaza.[1] Palestinian cities were further isolated when the Palestinian Territories were blockaded after the start of the Second Intifada in 2000, and Gaza was sieged in response to Hamas taking power in 2007. As for the Wall, on which construction commenced in 2002, it was clearly designed to further confine Palestinians in the West Bank. Israeli policies of blockading, controlling movement (checkpoints, work permits, and entry permits to Israel, etc.) and the construction of the Wall has progressively reduced physical contact between Israeli Palestinians and those in the OPT. These separation policies have affected most people, but have also led to a reorganization of clandestine travel and the balance of power among Palestinians (Parizot 2006, 2008), and fed affirmations of inter-Palestinian solidarity (*tadammun*). As a result, these security measures also fostered "a foliation of space and time" (Parizot 2009c), i.e., different timescales, perceptions, and especially experiences of separation among Israelis, Palestinian Arab citizens of Israel, and Palestinians in the

1 As of the beginning of the 1970s, Israel implemented a collective permit system enabling Palestinians from the Occupied Territories to move freely between the West Bank, Israel, and Gaza. This policy was reconsidered as of 1988 with the deployment of magnetic cards and then in 1991 with the issuing of individual permits (Mukh 2006; Abu Zahra 2007; Parizot 2008).

OPT. Even among Israeli Palestinians themselves, there are different "map perceptions" (Hamidi 2010)[2] and experiences depending on the geography of family networks, economic activities, or political convictions.

Many studies point to the awakening of nationalist policies and activism among Palestinian Arab citizens of Israel who publicly affirm their Palestinian identity. They describe "Palestinisation" and "politicisation" processes (Smooha 1984; Rekhess 1989; Rouhana 1997), partly encouraged by the occupation of the Palestinian Territories in 1967 that allowed contacts to resume between Palestinians who had been separated by the Green Line[3] since 1949.

Laurence Louër (2001a, 2003) takes this analysis further by pointing out that it is the bipolarization of Israeli politics at the end of the 1970s that allowed Arab political leadership to emerge in Israel. The consensus on the Palestinian identity is thus mainly a result of structural factors and an internal process of politicization. As for relations with the Palestinians in the OPT, there appear to be very different historic and political trajectories that preclude any notion of the Palestinians as "a single people from the Galilee to Hebron" (Louër, 2002). In addition, from a social and economic perspective, there are several studies of day-to-day interaction among so-called ordinary Palestinians, such as family networks, trips to markets in the West Bank, or relations between Palestinian employers in Israel and legal or illegal Palestinian workers (Forte 2001; Bornstein 2002; Parizot 2006a, 2006b, 2008a, 2008b, 2014). They all point out that the balance of power among Palestinians has changed as the Palestinians in the OPT become increasingly economically dependent on Israeli Palestinians. In this context, the dividing lines between Palestinians are shifting and polymorphic. On the one hand, they reveal multiple individual and collective trajectories that reflect different experiences and uses of the security barriers. On the other hand, they reflect complex identity adjustments causing the frontiers within and between communities to shift.

As a result, we need a new understanding of the links between Israeli Palestinians and those in the OPT, based on a non-linear process of interdependence and autonomization. This approach reminds us that identities are socially constructed. As highlighted by Riccardo Bocco and Daniel Meier (2005), borders in the Middle East are a component of identity, but they are not identities. Here in the Israeli-Palestinian context where the dividing lines are not actual borders, we should be even more aware that senses of belonging and self-attributed identities are fluid (Barth 1995). Next, as shown by studies done in various contexts (Vila 2003; Parizot 2008a), when dividing lines are crossed this can influence the balance of power, feed antagonism and fragment perceptions of space within the same community or ethnic group. Lastly, when analyzing inter-Palestinian relations we run into the question of how to qualify them. Are they inter- or intra-community relations? Are Israeli Palestinians and those in the Occupied Territories a single community? Using Giovanni Sartori's theory (1977), in 2000, Laurence Louër showed that community is founded on a consensus around common representations produced and maintained by a network of institutions. As such, if the structural criterion is fundamental, then there is

2 This formulation is credited to Michèle Lamont in her work on the construction of symbolic borders. It was taken up by Camille Hamidi (2010), who proposed the following definition: "These are instruments by which individuals and groups struggle and finally agree on the definition of the reality. They split up individuals in different groups and generate a feeling of similarity and a sense of belonging to a group."

3 The Green Line is the line that separates Israel from the West Bank following the ceasefire agreements signed in 1949 at the end of the first Arab-Israeli war.

indeed an Arab community in Israel, but not a "Palestinian community" in which Israeli Palestinians and those in the Occupied Territories are united around common institutions. Having said that, family networks, religious solidarity, and economic interaction between Palestinians have always been maintained, and there is a collective agreement on the idea of a common Palestinian people.

This chapter offers a top-down analysis of the new political and socioeconomic relations between Palestinians faced with the separation limits whether they be of a security, legal, ideological, geographical, or religious order. This approach allows day-to-day interactions and relations among activists to be studied at the same time and reveals how these relations oscillate between solidarity around aspects of identity and different ordinary experiences. In addition, this study focuses mainly on relations as perceived and experienced by Israeli Palestinians since 1949 until the present day, with a more specific look at the first decade of the twenty-first century. This period is marked by a turning point in security issues and stronger separation policies implemented by Israel on the one hand, and more intense solidarity among Palestinians expressed by Arab citizens of Israel and their institutions (political parties and associations) on the other.

Therefore, there are three parts to this chapter, in chronological order: Part one outlines the history of inter-Palestinian relations between 1949 and 2000. The other two parts focus on the recent period since the Second Intifada began in 2000, presenting first the reconfiguration of inter-Palestinian socioeconomic relations, and then addressing the political dimension of this solidarity that defies the dividing lines.

1949–2000: A Process of Differentiation between Palestinians in Israel and in the Occupied Territories

1949–1991: Inter-Palestinian Relations—From Rupture to Reconstruction

At the end of the first Arab–Israeli War of 1947–1949, Israeli Palestinians were separated from those left under Jordanian or Egyptian rule. The Palestinians who became Arab citizens of Israel were placed under military administration until 1966 and subjected to strict restrictions on their movements and on their economic, social, and political activities (Lustick 1980). However, links between the various Arab Palestinian populations were never entirely cut off.

On an economic level, although the regular trade that existed before 1948 broke down, there was still some contact through contraband networks and smuggling, especially between Hebron area, the north of the Negev (around Beersheba), and the Gaza Strip (Parizot 2001b). However, day-to-day relations did not resume until after 1967, when the occupation of the Palestinian Territories meant Palestinians could travel freely, allowing them to rebuild economic relations, as many Arab citizens of Israel would shop at the markets in Jenin and Hebron in the West Bank, where produce was cheaper than in Israel. Family relations were also reactivated, and Palestinians from Israel and the OPT soon began marrying each other (Parizot 2001b).

In the political arena, the activities of Arab parties were strictly controlled by the Israeli authorities until the end of the military government in 1966. Local leaders were massively co-opted by the Zionist parties, especially on Arab satellite lists close to Mapaï,[4] which

4 Predecessor of the Labour Party.

began standing for Parliamentary election in 1949. Apart from these occasional lists, the Israeli Communist Party (ICP) was the only party to represent the interests of Arab citizens of Israel. From its Judeo-Arabic foundations, the party split into two groups in 1965, with the Maki advocating a Zionist stance and Rakah supporting the Arab cause in Israel. Since then, the latter has always claimed to be a Judeo-Arabic party, criticizing Israeli policies in the OPT and advocating equality between Jewish and Arab citizens in a democratic Israel (Greilsammer 1978; Louër 2003a). In 1959, an exclusively Arab, nationalist party was founded called Al-Ard (The Land). The party was banned by the Israeli authorities in 1965 for its support of Nasser's pan-Arabic ideas.

As well as the lack of political parties, there were very few Arab associations during the military government era. The fertile network of Palestinian associations that existed prior to 1948 was wiped out after the war. Women's associations, for example, had been very active since the 1920s, doing charity work and organizing educational activities while supporting the resistance movement against the British authorities and the Zionist presence (Fleischmann 2003). There was cooperation between several women's associations based in large Palestinian cities like Jerusalem, Nablus, Bethlehem, Jaffa, Acre, Haifa, and Nazareth. However, the fragmentation of the Palestinian people in 1949 caused many of these organizations to close and left the Palestinians in Israel isolated. In the West Bank and Gaza, new associations were progressively formed but were also subject to strict control by the Jordanian and Egyptian authorities (Curmi 2002).

The Israeli military government's strict policy of control contained Arab citizens' political activities and also contributed to limiting relations between Palestinian activists on both sides of the Green Line. Any relations that were maintained were kept secret, since the young Israeli State saw them as a potential security threat. Arab communist activists would meet at gatherings organized in former Soviet bloc countries (Greilsammer 1978). But the borders were not hermetically sealed. Al Ard's leaders had links with Palestinian activists under Jordanian rule. Thus, contacts were never entirely cut off, even under the military government.

At the end of the 1960s, Palestinian mobilization was reorganized under the combined effect of a structured Palestinian national cause (the PLO was founded in 1964), the end of the military government in Arab zones of Israel (1966), and renewed contacts between Palestinians (1967). In addition to these political events, Israel's discriminatory policies against Arab citizens fostered their desire for protest and political organization. A new Arab nationalist party, Abna al-Balad (Children of the Nation), appeared in 1973 and openly admitted that it was close to the PLO. Meanwhile, the Palestinian resistance in exile progressively began addressing the case of Israeli Palestinians. Until the start of the 1970s, the PLO, under the leadership of Yasser Arafat, made no mention of Israeli Palestinians. But in 1972, the Palestinian National Council adopted a resolution calling to "support the Israeli Palestinian's struggle to reinstate their Arab identity and strengthen their links with the rest of the Palestinian people" (Rekhess 1989). In 1977, a Palestinian National Council resolution even encouraged contacts with the "progressive, democratic and anti-Zionist forces in Israel" (Rekhess 1989). It would seem that the 1976 Land Day helped change Palestinian nationalist leaders' representation of Arab citizens of Israel (Frisch 1996, p. 451). Six Arab demonstrators were killed during strikes and demonstrations organized in Galilee against the Israeli authorities' policy of expropriating Arab land. This protest movement was supported by several demonstrations in the OPT and marked a turning point in the public expression of Arab discontent in Israel.

Meanwhile, the ICP had links with communists in the OPT (Greilsammer 1978) and from the mid-1970s was organizing summer camps in Nazareth in partnership with the Palestinian Communist Party. The mayor of Nazareth, Tawfiq Ziyad, a communist whose election in 1975 was welcomed by the PLO, continued to organize these camps until the 1990s. The nationalist movement Abna al-Balad adopted a regional perspective and since its beginnings in 1973 had openly supported the PLO and advocated a single state solution. The party refused to take part in Israeli national elections, and had links to members of the Popular Front for the Liberation of Palestine (PFLP). However, nationalists reached a political watershed in the 1980s when the leaders of the Progressive List for Peace, set up for the 1984 elections, publicly declared their support for the PLO and even wrote in Fatah's official newspapers,[5] marking a turning point in their identification with the Palestinian cause (Frisch 1996). One of these leaders, Mohammad Miari, who was one of the founders of the Al-Ard nationalist party, was an internally-displaced Palestinian from Al Birweh, a village destroyed during the war. Active in student politics while attending the Hebrew University of Jerusalem, he later became a lawyer specializing in the cause of Arab citizens of Israel. He was placed under house arrest on several occasions and banned from the OPT, where Israeli security services suspected him of having contacts with "Palestinian terrorist groups" (Louër 2003a).

This affirmation of a Palestinian identity, which began in the 1960s and was taken up politically by Arab citizens of Israel, was perceived as symptomatic of the "Palestinisation"[6] of Arab citizens of Israel. Their open support for the First Intifada when it began in 1987 (Rekhess 1989) reinforced the Israeli authorities' fears and perception of them as the fifth column of the Palestinian struggle.

1991–2000: Separation Policies and the Politicization of a Common Palestinian Identity

Israel started to implement its policy of separation at the beginning of the 1990s by deploying the first checkpoints in the West Bank and Gaza Strip, along with a system of individual entry permits for Palestinians wishing to travel to Israel. The outbreak of the First Intifada in 1987 then the Gulf War in 1991 served to justify Israel's security policy aimed at pushing the Palestinian threat outside Israeli territory. The idea of a border remerged for all people living in Israeli-Palestinian space. According to Cédric Parizot (2008), "inter-Palestinian borders" even emerged from that point onwards. The first to be hit by restrictions on movements were Palestinian workers from the OPT who could no longer travel freely in Israel. Between 1992 and 1996, the number of Palestinian workers entering Israel dropped by 51 percent (Arnon et al. 1997). The application of the Oslo Accords (1993–2000) contributed to this process by dividing the West Bank into three zones (A, B, and C) and drawing new political and administrative lines along which were deployed mechanisms of filtering such as checkpoints and barriers run by the Israeli army. With Gaza surrounded by a barrier, social and economic relations among Palestinians became concentrated in the West Bank, not without some difficulty. Moreover, these restrictions on movements shifted the balance of power among Palestinians as workers from the West Bank found themselves

5 Fatah is the acronym in reverse of the "Palestinian National Liberation Movement" founded in 1959 by Yasser Arafat.

6 An abundance of studies were carried out on the topic of the "Palestinisation" of Arab citizens. See Louër (2001a, 2003a) for a review of this literature.

increasingly dependent on Israeli Arab employers, and illegal workers were forced to rely on smugglers (Parizot 2006a, 2006b).

At the same time, Palestinians who were Israeli citizens had easy access to Palestinian towns, and would often frequent the markets in Jenin and Hebron or go out in Ramallah. The city became the Palestinian economic and intellectual capital in 1994 when the creation of the Palestinian Authority brought returnees and Western expats to Ramallah. The urban center became very popular for young Palestinian couples coming from Israel. Those who would go there regularly recall a dynamic city with a choice of bars and restaurants, and shows and concerts in Arabic. They also say that they enjoyed being in an exclusively Arab environment where they heard their language being spoken and could be with people who listened to the same music. They felt quite at home there, in a cultural setting that was familiar to them.

Yet these outings were also an opportunity to realize their differences. Tania Forte (2001) described Israeli Palestinians' trips to the market in Jenin, outings that confirmed Arab citizens of Israel's preconceived ideas that they themselves were more "independent," "modern," and "affluent" than Palestinians in the OPT. Tania Forte described these multiple identities as follows: "The women … show themselves to be, through shopping and home making, at times female and at times Palestinian, at times Palestinian and at times Israeli, at times Palestinian-Israeli middleclass women, and at times economic persons making 'rational' choices" (2001, p. 212).

The same complexity of individual and collective behaviors existed in the political arena. Arab parties continued to affirm a Palestinian identity, setting up the Arab Democratic Party in 1988 then the National Democratic Assembly (Tajammu') in 1996. The Islamic Movement also made its political debut in municipal elections at the end of the 1980s, before splitting into two branches for the 1996 elections. The "southern" branch took part in parliamentary elections while the "northern" branch boycotted any idea of representation in the Knesset (Israeli parliament). The Islamic Movement built close relations with preachers and Islamic activists in the OPT (Rekhess 1997; Tal 2000), particularly by providing training to Imams and to the movement's senior party officials (Louër 2003a). Its founder Abdallah Nimr Darwish studied Sharia in Nablus from 1969 to 1972, where he met Ahmad Yassine and was introduced to the ideas of the Muslim Brotherhood. Ra'ed Salah, the current leader of the hard-line northern branch, studied at the University of Hebron. The Islamic Movement of Israel also organized assistance to Palestinians in need. In 2000, the northern branch's Humanitarian Aid Committee sponsored 8,000 orphans in the West Bank and Gaza, who each received a monthly stipend of 30 dollars (Louër 2003a, p. 103).

Just before the Second Intifada began in September 2000, there was consensus among all Arab parties on equality between Jewish and Arab citizens and the transformation of Israel into a "State of all its citizens" living in peace alongside an independent Palestinian State within 1967 borders (Louër 2003a). To varying degrees, they all had links to political formations in the OPT. Several Israeli Arabs emerged as intermediaries between Israel and the Palestinian leadership. Ahmad Tibi, for example, took part in the secret negotiations in Oslo before becoming official advisor to Yasser Arafat (Bligh 1999, p. 157). Moreover, the PLO and Yasser Arafat in particular intervened on several occasions to call on Israeli Arab leaders to unite for parliamentary elections (Frisch 1996, p. 458), something that the Arab parties always refused to do, preferring to put forward several lists of candidates for the Knesset. Political orders from Palestinian leaders have never really had an impact on Arab politics in Israel. While it was important for the Arab parties to be close to the PLO and affirm a Palestinian identity, the PLO and its various rival factions never had any real influence on political life in Israel and were never viewed as representative organizations (Amara 2000; Ghanem and Ozacki-Lazar 2003). Yasser Arafat supported the Intifada by Palestinians in the OPT but never directly called for an uprising or violence against Israel on the part of Arab citizens (Frisch 1996). Moreover, Israeli Palestinians were absent from the Oslo peace negotiations, a further demonstration of the political differences between Arab citizens of Israel and the rest of the Palestinian population. Palestinian politics and territorial matters became firmly centered within the OPT in 1994, with the territorialization of the Palestinian Authority then its formal election in 1996 (Radi 2002, p. 206). The West Bank and Gaza thus became the center of the Palestinian cause and the representative territory and political structure for refugee Palestinians. At the end of the 1990s, Israeli Palestinians and those in the OPT therefore had two distinct political arenas.

2000–2010: Palestinians' Socioeconomic Relations under the Constraints of the Israeli Separation Policy

With the start of the Second Intifada (2000), Palestinian enclaves within the OPT soon found themselves under siege. They were repeatedly blockaded, curfews lasted months at a time, and Gaza remained cut off from the West Bank (Mukh 2006). It was practically impossible to enter or leave Gaza, there were tighter controls at crossing points into and out of the West Bank, and the number of work permits for Palestinians was drastically reduced. As a result of these travel restrictions, the number of Palestinian workers in Israel dropped from 107,630 in 2000 to 59,000 in 2001, well before the Wall was built (Parizot 2009a, p. 65), meaning that contacts between Israeli Palestinians and those of the OPT were limited. Arab citizens of Israel also progressively stopped entering the OPT. Not only did they view the West Bank as dangerous, but the Israeli authorities banned all Israelis from travelling to the OPT.

Even within Israel itself, the events of October 2000 reactivated the split between Jewish and Arab communities. The military crackdown on pro-Intifada demonstrations in the north killed 13 Palestinians (12 Israeli citizens and 1 Palestinian from Gaza). The city of Nazareth also saw violent clashes between its Arab and Jewish residents. Stones were thrown, vehicles and shops set alight, and young demonstrators lynched, seriously damaging relations between Jews and Arabs in the long term. The immediate result of these clashes was that Jewish citizens deserted Arab towns. Until 2005, few Jews dared to walk around the souk in Nazareth on a Saturday, go shopping, or have lunch in the town's restaurants. An even longer-term effect was that the Israeli authorities became more suspicious of Arab citizens, whom they deemed to be potentially disloyal, feeding fears of an "Israeli Arab Intifada" (Louër 2001b, 2003b).

When Sharon's government launched the construction of the Wall in 2002, Israel's policy of separating the West Bank from Israeli territory became more concrete. However, because the Wall was erected much later in southern Hebron region, some Israeli Bedouins continued to frequent the Dahriyya market, where goods were cheaper. These trips depended on the condition of roads circumventing the Israeli checkpoints. As a result, these trips no longer resembled the family outings seen before 2000, where entire families would go to the West Bank to buy cheap wedding dresses, food, or furniture. Palestinian workers also continued to enter the Negev to work in construction or in local shops. Both legal and illegal workers found that the balance of power had shifted further still and that their Bedouin employers had the upper hand (Parizot 2006b, 2008b). Smuggling and trafficking of Palestinian workers took place in all Arab zones, such as the Triangle, Galilee (Agbarieh 2004), and the northern Negev (around Beersheba) (Parizot 2014).

In addition, in recent years, Arab zones in Israel have witnessed the arrival of Palestinian women from the West Bank working as cleaners, or begging from door to door in the Negev. The rise in female Palestinian migration is mainly due to deteriorating economic conditions in the Palestinian Territories. In the Negev, for example, these women are generally married with children but their husbands and sons are unemployed.[7] They come to work in Bedouin towns and send their earnings back to their families in the Hebron area. Some women spend several years in the Negev and return home once a month, a trip that is becoming more costly as the number of crossing points is progressively limited. This same situation is seen in Jerusalem, where local Palestinians or expats employ cleaners from the West Bank. Since most of these Palestinian workers, male and female, enter without a work permit, they are dependent on the good will of their employers, and often endure difficult conditions. They live on the margins of Arab society, with an inferior status, respected as Palestinian "brothers" yet at times excluded as undocumented migrants. Very few Israeli social organizations, including the Arab ones, help them. Only the workers' rights association Kav LaOved and the Arab trade union Sawt el Amal in Nazareth[8] are struggling to handle the cases of illegal Palestinian workers, especially women. Abuses of power, harassment and violence are hardly reported, since these workers are unable to protest as they do not fall within any legal framework.

Israel's separation policies have also affected marriages. From 2000 onwards it became more complicated for Israeli Palestinian men and women to find a spouse in the OPT. Not only was travel to Israel restricted, but family reunification between Palestinian spouses was strictly prohibited in 2003. Under the Citizenship and Entry into Israel Law, passed in July 2003 as a temporary provision and later renewed, a Palestinian marrying an Israeli citizen does not have the right to Israeli residency or citizenship. Some of these marriages are polygamous (prohibited by Israeli law) or contracted solely for convenience, to allow Palestinians to live and work in Israel. Since 2003, such marriage strategies have come up against the Israeli ban.[9] Officially, the law affected 15,000 couples, but this figure is much higher when undocumented residents are factored in. These measures were adopted on security grounds, but also for demographic reasons due to Israeli fears that Arabs will outnumber them. Today, inter-Palestinian marriages are still restricted, and Palestinian spouses living illegally in Israel have no rights and no social or economic assistance. Many Palestinian women, particularly those engaged in polygamous marriages, are living in appalling conditions. The women I met in the Negev told of a day-to-day existence of fear and uncertainty. They mostly stay within the confines of the Bedouin villages out of fear of being arrested if they venture further afield. Their children's status is not clear. The law

7 A World Bank report from February 2010 stresses the growing participation of married, middle-aged Palestinian women in the labor market since the start of the Second Intifada. The deterioration of economic conditions in the West Bank and Gaza, notably due to the closure and movement restrictions, has led these women to accept precarious often underpaid jobs. While these women enjoy a greater freedom of movement, younger graduate women are subject to a more stringent social control and find less skilled jobs.

8 Interview, July 2010.

9 As early as 2000, Israel suspended the family grouping procedures. But it is only as of 1991 and the obligation to hold a residence permit that the notion of family grouping really became relevant since, until then, the free movement of Palestinians prevailed (Conte 2005).

stipulates that they are not entitled to permanent residence in Israel and must leave the country when they turn 12 (Conte 2005).

Arab citizens of Israel are also affected by these Israeli restrictions, although they have recently been re-authorized to enter Palestinian towns. Palestinian enclaves were reopened to Israeli Palestinians only in 2009, officially to stimulate the West Bank's economy. The Gaza Strip remains cut off. Hebron was the first city to which Israeli Palestinians returned, way before the checkpoints were officially opened. As early as 2008, chartered buses would run every Saturday between Israel's various Arab regions and Hebron. These trips were mainly organized by travel agencies and allowed Israeli Palestinians access to cheaper goods whilst boosting business for local shopkeepers. Other Palestinian cities like Jenin and Nablus became accessible one day a week in early 2009. The first buses from the Bedouin village of Lakiya to Jenin began operating in May 2009, and in June they also began services to Nablus and Ramallah for some 10 shekels.[10] Since mid-2010, Palestinian towns have become freely accessible to Arab citizens of Israel. As well as these organized tours, a regular flow of individual travelers and Israeli Arab families can now be found travelling to the West Bank. Israeli Palestinian women see these trips as opportunities to find bargains. They also often say that the food, such as bread, pastries, and fruits and vegetables, is better quality. They appreciate being able to bring back the "genuine" *knafe* from Nablus, CDs of the latest Lebanese and Egyptian music, or finding a greater selection of headscarves and tunics (made in China). The women are very proud to have shopped in Palestinian towns, and mention their shared culture and how they enjoy spending a few hours in an exclusively Arab environment. At the same time, they protest against the Israeli occupation, the blockade of Gaza and their own marginalization within the state of Israel. These trips reinforce the Palestinian identity of Arab citizens of Israel, who find themselves briefly immersed in a familiar atmosphere with a chance to affirm themselves as full members of the Palestinian people (*sha'ab*).

While we cannot talk of separation between Palestinians, there prevail differences in their daily lives and economic and social relations, including marriage. Firstly, not all Israeli Palestinians have relations with Palestinians in the OPT. Some have no family there, have not returned to the West Bank since the 1990s and have never been to Gaza. People's experience of these areas varies depending on whether a family network or economic interests exist, or whether they are geographically close to West Bank cities (like the Triangle and northern Negev). Thus it is not surprising to hear a variety of different attitudes to links with the OPT. Secondly, Israeli Palestinians, including those who frequent the West Bank and for whom such trips and day-to-day contact are important, are also aware of the social, economic and political differences between themselves and the "*dhaffawiyin*" (West Bank Palestinians).

Furthermore, the absence of a Palestinian community transcending the dividing lines cannot be explained solely by the differing ordinary experiences of Israeli Palestinians and those in the OPT. Even the Palestinian Arab community within Israel itself is divided along religious (Christians/Muslims), socioeconomic and geographic (Northern Palestinians/ Negev Bedouins) lines. There is prejudice between these different social groups. For example, the northern Arabs often view the Bedouins of the Negev as conservative, traditional nomads. Very few Palestinians from Galilee and the Triangle have ever set foot in a Bedouin village in the Negev. Israel has in fact made great efforts to manipulate and "ethnicize" the Bedouins by distinguishing them from other Arab citizens: co-opting the elite, conscripting

10 Five shekels correspond to about one euro.

them for military service, etc. (Parizot 2001a; Yonah, Saadi and Kaplan 2004). Marriage between Christians and Muslims is not the norm, due to the existence of strong religious identities that at times provoke political tension, as in the case of Shihab ed-Din[11] in 1997 in Nazareth (Louër 2003a). As a result, if we are to explain the absence of a Palestinian community transcending the dividing lines, it is not enough to recognize the imbalance of power between Israeli Palestinians and those in the OPT and the fact that they live under different socioeconomic conditions and perceive the separation lines differently. The only way to understand how the separation policies have created two distinct groups among those who all claim the same Palestinian identity is to take a look at politics.

Palestinian Solidarity across the Dividing Lines: Symbolism and Political Reality

The start of the Second Intifada in 2000 and especially the violence in Galilee in October certainly intensified Arab citizens of Israel's identification with the Palestinian cause. While the assertion of a Palestinian identity is not new (Rekhess 1989; Smooha 1989), nor is its routinization and status as a consensus among the Israeli Arab population (Louër 2001a). Today most members of the population—including non-activists—accept that identity as theirs and express it through symbolic actions (like wearing the keffiyeh or Handala pendants,[12] etc.) and concrete activities like charity work and demonstrations of support. Since the start of the Second Intifada, and especially after the Israeli army attack on a refugee camp in Jenin in 2002, inter-Palestinian solidarity has become widespread and structured, with all political parties, non-profit organizations, and Arab municipalities collecting money, food, and school supplies for Palestinians. Women's charities collect clothing all year round for Palestinian children and visit Palestinian prisoners in Israel. Religious networks are also very active. Through mosques, charities, and women's associations, the Islamic Movement collects money year round and meet during Ramadan, and sponsors Palestinian children. Meanwhile, Israeli Christians are also in contact with their counterparts in the OPT. These long-standing networks exists within every Christian community, be it Orthodox or Catholic. Churches and charities collect funds and goods for destitute Christians and for orphanages in the West Bank, especially Bethlehem. Prayers and sermons expressing support for Palestinians are regularly heard during mass in Israel.

Demonstrating has also become a regular part of Arab citizens of Israel's repertoire and their preferred form of action. They regularly hold protest demonstrations, as well as annual marches commemorating the *Nakba* (the "disaster" the Palestinians lived through at the occasion of Israel's declaration of independence in 1948), Land Day and the October events, and are quick to take to the streets to protest against Israeli military intervention in the West Bank and Gaza. Recently, in 2008 and 2009, Israel's military intervention in

11 Tension between Christians and Muslims broke out in 1997 around plans to build a mosque a few meters from the Basilica of the Annunciation in Nazareth on land that had been designated to become a square for the Pope's visit in 2000. The Islamic Movement and all the Arab political parties as well as the Zionist parties spoke out on the issue. There was so much tension that both the Vatican and the Palestinian Authority spoke out against building a mosque on the Shihab ed-Din site. In the end the mosque was never built, but a place of prayer was installed just next to the square.

12 The Handala character is a child who is 10 years old, the age of the artist when he went into exile. The character will remain a child and always have his back turned until Palestine is liberated. He was drawn by the Palestinian Naji Salim al Aali, and has been the symbol of Palestinian identity and resistance since the 1970s.

Gaza provoked numerous demonstrations in Israel's Arab cities. One of the biggest Arab demonstrations ever took place in Sakhnin on March 1, 2009, mobilizing tens of thousands of demonstrators. Tajammu' Knesset member Wasil Taha offered this explanation for the level of mobilization: "This is one of the biggest demonstrations we have ever seen because it affects every family. People want to express their suffering by showing solidarity with the members of our nation."[13]

Acts of solidarity are thus linked to a shared reference, that of being Palestinian. Two examples of the expression of this solidarity by political parties and professional associations are described below.

Firstly, solidarity exists between all Arab political parties in Israel. Since the end of the 1990s, all Arab parties have been aligned on the notion of "a State of all its citizens," calling for an overhaul of the Israeli political system, an end to its Jewish nature and for Arab citizens to be recognized as a national minority. This consensus advocates a hybrid political system, combining the foundations of a liberal democracy and a binational state (Ghanem and Ozacki Lazar 2003). It is also the sign of a conflicting integration of an Arab minority that, despite their Israeli citizenship, have never been integrated, and are today demanding to be recognized as a Palestinian minority. The binational State solution is not unanimously advocated but appears to have the support of many Arab activists. The nationalist right—but also many left-wing Jewish activists—see the fate of these Arabs as linked to that of Palestinians in the OPT. This is reflected in Avigdor Lieberman's idea of transferring sovereignty over Israeli Arab villages in the center district of the Triangle to the Palestinians.[14] Pursuant to its objective of securing the Jewish nature of the State of Israel, the nationalist right offers Arab citizens two choices: integrate in silence or leave to one of the Palestinian Territories, which would become the Palestinians' nation state, while the Jews would have Israel as theirs. However, despite their sense of a common Palestinian identity, most Arab citizens want to remain in Israel and have no desire to move to the Palestinian Territories, even if an independent state were created.[15]

Arab citizens' mobilization is very much anchored within the Israeli state, their protests and calls for greater rights being principally addressed at the Israeli authorities. In 2002, Elie Rekhess wrote of the "localization of the national struggle," based on three points. Firstly, Palestinian Arabs in Israel criticize the country's Jewish, democratic nature and advocate alternative state models. Next, they perceive themselves as a national minority and call for collective rights. And, thirdly, they have put sensitive issues dating from 1948 on the table, such as land, the right of return, and the commemoration of the *Nakba*. These issues are indeed an integral part of the Palestinian national struggle and Arab activists within Israel are campaigning on all three.

Moreover, all Arab parties officially advocate the creation of an independent Palestinian state with Jerusalem as its capital, and all of them have relations with Palestinian political forces. The Israeli authorities, who keep them under close surveillance, have accused some Arab political leaders of plotting with Palestinian nationalist organizations or with neighboring countries in conflict with Israel. Mohammad Kana'neh, leader of Abna al-

13 Sharon Roffe-Ofir, 2009. "Sakhnin protest: IDF op a war crime." *Ynet* [Online], http://www.ynetnews.com/articles/0,7340,L-3649584,00.html, accessed November 21, 2014.

14 Israel Beiteinu ("Israel is our home") is an ultra-nationalist right-wing party founded by Avigdor Lieberman in 1999.

15 In a 2001 survey by the Givat Haviva Institute for Peace, only 4.7 percent of Arab citizens of Israel surveyed said they wanted to resettle in an independent Palestinian state.

Balad, was accused of contacts with a foreign agent, particularly with George Habash, founder of the Popular Front for the Liberation of Palestine (PFLP), and its general secretary Ahmad Saadat. He was arrested in 2004 and sentenced to 30 months in jail. Other organizations close to the movement have also been raided, and Amir Makhoul, director of the organization Ittijah, was arrested on May 6, 2010 and accused of spying for Hezbollah. In 2007, Azmi Bishara, the former leader of Tajammu‘,[16] was also prosecuted for his alleged relations with Syria and Hezbollah during the 2006 Lebanon War. He left the country and went into exile to escape the proceedings.

The 2006 Lebanon War marked a turning point in Arab activism in Israel, not least because it mobilized Arab citizens to such an extent that accusations of treason and terrorism were reactivated with a vengeance. In January 2008, Avigdor Lieberman stated: "Our problem is not Judea and Samaria but the extreme fundamentalist leadership that is in the Knesset ... Our problem is Ahmed Tibi and Barakeh—they are more dangerous than Khaled Mash'al and Nassrallah, because they work from the inside; they operate methodically to destroy the State of Israel as a Jewish state" (Hofman and Jerusalempost.com staff, 2008). Similar accusations were made more recently against Tajammu‘ MP Hanin Zoubi for her presence on board the MV *Mavi Marmara*, part of the humanitarian aid fleet sent to Gaza on May 31, 2010 and boarded by the Israeli army with tragic consequences. In the Knesset, she later condemned the "pirate military operation" by Israel and called for the blockade of Gaza to be lifted. She came under vicious attack from some right-wing MPs, particularly Miri Regev of Likud,[17] who shouted at her: "Go to Gaza, traitor!"[18]

Hanin Zoubi was also stripped of her parliamentary privileges (diplomatic passport, financial assistance, and the right to visit countries that have no diplomatic relations with Israel). This perception of Arab members of parliament as the Palestinian struggle's fifth column and Trojan horse is growing, as they are more mobilized than ever within the Israeli political system. They are fighting for their differentiated integration, using the legal machinery of the Israeli State and acting as citizens who protest but are willing to negotiate. In recent years, Sheikh Ra'ed Salah, leader of the Islamic Movement's northern branch, has come to personify Arab resistance in Israel and is kept under close surveillance by Israeli intelligence. He is one of the rare political leaders who is regularly seen demonstrating and engaging in acts of resistance in public while refusing to be part of the Israeli political system. He focuses on the holy sites in Jerusalem, claiming that only the Palestinians should be entrusted with protecting Muslim property and that they and not Israel should have sovereignty of the city. This nationalist-Islamist position is clearly aimed at blurring the frontiers between Israel and the Palestinian Territories by invoking religious solidarity that should transcend territorial boundaries. Sheikh Ra'ed Salah has already been prosecuted. He was jailed for 26 months for conspiring with the enemy, aiding and abetting terrorist organizations, illegally leaving the country, and for meeting with Ahmad Jibril of the PFLP. When he was released from prison in 2005, Sheikh Ra'ed Salah returned to politics more credible than ever and is one of the Israeli Arab population's major political forces today. More recently he was charged with violence and racist hate speech after the 2007

16 *Al Tajammu' al Watani al Dimuqrati* (National Democratic Assembly) was established in 1996 by Azmi Bishara.

17 The Likud ("consolidation") is the main Israeli right-wing political party founded in 1973 by Menahem Begin.

18 Barak Ravid, 2010. "Israeli Arab MK who joined Gaza flotilla: IDF raid was a 'pirate' operation." *Haaretz* [Online], http://www.haaretz.com/news/national/israeli-arab-mk-who-joined-gaza-flotilla-idf-raid-was-a-pirate-operation-1.293769, accessed November 21, 2014.

protests against archaeological excavations launched as part of the Mugrabi Gate ramp renovation at the Temple Mount/Al-Aqsa Mosque.

While some see him as an independent player on the Arab political scene, others view him as close to the Palestinian Hamas.[19] In 2008, police shut down the offices of the Al-Aqsa Institute, founded in 2000 in Umm el-Fahm. The institute was accused of operating illegally and of logistic and financial links to Hamas and Al-Dawa in East Jerusalem. Links to Hamas do indeed exist, but the Islamic Movement of Israel remains independent. It has nothing to gain from being seen as close to a political movement criticized in the OPT themselves and liable to force it underground in Israel. By protecting its independence, the Islamic Movement of Israel has positioned itself as a charismatic leader that has not strayed into any form of political representation.

In addition, in recent years, Arab human rights organizations have become a key channel through which concrete support for the Palestinians in the OPT is expressed (Adalah,[20] Mossawa,[21] the Human Rights Association of Nazareth, or Ittijah[22]). These associations mainly developed in the 1990s as they fought for the rights of Arab citizens of Israel. But today they are all developing activities to support Palestinians in the OPT. They have all published reports criticizing Israel's occupation policy, and organize pro-Palestinian activities (conferences, documentary screenings, demonstrations, etc.) and seminars with their colleagues from the OPT. Adalah is an Israeli Arab legal aid organization founded in 1996, which focuses on Arab citizens of Israel's land, civil, political, and economic rights. It also defends the rights of Palestinian prisoners and challenges the Israeli army's practice of demolishing homes in the OPT. As well as the cases it takes to the Israel Supreme Court, it also conducts international advocacy activities. It was one of the NGOs that provided the information behind the Goldstone report on Israeli military intervention in Gaza in 2008–2009. The report caused a scandal when it was published in Israel in 2010, with some Arab associations being accused of betrayal. The New Israel Fund (NIF),[23] which funds most Israeli NGOs including Arab ones, was severely criticized by politicians, the media, and also by its Jewish donors in the US, some of whom threatened to suspend donations if the NIF continued to finance organizations like Adalah, whom they labelled as "terrorists."

Today, Arab NGOs in Israel are actors of the Arab political struggle in the country. More so than in the past, they are taking a clear stand on reforming the Israeli political system and are clear about their Palestinian identity. There were four publications in 2006–2007 that drew attention to them on this matter: *The Future Vision of the Palestinian Arabs in Israel*, published in December 2006 by the National Committee of the Heads of the Arab Local Authorities; *The Haïfa Declaration* published in 2007 by the Mada al-Carmel research center; the "Ten Points" of the Mossawa Center; and *The Democratic Constitution* published by Adalah in 2007. These documents all insist on the historic roots and Palestinian identity of Arab citizens and call for the abolition of the definition of Israel as a Jewish state, resulting in a democratic state for all. They all call for Arab citizens to be recognized as a

19 Hamas ("enthusiasm") is the acronym of the Islamic Resistance Movement founded in 1987 by Sheikh Ahmad Yassin, Abdel Aziz al-Rantissi, and Mohammed Taha, members of the Muslim Brotherhood.

20 Adalah ("justice") is a legal aid non-profit organization for Arab citizens in Israel.

21 Mossawa ("equality") is an advocacy center to promote equality for Arab citizens in Israel.

22 Ittijah ("direction") is the Union of Arab Community-Based Associations in Israel.

23 The New Israel Fund was created in 1979 to promote "social justice and equality" in Israel, notably through the funding of community projects.

national minority with the autonomy to manage their cultural and educational affairs. These documents all focus solely on Israeli Palestinians and only address the conflict in terms of their support for the creation of an independent Palestinian state.

As a result, despite their regular activities supporting Palestinians in the OPT, these NGOs and the political parties act mainly within the Israeli political system. One conclusive example is the work done by women's organizations, although they do have links with their counterparts in the OPT (Marteu 2009b). The General Union of Palestinian Women, founded in 1965 as part of the PLO, has structures among the different Palestinian populations in the Middle East but has never had a representative within Israel's Arab community. Cooperation does exist between Palestinian women's organizations in Israel and in the OPT, especially on issues like domestic violence, young children, or advocacy targeted at the United Nations (they jointly draft alternative reports to the CEDAW committee).[24] There are also occasional contacts, like in 2010 when Bedouin women's associations from Lakiya visited embroidery associations in Ramallah, or the Nazareth based Movement of Democratic Women visited a charity in Bethlehem. Ma'an, a women's organization in Beersheba, also organized a trip to Ramallah for young Bedouin men and women, the goal of which was to stimulate their Palestinian identity. Yet these relations have not yet led to any more significant forms of cooperation. So far, the Israeli and Palestinian non-profit sector remains structured by different legal realities, priorities, and constraints (Tamari 1999). The blockade of the territories and geographic constraints alone are not enough to explain why relations between Arab and Palestinian activists are not more mainstream. Jerusalem is a case in point, with no more inter-Palestinian cooperation than elsewhere, despite the absence of geographic constraints or military barricades. Thus, the differences run deep and stem from modes of differentiation that are at once territorial, socioeconomic, and political.

Signs of inter-Palestinian solidarity are clearly expressed more often by Israeli Palestinians than by those in the OPT. As well as frequent demonstrations, they also support en masse the BDS campaign to boycott Israel (Boycott, Disinvestment and Sanctions) launched by the Palestinians in 2005. Their support has its own name, the "Palestinian BDS Call from Within," and is currently backed by Jewish and Arab citizens in Israel. However, when asked about their links with Israeli Palestinians, those in charge of Palestinian NGOs in the West Bank are unanimous. They recognize that links exist, especially within political parties and human rights organizations, but, as one feminist activist in Ramallah put it: "Each of us must act within our own State and put pressure on our own governments."[25]

This is precisely what Israeli Arab militants are trying to do, within the parties and non-profit organizations, by putting pressure on the Israeli government. But they are increasingly rejected by Zionist parties and by public opinion in Israel. As long as their Palestinian identity was based on ethnic and cultural differentiation it was accepted. Now that it has become a resource for political action and protest, it is perceived as radical

24 All the signatory States of the Convention for the Elimination of All Forms of Discrimination against Women (CEDAW) must submit, on a regular basis, a report on the state of progress of women's rights to the UN CEDAW committee. In turn, women's organizations also draft an alternative report, also known as the shadow report, aimed at bringing out elements "glossed over" in the official reports. The workgroup on Palestinian women in Israel drafts its own report, in collaboration with the Palestinian organization Women's Center for Legal Aid and Counselling (WCLAC) in Ramallah, which stresses the impact of Israeli military occupation on the conditions of women in the West Bank and Gaza.

25 Interview, Ramallah, 2008.

and subversive. The growing success of Palestinian nationalist arguments (from Israel and from the OPT) in favor of a binational Israeli State, or even of a single state from the Mediterranean to the Jordan River, further stokes the Zionist parties' fears. This is a context that the far right has exploited to reiterate its plans to transfer the Arab population outside Israeli borders. Accusation of treason and disloyalty abound, proving that while seeking to fragment the Palestinians, Israel has always managed to exploit the amalgamation between Israeli Palestinians and those in the OPT to its own ends.

Chapter 11
From a "Gay Paradise" to a Pioneer Frontier: Constructs of the "Frontier" in the Activist Struggle and Activist Discourse of LGBTQs[1] in Israel and Palestine, 1988–2012

Valérie Pouzol

Introduction

In May 1998, Dana International, an Israeli transsexual songstress, won the Eurovision song contest. Despite strong opposition from religious circles in the country, the victory demonstrated that Israel could be represented by a person with an ambivalent sexual identity who plays with different trajectories and affiliations: a man who became a woman, a Sephardi Jew from a lower middle class background, an icon of a counterculture. In 2007, in Haifa, members of Aswat,[2] a group of Israeli Palestinian lesbians organized their first conference, which revolved around the presentation of a work on lesbianism and which drew an audience of more than 300.[3]

Although separated by an interval of 10 years, these two events mark the affirmation in both the Israeli and Palestinian public spheres in Israel of LGBTQ identities. The chief protagonists in the two events had one thing in common: They were defending the rights of sexual minorities and were emphasizing the complexity of the struggle in which sexual oppression is allied with national oppression. In the 1990s, Dana International, who sings in both Hebrew and Arabic, already made the question of sexual, ethnic, social and national borders a central issue in her songs, which were sometimes takeoffs on songs that were symbolic of Israeli nationalism.[4]

Her emblematic and polemical success in the Israeli public sphere was closely linked to a dramatic change that has been taking place in Israeli society since the 1980s. This major change has made the country's sexual minorities visible and has reinforced them judicially, culturally, and politically. It has given rise to a activist LGBTQ network that has positioned itself, on the one hand, as an offensive player in the negotiations over the rights, and total social integration of, sexual minorities in Israel and, on the other, as a committed movement that seeks not only to promote community-oriented demands but also aspires to the formation of much broader political options. Ten years later, Palestinian LGBTQ

1 I shall use the abbreviation of the generic term in English, LGBTQ (Lesbian, Gay, Bisexual, Transgender, Queer) to designate sexual minorities that, however, do not together form a unified community.

2 "The Voices" in Arabic.

3 *Home and Exile in Queer Experience: A Collection of Articles about Lesbians and Homosexual Identity* (Heinrich Böll Foundation, Open Society Institute and Soros Foundation Network, 2006).

4 For example, the song "Nosa'at L'Petra" (whose title means "Going to Petra") parodies the famous nationalistic song of the 1950s that was sung at the time by Arik Lavi, "Hasela Ha'adom" ("The Red Rock").

activism was recognized first in Israeli Arab society and then clandestinely in the Israeli-occupied Palestinian territories.

Initially, some Israeli and Palestinian LGBTQ activists in Israel worked together in centers for the defense of minorities throughout the country. However, they went their separate ways in the first decade of the twenty-first century. In the context of both the Second Intifada and the reinforcement of the mechanisms of separation, joint activism was no longer considered acceptable for either Palestinians living in Israel or for those living in the Occupied Territories. The rift became more visible when Israel chose to promote, at the national and international levels, the image of a country that protects its sexual minorities and to thereby emphasize the uniqueness of Israeli democracy in a homophobic Middle East.

This chapter traces, first of all, the history of the LGBTQ struggle in terms of both its quest for recognition and its complex approach to politics and to the question of the Palestinian-Israeli conflict. We will then show how, early in the first decade of the twenty-first century, this struggle came face to face with a new national and international context that led the State of Israel to make the issue of the defense of sexual minorities part of its effort to promote the image of an all-embracing democracy. In the third part of the chapter, we will demonstrate how a Palestinian LGBTQ activism first established itself in the shadow of the leading Israeli LGBTQ groups, then progressively became an autonomous entity opposed to the vast majority of these groups not only over political questions but also over activist strategy and activist discourse.

The LGBTQ Community in Israel: A Spectacular Case of Recognition

Since the late 1980s, sexual minorities in Israel have discarded their previous clandestine existence and have achieved considerable judicial successes with regard to recognition of their status. This recognition, symbolized by the gay pride parade, which has become an annual event since 1998, is not something that can be taken for granted—in view of the fact that Israel is a country that lives with a tense security situation where the national narrative regards the fruitful heterosexual family as a guarantee of the survival of a nation that feels threatened by Arab demography. Zionism, which is both a political utopia and an erotic one, dreamed of the heroic, invincible masculinity of the new Jew reestablished in his native homeland and partnered with a new Jewish woman whose fertile womb would assure the survival of a new Jewish state (Biale 1997); moreover, Zionism has incorporated that dream in its discourse, literature, and initial films. Eager for recognition, sexual minorities in Israel *ipso facto* contradict this national project.

The increased clout of the LGBTQ community in Israel can be attributed to the profound ideological and political changes that have occurred in Israel from the time of the First War in Lebanon in 1982 and which plunged the country into a crisis situation over its collective values. One factor that has greatly contributed to the reinforcement of the discourse on the right to be different has been the emergence of post-Zionism, which attaches great importance to the defense of the individual's rights in a society that, up until now, has been modeled on the principle of the fulfillment of only collective needs. Another factor that has promoted the "gay cause" has been the growing tension between secular and religious Israeli Jews. While Orthodox Jews are openly homophobic, secular Zionists have discovered that the defense of sexual minorities is an especially effective lever in the battle against religious fundamentalism in Israel (Solomon 2003, pp. 149–65).

The gay community in Israel has gradually developed a activist network that operates on two fronts. The first front is led by independent activists who have chosen to wage a judicial battle for gay rights by submitting every case of anti-gay discrimination to the Israel Supreme Court; the second is led by gay-friendly political figures who fight for gay rights in various parliamentary and extra-parliamentary institutions. In July 1975, in Tel Aviv, a small group of gays—homosexuals and lesbians—founded the Society for the Protection of Personal Rights (SPPR). Without explicitly declaring that their struggle is aimed against discrimination based on sexual orientation, they have consistently promoted gay rights and were instrumental in the removal of sexual relations between men (referred to as the criminal act of "sodomy") from Israel's criminal code in 1988. In 1992, the formation of a Labor government provided additional momentum to the struggle for gay rights. It was decided that a subcommittee of Israel's parliament, the Knesset, on gay affairs would be headed by a member of Knesset Yael Dayan, who has done much to battle discrimination on the grounds of sexual orientation in both the workplace and the army. As of 1993, the Israel Defense Forces (IDF) openly permits homosexuals and lesbians to serve in its ranks. Important achievements have also been made in Israeli legislation with regard to the issue of personal (or family) status. In 2006, the Supreme Court ordered the Ministry of the Interior to register same-sex marriages that are performed abroad; as a result, partners in same-sex marriages have the same rights as partners in heterosexual marriages with regard to benefits, tax concessions, and inheritance rights. Concerning gay parenthood, significant progress has also been made. In 1999, the Israel Supreme Court ruled that the Interior Ministry must recognize the non-biological mother in a lesbian couple as the second legal mother. On February 12, 2008, the Israeli government decided to grant to same-sex couples the same rights regarding the adoption of children as heterosexual couples.

In recent years, figures in civic society and in the political world in Israel have "come out of the closet" and have linked their identity as gays with their public commitments. At the level of municipal politics, Michal Eden was elected in 1998 to the Tel Aviv-Jaffa municipal council, becoming the first lesbian in Israel to serve as a member of a municipal council. In 2003, Saar Netanel followed in her footsteps when she was elected to the Jerusalem municipal council. At the level of federal politics, Uzi Even became in 2002 the first self-declared gay to be elected a member of Knesset. This emergence of gays "from the closet" can be regarded as important as the fact that, as a whole, Israeli politics continues to be characterized by widespread homophobia. It should be noted that the election of gays in municipal elections has led to the creation of committees on gay affairs in those cities where gays have won seats on the municipal council, such as in Tel Aviv-Jaffa.

The judicial struggles have been accompanied by the formation of, on the one hand, organizations committed to the advancement of the welfare and rights of gays and, on the other, community centers catering to the gay population; these organizations and community centers are part of an umbrella organization, the Israeli National LGBT Task Force, or the Aguda,[5] the successor to the SPPR. The Aguda, meant to stand for the rights of all sexual minorities, was first headed by gay men. It was hard during this period for Israeli lesbians to find a place for themselves in the sun: They did not feel they could express themselves sufficiently in contexts that served both homosexuals and lesbians or in Israeli feminist organizations that, in 1976, rejected the inclusion of lesbianism in the feminist agenda. In 1987, the first lesbian group was founded in Israel: Klaf (Hebrew acronym for *Kehila Lesbit Feministit*—the Lesbian Feminist Community), which identified itself as a pressure

5 The Hebrew word *aguda* means association.

group dedicated to the advancement of lesbian rights and to the attainment of public and social recognition for lesbians.

During the First Intifada (1987–1993), lesbians in Israel found opportunities for self-expression in the context of radical women's peace groups, such as Women in Black (Pouzol 2007, pp. 75–87). Community centers, financed largely by the New Israel Fund and international NGOs, began to spring up in main urban centers throughout the country; these centers, which provided information and documentation, were not only contexts where debates could be held but also places of refuge for gays facing personal, family, or social distress. Whereas the Tel Aviv-Jaffa municipality was quick to fund this initiative, the same could not be said for Jerusalem, where the Jerusalem Open House, which was opened in 1999 to serve the city's gay population, had to wage numerous battles in order to secure funding for its activities.

Israel's LGBTQ community, which is not homogeneous, has found it difficult to structure itself. It is crisscrossed by various fracture lines[6] that separate homosexuals, lesbians, transgenders, and bisexuals. It is also divided over the perception of activist action and over the application of such activist action in political and national issues. The surveys conducted by Erez Levon within the Aguda showed that many Israeli gays were demanding full adherence to collective and national values. These gays demanded inclusion in Israeli society, focusing their activism on the demand for equal rights for gays in the army and on the demand for the right of gays to be parents (Levon 2010).

Only a minority of these activists stood out with their radical positions and with their rejection of both the dominant perception of the state and the norms of gender and sexuality linked to that perception. Rather than fighting for the inclusion of gays in Israeli society in its present state, they wanted to abolish the systems of control and marginalization. Their working assumption was the existence of a link between different forms of oppression and discrimination (whether sexual, ethnic, or national). The ephemeral group Kvisa Shchora (Black Laundry),[7] an LGBTQ group that was founded in 2001 during the Second Intifada, and which consisted of some 40 Jewish and Arab activists (specifically Jews and Palestinian Arabs from Israel), marched to Tel Aviv during the annual gay pride parade; they carried placards demanding the withdrawal of Israeli troops from the occupied Palestinian territories. Using humor, scorn, and the tactics of performative demonstration, the group tried to show a connection between the various forms of oppression, during the demonstrations that the group held in different cities throughout Israel (Ziv 2010, pp. 538–56). The most radical wing of Israel LGBTQs activists strongly protested the representation of Israel as a country that boasts of its tolerance toward sexual minorities. Since 2006, small groups—often led by "marginal members" of the LGBTQ community, such as bisexuals and transgenders—have been organizing their own alternative gay pride parades. Associating their struggle with that of the Palestinians, they try to disrupt public events aimed at promoting the image of Israel as a "gay paradise." On June 11, 2009, a small number of radical Israeli activists tried to interrupt the proceedings of a public meeting that was organized by activist Israeli LGBTQs and the American pro-Israel group Stand with Us and which was intended to defend and advance Israel's image as a "gay-friendly" country. In the Stand with Us promotional campaign, which took pains to stress the contrast

6 In Israel, use is made of North American categorizations: Lesbians, Gays, Bisexuals, Transgenders, Queers. These terms have been Hebraized as Lesbiot, Gayim, Transgenderim, and Queer. On the categorizations and their incorporation into the Hebrew language, see Levon 2010.

7 Dirty, literally black, laundry—that is, black sheep or ostracized people.

between a democratic Israel and a backward, authoritarian Arab world, a central topic was the protection of gay rights. Radical activists, with the support of the Coalition of Women for Peace and radical feminists,[8] distributed leaflets and heckled the organizers of this event for their promotion of the concept of Israel as a gay paradise and for their representation of Palestinian gays as eternal victims of a backward, homophobic Palestinian society. A., a transgender who was an activist with Anarchists Against the Wall,[9] had since 2007 been a member of the radical queer group Queer It Up, which tries to join together all of the various struggles conducted by diverse segments of the LGBTQ community. At the Stand with Us meeting, she distributed leaflets at the entrance to the auditorium and heckled the participants.

> Stand with Us was an extreme right-wing pro-Israeli organization that had no compunctions about manipulating Israelis students. When the Aguda organized its first gay pride parade in 2009,[10] we asked the Aguda's leader, Mike Hamel, a number of questions about its activities vis-à-vis Palestinians. In his response, Hamel declared that the Aguda was trying to "save" them. They defend the idea of a gay paradise without condemning the reality of the occupation. Aguda is dominated by white Ashkenazi males, carefully chooses its issues and is a group that lacks political independence because it is funded by the Tel Aviv-Jaffa Municipality. For example, the municipality wanted to limit access to Independence Park, a favorite gay site for "cruising," and to turn it into a facility for the exclusive use of families. Aguda should have protested the municipality's decision but it failed to do so. The municipality paid its inspectors to chase gays out of the park.[11]

The LGBTQ community scored decisive victories in its negotiations over the recognition of the rights of its members. However, it has remained divided over the issue of political orientation and over the analysis of the "control relationship" in the Israeli-occupied Palestinian territories. Within only a few years, part of the LGBTQ community has transformed itself from a movement whose principal activity is to debate issues into a consensual organization seeking total inclusion in the collective national project.[12] Far from

---

8 Founded soon after the outbreak of the Second Intifada in 2000, the Coalition of Women for Peace is an Israeli NGO that is an umbrella organization for groups and individual women that are strongly committed in their opposition to the Israeli occupation of Palestinian territories. It is politically radical in its total opposition to all instances where one nation occupies another's land, in its promotion of a two-state solution with East Jerusalem as the capital of the Palestinian State and in its support for the BDS campaign against Israel. In addition, the Coalition of Women for Peace openly identifies itself as a feminist organization and links the battle against gender-based oppression with the battle against the national oppression of the Palestinians. See http://www.coalitionofwomen.org/, *accessed June 19, 2013.*

9 Anarchists Against the Wall is an activist group that was founded in 2003 by Israelis engaged in organizing demonstrations and direct action in the Israeli-occupied Palestinian territories.

10 Stand with Us was at the time spearheading a program entitled IPride (that is, Israeli pride) that was aimed at promoting a new image for Israel that placed less stress on violence and conflict and which instead highlighted fields of Israeli excellence, especially in the economic sphere. The project also attempted to emphasize Israel's tolerance and avant-gardism in the defense of the rights of sexual minorities.

11 Conversation with A., a transgender and a member of Queer It Up and Anarchists Against the Wall, Gan Meir, Tel Aviv, April 2010.

12 For details on the specific demands of the LGBTQ community (the right to start a family, protection from hate crimes, the right to a unique sexual identity, the right to access to comprehensive

being a phenomenon of social deviance, the struggle for the recognition of gay rights has now become one of the symbols of a democracy seeking to promote its own image.

A Gay Paradise, the Pioneer Frontier and the Erasure of Frontiers

The activist history of the LGBTQ community must also be examined in light of both state-sponsored public diplomacy campaigns and the promotion of Israel's image in the international arena. A portion of the political class in Israel has become strongly committed to the advancement of the issue of respect for the rights of sexual minorities in order to reinforce Israel's image as a democratic state and in order to include them in the national collective citizens who have finally become legitimate through their adoption of the values of the Israeli collective ethos. The political class in Israel has played a major role in the creation of the image of a "gay paradise": a paradise whose porous borders offer asylum to gays who feel threatened in a hostile Middle East.

National marketing strategy, which links opinion polls and re-evaluations of the altered image of a country through sophisticated public diplomacy campaigns, is a field that is taught and whose theory is constantly being developed at various universities in North America (Kotler, Jatusripitak and Maesincee 1997). Since 2003, American marketing groups, very skilled in these techniques of promotion, have carried out public opinion surveys in the United States and subsequently in Israel for the purpose of advising Israeli political leaders on the political interests to be gained from a re-evaluation of Israel's image. Since 2005, a public relations campaign aimed at winning back the heart of the American public is being conducted by a group of pollsters (BIG—the Brand Israel Group) in order to repair, and then to correct, negative images being disseminated about Israel. In October 2005, BIG's marketing directors submitted the findings of their research to a number of Israeli cabinet ministers in order to produce a common strategy and in order to launch a public relations project in Israel: the Brand Israel Project. Officially inaugurated in Tel Aviv in 2007 by then-foreign minister Tzipi Livni, the project, with a four-million-dollar budget, was subsequently publicized in the US by various pro-Israel lobby groups working together to improve Israel's image.[13] The public diplomacy campaign was intended to promote the qualities of Israeli democracy and to showcase the country's economic and technological achievements in order to transform negative views of Israel at both the national and international levels. The themes of liberty, especially for Israeli women, and the protection of sexual minorities were given considerable attention in this public diplomacy campaign (Schulmann 2013).

In 2007, Israel's electronic media (Newkey Burden 2007) disseminated a report compiled in Israel by a British journalist, who described the country as the "final gay frontier" and who contrasted the privileged position of Israeli gays with the precarious situation of their counterparts in neighboring Arab states. In 2008, to mark Israel's 60th year of independence, an American-Israeli lobby group, The Israel Project (TIP), publicized on its website an article on gay rights that lauded Israel for its avant-garde position in the defense of sexual minorities:

---

health services, the right to benefits from the state, etc.), see the LGBTQ Bill of Rights, which was presented to the Knesset in 2010.

13 By groups such as The Israel Project (TIP) and Israel21c. See: www:http://theisraelproject.org/ and www//http://israel21c.org/, accessed October 10, 2013.

> Israel has become a haven for gay Palestinians who flee persecution in the West Bank and Gaza Strip, where they are subject to severe abuse by their families, communities, Hamas and the Palestinian Authority. Compared to neighboring Arab countries, Israel is far more tolerant and accepting of gays. While homosexuality is legal and there is legal protection for gays in Israel, homosexuality is illegal in countries such as Lebanon, Libya, Syria, Iran and Saudi Arabia.[14]

The emergence of this inclusive discourse should be placed in a new international ideological context that, following 9/11, has associated the clash of civilizations with the clash of sexualities. From that date, the liberal and tolerant West is at war with a violent, sexist and homophobic Muslim world. This ideological argument has had a major impact on the respective agendas of certain gay movements. In the wake of 9/11, many of these movements have begun to support the position of their respective countries with regard to the international arena and to echo their country's demands—even the most bellicose ones—in that connection (Puar 2007). Similarly, in the U.S., a large segment of the gay movement is committed to an agenda of "homonationalist" normalization that supports the various wars being conducted against terrorism. The recruitment of Western gay groups in the support of a war against the Muslim world has had a significant effect on Israel where gays once again find themselves confronted *de facto* with the problem of the representation of the frontier between sexualities, seeing that they consider their country a pioneer frontier (Turner 1921) in a hostile regional environment.

Feature- and documentary- film production in Israel has taken part in this project of imagining and representing the frontier (Pouzol 2013). Eytan Fox's film, *The Bubble* (2006), presented a Tel Aviv that was a city in suspended animation where two homosexual lovers, an Israeli Jew and a West Bank Palestinian, can fall in love before the logic of both separation (of Jews and Palestinians) and violence cancels their rights. This erotic fiction unfolded on the screen when the politics of separation had already led to the disappearance of Palestinians from Israeli streets, as if the presentation of this story was only possible when the factual probability of such a situation no longer existed (Stein 2010). The director creates a frontier where the checkpoints are simultaneously places of control and violence and potential places of seduction and interaction. It should be noted that the two lovers meet at a checkpoint. In 2012, although operating from a different perspective, Yariv Mozer, an Israeli director, similarly took up the theme of flight to, and asylum in, Israel for Palestinian gays who feel threatened in their own indigenous society. In his documentary, *The Invisible Men*,[15] he presents conversations with Palestinian gays, and uses these conversations to describe a totally different version of the "gay paradise": Once they arrive in Tel Aviv, these men must once more live clandestinely and must conceal their Palestinian identity. He approaches the theme of "invisibilization," which, on the one hand, is experienced within a homophobic Palestinian society and which, on the other hand, is imposed by an Israeli society that condemns any person who comes in contact with, or provides lodging to, Palestinians from the Israeli Occupied Territories. Whether fictional or factual, these images contribute to the representation of a porous, open frontier, even if the latter form of invisibilization, once the frontier is crossed, entails the invisibilization of Palestinian identity.

---

14 The Israel Project, 2010. See: https://www.kintera.org/site/apps/nlnet/content2.aspx?c=hsJPK0PIJpH&b=689705&ct=8855675, accessed January 7, 2015.

15 See the film website: www.http//.theinvisiblemenfilm.com, accessed May 23, 2014.

There is, however, a gap between the illusion of a gay paradise in the midst of a hostile regional environment and the reality of the acceptance and protection of sexual minorities in Israel. Deep inside Israel and despite its military victories, homophobia is still widespread, and Tel Aviv is often presented as the only city in the country that is especially liberal. LGBTQ defense organizations point in particular to the homophobia of the Israel Police and to its arbitrary arrests in "pickup joints" and bars frequented by members of the gay community. In Israel, the taut political context and the conducting of frequent military operations often produce homophobic messages and violent actions such as those that were evident during the gay pride parade in Jerusalem in 2006 (*Ynetnews* 2006). Despite the gay community's victories, the murderous attack carried out on July 31, 2009 against a LGBTQ community center in Tel Aviv has again raised the question of the extent of tolerance and the extent of the acceptance of sexual minorities in Israel.

The tightening of movement restrictions that Israel has imposed on Palestinians since the 1990s places a large question mark over the idea that Israel can function as a refuge for Palestinian sexual minorities. Since the Six-Day War of June 1967, gay Palestinians from the West Bank and Gaza have crossed over into Israel in order to escape social pressures, specifically, the possible threat to their very lives. Besides seeking odd jobs, they have tried to benefit from the immense anonymity that Israeli cities offer them and from the possibility of meeting gays there. It is difficult to estimate their number because the overwhelming majority of these homosexuals have entered the country clandestinely. The few statistics that are available are furnished by the Israeli branch of the Aguda that works on behalf of "Arab minorities." According to Shaul Ganon, who has been the Aguda official responsible over the past several years for Arab sexual minorities, there are in Israel approximately 500 homosexuals who have found refuge of some sort in Israel and who have entered the country from the West Bank, the Gaza Strip, and the Kingdom of Jordan (men, for the vast majority of them):

> I began in 1995 with young gays who were at the time living in the street. Many of them were Palestinians and had entered Israel from the West Bank looking for work. There was an "organizational vacuum" because no one in Israel wanted to help these young people. These young gays needed, first and foremost, a place where they could sleep at night and also required papers that could enable them to remain in Israel because many were afraid to return to their homes. Some of these gays live under very difficult conditions, are drug addicts and work as prostitutes. On both sides, Israeli and Palestinian, Palestinian gays are subjected to harassment and blackmail in an attempt to recruit them as informers. Our efforts have been aimed at obtaining for these individuals temporary resident permits and at providing them with further help once the permit has been issued.[16]

From June 1967 to the outbreak of the Second Intifada in 2000, some 50 ethnically mixed (that is, Jewish-Arab) gay couples took up residence in Israel; however, when the popular uprising ended in 2004, only 15 couples remained, the majority having emigrated to Europe.[17] Since 2000, the Israeli authorities have neither granted nor renewed temporary resident permits for homosexuals who have entered the country because they feared for their lives. Although persons fleeing homophobic violence who need governmental protection are

16 The conversation was held in Tel Aviv on October 10, 2010.

17 Berlin appears to be a popular choice for Israeli-Palestinian gay couples who decide to emigrate; many young Israelis (approximately 12,000) have opted to take up residence there.

eligible for international protection under the United Nations Convention on the Status of Refugees, which Israel ratified in 1954, Israel has consistently refused to grant Palestinians the right to file a demand for protection on the grounds that Palestinians seeking asylum pose a possible national security threat. In refusing to grant a temporary resident permit to such persons, the Israeli authorities are not acting in compliance with the principle of "non-refoulement" which forbids the escorting back to the border of a person who has sought refuge and whose life is threatened in his or her country of origin. In April 2008, two Israeli lawyers attached to Tel Aviv University, Anat Ben-Dor and Michael Kagan, drew the attention of the Israeli authorities to these practices, demanding in a report (Ben-Dor et al. 2008) a relaxation of such measures and the creation of a mechanism to aid in the international relocalization of gays whose lives are in danger. Similarly, they protested the precarious situation of Palestinian gays, who, because of their illegal status, are often victims of blackmail and who, in order to survive, are prepared to work for starvation wages or to become prostitutes. Palestinian gays are prime targets for the Israeli security services who seek to turn them into informers.

As a counterpoint to its inclusive slogans, Palestinian LGBTQs with Israeli citizenship have made their voices heard and have raised the question of frontiers and national affiliation in these identitarian struggles, thereby leading Palestinian activists to reorient their own commitments.

Queer Activists in Palestine: The Mapping Out of the Intimacy of a Nation

There is a wide gap between the recognition of activist LGBTQ groups in Israeli society and the confidential, clandestine recognition of small groups that are working on behalf of Palestinian sexual minorities in Israel, the West Bank and the Gaza Strip. In Israel, activism, even when it is poorly structured at the national level, is visible and it can rely on the activities of influential public figures, on a network of organizations and on a political lobby operating in the Knesset and in certain municipalities. Inversely, on the Palestinian side, the pursuit of a national struggle against the Israeli occupation of Palestinian territories has strongly reinforced the expectations regarding individuals in terms of social cohesion and resistance to the occupation. In the context of this resistance, the human body, sexuality and the preservation of honor are clearly invested with national aspirations (Amireh 2003).

The status of sexual minorities in Palestinian society is difficult to define because of the parcelization of the Palestinian territories and the heritage of different judicial cultures in the West Bank and in Gaza. Whereas the fundamental laws of the Palestinian constitution, which was amended in 2003, guarantee to all Palestinians the principle of non-discrimination on the grounds of race or sex, no civil legislation has yet been passed to protect the rights of sexual minorities. Since 1994, in the territories placed under the control of the Palestinian Authority under the terms of the Oslo Accords, the principle has not been given concrete expression in Palestinian law books; instead, Jordanian law, which was in effect in the West Bank before June 1967, is applied by the PA by default. Under Jordanian law, homosexuality has been decriminalized ever since 1951. By contrast, in the Gaza Strip, Ordinance No. 14 of the British Mandate's Criminal Code of 1936 is still in effect. According to that ordinance, sexual relations between men are defined as being contrary to nature, illegal and punishable by 10 years in prison, while the absence of any reference to lesbian relations renders such relations immune to legal prosecution.

On the Palestinian side, the choice of a sexual orientation that goes against the current poses a problem because abnormal sexuality is associated with a threat to the cohesion of the national group: Homosexuality is connected to the Westernization or Israelization of Palestinian society. A breach of sexual morality is regarded as the accelerator of the moral disintegration of Arab society (Latte Abdallah 2006). Research studies conducted in the mid-1990s among young Palestinians in Israel have shown that they consider sexual morality as the ultimate defense against the process of Israelization (Louër 2003a). Similarly, homosexuality is associated in Palestinian society with the specter of collaboration with the Israeli security forces.

During the 1980s and early 1990s, especially in the course of the First Intifada (1987–1993), the threat to social integrity and sexual morality was associated with the specter of collaboration. The suspicion of "horizontal collaboration"[18] was initially linked with women and then with individuals considered to be "deviant" (drug addicts, alcoholics, consumers of pornography, and homosexuals). In Palestine, sexual minorities, like women, were suspected of collaboration; they were thought to be potential targets of the Israeli and Palestinian security services, which might be tempted to utilize the "fragility" of these sexual minorities, to blackmail them and to turn them into informers (B'Tselem 1994, p. 89).

> First of all, the situation of homosexuals and lesbians, where you have no civil protection whatsoever, is very bad. We are the weakest link in society and, for that reason, we can be used by anyone in the worst possible manner, such as by a police officer, your neighbor, and so on. When you are weak and no one is protecting you, this can be really bad. Then just imagine that you live in Palestine, trying to survive from one day to the next under the [Israeli] occupation [of Palestinian territories], the situation is even more difficult for homosexuals and lesbians. [Palestinian] gay men are also victims of both the Israeli security forces and the Palestinian security services because both entities, which know that gay men are the weakest members of society, want to exploit them. The Israelis try to lead more and more people to collaborate and they say to them [Palestinian gay men], "We'll tell your family, we'll tell your friends, and that will really harm you." The Palestinians [that is, the Palestinian security services] also exploit them in order to prevent gays from collaborating [with the Israeli security forces]. [Palestinian] gay men are frequently suspected of being collaborators. Palestinian gay men can be killed not because of their sexuality but because of their suspected collaboration. Palestinian gay men are victims of both sides [i.e., the Israeli and Palestinian security services]. (Beaujouan 2010)

Thus, it was on Israeli territory and in the midst of mixed activist groups (which bring together Jews and Arabs) that some Palestinian homosexuals and lesbians were able to explore the theme of sexual diversity, their minority status and the issue of social exclusion. This period of activism "under an Israeli cover" was short-lived, demonstrating the limits of the Palestinian LGBTQ community, the rights of whose members are not even recognized in the context of the struggle spearheaded by Israeli LGBTQ organizations. After a brief period of joint activist action, Palestinian LGBTQs, most of them members of Israeli Palestinian society, decided to create their own activist groups.

18 "Sexual collaboration."

Aswat and Al Qaws,[19] two Palestinian groups working openly for the promotion of sexual diversity in Palestinian society were founded in 2001 and 2007 respectively by two Israeli Palestinian women. This autonomy is not limited to LGBTQ groups and, since the beginning of the present century, has been reflected in the affirmation of questions of identity among Israeli Palestinians and in their solid commitment to the Palestinians in the Israeli Occupied Territories during the Second Intifada (Marteu, this volume). In this context, it is important to note here that two lesbians have openly assumed the leadership of two LGBTQ organizations aimed at Palestinians. Both of them are Israeli citizens, have attended Israeli universities and have the experience of feminist activism. Haneen Maikey, founder of Al-Qaws (The Arc), an independent NGO aimed at Israeli Palestinians, participated for several years in the activities of Jerusalem's only LGBTQ center, the Jerusalem Open House (JOH). At the time of its creation, Al-Qaws contributed, with the support of the JOH's director, to the mounting of a project on behalf of Palestinians in Israel and in the Israeli-occupied Palestinian territories. Initially intended to serve the needs of the LGBTQ community in East Jerusalem, Al-Qaws program functioned under the JOH's aegis, which principally expressed itself in the granting of funds for the program's operation. Like Aswat, Al-Qaws separated from the JOH in November 2007, becoming an independent Palestinian organization registered in Israel. Although Al-Qaws is still located in the JOH office in West Jerusalem, the Palestinian NGO determines the orientation of its programs and conducts its own fundraising activities. For Al-Qaws's director, this stage is the result of the mapping out of a complex course of action aimed simultaneously at centering its activities on the Palestinian LGBTQ and at demanding autonomy in the face of the patronage of Israeli activist gays:

> Our organization conducted its site in Arabic and that it aimed its activities at Palestinian LGBTQs. It has offices in Haifa and Jaffa because it is easier to work in the cities than in rural areas. Al-Qaws has a clandestine presence in Ramallah where it has launched a discussion group for young gays. The organization helps young homosexuals in East Jerusalem who are encountering great difficulties in finding lodgings and work. In 2005, I grappled with the question of Palestinian identity in the context of the JOH and came to the conclusion that there was a need for focusing on an identity beyond the limits of a gay identity. For a considerable period of time, Palestinian gays in Israel sensed that they had to hide behind Israeli gays. Although I did feel a certain degree of solidarity with Israeli gays, I was also aware that the needs of the Palestinian LGBTQ community were very different from those of the Israeli LGBTQ community.[20]

In the activist discourse and activist activities of these two groups, the issue of frontiers—sexual, political-national, social, and ethnic—is central. After a period of clandestine existence in Palestinian society and transparent existence in the midst of gay organizations in Israel, it became clear to the two groups that their priority should be their ability to function autonomously. The designation of national frontiers has been an important stage in this process of affirmation and emancipation.

In order to underline their uniqueness, the two groups place particular emphasis on the use of their mother tongue, Arabic. The activists in Aswat and Al-Qaws have dedicated

19 *Al Qaws lel taadodiya al jinsiya wal genderia fi mustama el filastini*. Literally, The Arc for Sexual Diversity and Gender Diversity in Palestinian Society.

20 Conversation with the director of Al-Qaws, Jerusalem, August 9, 2008.

their energies to a thorough task of linguistic adaptation and linguistic research in order to define sexual difference and to unearth words from Arab culture to express terms related to homosexuality and lesbianism. Ancient expressions from medieval Arabic (such as *suhakiya* for lesbian) and the pejorative term "children of Lot" (*luwatiyat, luwatiyun*) that associate single-sex relations with the deviant behavior of the children of Lot have been rejected by activists who denounce the religious and clinical implications of such terms. Other expressions commonly used in Arabic such as *shadh*, which literally means abnormal, have also been rejected because they are associated with deviant or perverted behavior. Activists in Aswat prefer the expression *mithliyun/mithliyat* or simply *methliya jensiya*, "those who have the same sexuality." The Western term Queer, which was initially discarded, is now increasingly being used by activists who find its performative and political usage to be particularly effective. Aswat has also published several collections of feminist and lesbian articles in a series of collections that were launched at its first public conference in Haifa in 2007. This event, which took place under massive police protection, brought together some 300 persons despite warnings issued by Islamic movements. The conference constituted an important first step in the history of lesbian struggles in the Arab world (Aswat 2006, 2007, 2010).

The internet has allowed these organizations to position themselves within a rubric of regional activism that transcends borders and which enables Palestinian gays to link up to discussion forums and to open themselves up to a activist regional Arab-speaking community. Art, dance, and photography have been the media of affirmation for Palestinian queer culture and, in this context, Aswat organizes, in addition to its lectures, a monthly evening program devoted to oriental dancing. A young Palestinian photographer, Ahlam Shibli, is familiar with the themes of the quest for "a place where you can feel at home," trauma and the expulsion. Between 2004 and 2006 she dedicated a series of photographs to sexual minorities forced to lead lives characterized by self-concealment and exile. Her series of photographs, entitled "Eastern LGBT," showed Eastern gays, transsexuals, and transvestites in different European capitals as well as in Tel Aviv. These photographs are part of a global project that audaciously questions the issue of dispossession, whether of one's destiny, land or home, as well as the dispossession of one's right to be master or mistress of his or her own body and intimacy.[21] In her photographs, Shibli presents the human body as a potential metaphor of one's home and one's country (Lagnado 2007).

The activists in both Palestinian gay groups have tried to include their struggle for emancipation in the context of clearly defined political and national frontiers in order to connect their struggle against sexual oppression to their own national struggle.[22] They want to disassociate themselves from Israel's public diplomacy campaign promoting the inclusive theme of a gay paradise so that they can prevent themselves from being exploited. In their view, the illusion of an Israeli gay paradise ignores the reality of the Palestinians' national oppression: The prioritization of the defense of sexual liberty is a clear example of "pinkwashing,"[23] where the question of national borders is confused with one's position on territorial issues.

---

21 See the photographs at: http://www.ahlamshibli.com/Work/LGBT.htm and http://www.ahlamshibli.com/installation/LGBT.htm, accessed October 27, 2013.

22 This is the principle of intersectionality, introduced by Crenshaw (1991).

23 This process, in which the LGBTQ struggle for gay rights is recruited in order to promote Israeli democracy, has been condemned by LGBTQ activists opposed to the process as an example of "pinkwashing." Queer radical organizations, operating as a group calling itself QUIT, and opposed to the Israeli occupation of Palestinian territories, began using this term in 2010. The struggle against

This preoccupation has other repercussions for their activist strategies: The "emergence from the closet" has not been prioritized because it is too closely associated with the desire for normalization and for integration with a political reality rejected by Palestinians. Gay activists have become very sensitive to the fantasy of an erasure of frontiers, an idea presented by the Israeli LGBTQ community when the World Pride event in Jerusalem in 2006 was organized around the theme of "Love without Borders." For Israeli gays, emergence from the closet is associated with being recognized as full-fledged citizens (the right to marry, adopt children, serve in the army, etc.) and with the attainment of normalization without any questioning of national values (Ritchie 2010). Activist Palestinian gays reject this object of visibility:

> And we are very disappointed that the Open House was supporting this event and also (that it is) called love without borders. We want love with borders. We want love. We want our borders as a Palestinian community, we want our borders as women in our society, we want our borders as gays. (Krahulik 2005, p. 517)

The two Palestinian gay groups have strongly politicized their activist discourse and have proposed a mapping out of the parameters of an "intimate" Palestinian nation. They want to be regarded as a minority within the Palestinian people; in addition to demanding a frontier between gays and heterosexuals as well as between gay men and lesbians, they demand a frontier between Israelis and Palestinians. These activists have chosen to question the national narrative by rendering the question of the diversity of sexual identities and the question of sexuality visible and legitimate. They have clearly shown their support for the Palestinian cause by rendering that cause compatible with, and linked to, the struggle against sexual oppression. Members of Aswat and Al-Qaws and other Palestinian queers have created a group called Palestinian Queers for BDS (PQBDS) in order to join the initiative led by Palestinian civic society as they clearly affirm that their struggle as sexual minorities is linked with the Palestinian struggle for liberty (Moussa 2011, p. 87).

The establishment of clearly defined borders between Israeli and Palestinian activists and the determination to reintroduce the question of national oppression into the activist LGBTQ agenda are part of a more general movement. Like the boycott of Israeli universities, the establishment of borders, and the inclusion of national oppression in the LGBTQ agenda are aimed at opposing any process of normalization in a context where the Israeli occupation of the West Bank and the Gaza Strip is allowed to continue and where the radical asymmetry of the power relations between Israelis and Palestinians remains in place. Some Palestinian LGBTQ activists are calling on Palestinian gays to remain in the Palestinian zones (that is, in Israel or the Israeli-occupied territories). Like the refusal to participate in gay pride parades, the refusal of both "flight" and mobility has now become an integral part of activist gay objectives. This position is regarded as a strategy of integration within the Palestinian national community (Moussa 2011, p. 99).

---

pinkwashing has been organized at the international level as part of other anti-Israel campaigns, such as the BDS movement.

Conclusion

Since the 1990s, the struggle of LGBTQs has considerably evolved in Israel and in the Israeli-occupied Palestinian territories. Israeli gays have won numerous judicial battles and have established their presence in the "public space" through their demand that their full social integration be linked to their being recognized as full-fledged citizens. This inclusive position and the proclamation, made by most Israeli gays, of a total allegiance to the State and its national values have enabled Israel's LGBTQ community to end its clandestine existence and to shed its status as an ostracized entity. In this new activist landscape, the emerging Palestinian LGBTQs initially experienced considerable difficulties in their attempt to find their own space for action and to formulate their strategies. The struggle of the queers has often been forced to advance with the use of camouflage—whether in the context of Israeli organizations or in the context of Palestinian ones.

The emergence of a strong homonationalist trend in Israeli society, soon after the outbreak of the Second Intifada in 2000 and the multi-pronged terrorist attack on American soil on September 11, 2001 have dramatically affected this process of mobilization. A considerable number of Israeli LGBTQ activists have shown their desire to be integrated into an envisaged national community, namely, a gay paradise and its system of representation of frontiers. As a result, Palestinian LGBTQ activists who have already shown their desire to actively participate in autonomous structures and to defend their unique approach to the issue of gay rights have accelerated the process of rupture. The objective was to avoid any possibility that the issue of normalization could be turned into a tool for exploitation. Priority was thus given to the development of independent organizations, to the search for diversified international funding and to the development of programs and strategies for struggles adapted to the diversity of Palestinian society—in Israel and in the Israeli-occupied Palestinian territories.

PART IV
Political Crossings

Chapter 12
Activists without Borders? Tours to Israel and the Palestinian Territories Organized from France

Marc Hecker

Introduction

Sometimes conflicts overflow their immediate field of operations. Such geographical spreading can lead to the exportation of violence to a neighbouring or to a more distant region. In the 1970s, for example, the Israeli–Palestinian conflict had violent repercussions in some European countries. This internationalization led to the highjacking of planes, to terrorist attacks, to the taking of hostages by various Palestinian groups, and to targeted assassinations committed by the Israeli secret services.

However, the effects of war on other countries can also take much more peaceful forms. The existence of associations which support Israel or Palestine are a good example of this. These associations, which have greatly increased in number in France since the 1967 War, use many different "repertoires" (Tilly 1984), such as demonstrations, distribution of leaflets, organization of conferences, meetings with politicians or journalists, etc. Another very widespread form of action is the organization of "activist" tours to Israel and/or to the Palestinian Territories. Such tours are not new but have increased considerably since the beginning of the Second Intifada (2000).

The various associations which organize trips to Israel and/or to the Palestinian Territories have differing conceptions of travelling. Someone who has no previous knowledge of the Near East and who takes part in a journey organized by a pro-Israeli group, would return to France with a very different vision of the Israeli–Palestinian conflict to that of someone who had travelled with a pro-Palestinian group. In other words, the people who organize these visits create particular conditions in order to produce a certain type of knowledge which reflects the dominant interpretation of the Israeli–Palestinian conflict found in pro-Israeli or pro-Palestinian circles. In pro-Palestinian circles the colonial interpretation is very widespread. Israelis are thus presented as the colonizers and Palestinians as the colonized. In pro-Israeli circles the democratic argument is widely held: Israel is seen as a democracy struggling against tyrannical Arabic regimes and against Islamist groups—supported by Iran—rejecting the Western model.

This chapter was written while I was finishing my PhD thesis in political science at the University of Paris 1 Panthéon-Sorbonne. This thesis, entitled, "Transnational actors and the State: the example of activism in France with reference to the Israeli-Palestinian conflict," is based on an in-depth study of pro-Israeli and pro-Palestinian circles. It draws attention to the diversity of the associations within these two spheres and to their important political differences. In this chapter, because of lack of space, these political divides will not be examined in detail, but it is important to bear in mind that the pro-Israeli and pro-Palestinian camps are far from being monolithic.

This chapter is largely based on interviews carried out with activists and on documents produced by associations. There is no question of assessing the effectiveness—and even less the legitimacy—of the activist practices of either faction, but the aim is to study in as dispassionate a way as possible certain kinds of mobilization—in this instance the organization of tours—brought about by a distant conflict.

To this end, we will first establish a "typology" in order to distinguish between "fact-finding tours," "solidarity tours," and "political-spiritual tours." Subsequently, we will explore the manner in which the participants in these tours envisage the question of borders within the Israeli Palestinian space.

"Fact-Finding Tours": Proclaimed Objectivity

Pro-Israeli and pro-Palestinian associations have one point in common: they both denounce the way the media present the conflict. In order to counter this alleged misinformation, they try to present information which they describe as "objective." This usually takes the form of magazines, bulletins, websites, or e-mails. The leaders of the associations all agree that the best way to obtain and convey information is to travel to Israel or to the Palestinian Territories.[1] As a result, they organize "fact-finding tours" for people whom they consider influential, such as politicians and journalists.

The composition of the groups is important. It serves no purpose to organize a tour for people who are already totally convinced of the rightfulness of the cause they are supporting. On the other hand, it can be useful to add a "friendly" politician or journalist, who will help to communicate a positive image to the other participants. It may also be relevant to include people who may prove useful in the future, such as young members of parliament, likely to become ministers.[2] Tours have also been organized for students in journalism schools. For instance, in February 2006, the journalism school of the Paris Institute of Political Studies (Sciences Po) went on a tour which was supported by the CRIF (Conseil Représentatif des organisations juives de France) and partly funded by the American Jewish Committee (AJC) (CRIF 2006).

The size of the groups varies: some organizers favor groups of a dozen people—like the France-Israel Association or Medbridge—while others prefer bigger groups, like Siona, which, to quote its president, invited about "350 members of parliament during the course of five consecutive tours … to show them the real Israel."[3]

Pro-Israeli associations organize more tours than pro-Palestinian associations, probably because they have more funding and so can more easily afford the transport and accommodation costs. For all the pro-Israeli groups there are at most 10 tours organized each year.[4] However, the financial argument does not explain everything. Some pro-Palestinian groups also have reasonable budgets at their disposal. The Committee for Charity and Support for the Palestinians (Comité de Bienfaisance et de Secours aux Palestiniens, CBSP) has an annual budget of more than six million euros (CBSP 2007). The France Palestine Solidarity Association (Association France Palestine Solidarité, AFPS),

1 Interview with Yves-Victor Kamami, June 16, 2004.

2 Interview with Valérie Hoffenberg, October 2, 2008.

3 Interview with Roger Pinto, president of Siona, November 8, 2007.

4 Roger Benarrosh, honorary vice-president of the CRIF, estimates that five or six tours of this kind are organized each year by the CRIF. Interview with Roger Benarrosh, June 28, 2004.

for its part, transfers tens of thousands of euros each month to the Palestinian Territories.[5] Such amounts could be used to organize several "fact-finding Tours" each year. However the former president of the AFPS Bernard Ravenel believes that "it is not possible to put money into such projects."[6] Faced with the urgency of the situation in the Palestinian Territories, activists would question the legitimacy of spending money on plane tickets and hotel accommodation for French politicians and journalists.

"Fact-finding tours" are usually quite short—between three and seven days—for two main reasons: the high costs involved and the busy schedules of the "decision-makers" who take part and who cannot allow themselves to be away from France for very long. In such a short time, it is impossible to make an exhaustive tour of the region. Consequently, the "facts" provided depend on the places—carefully selected by the organizers—which are visited and on the people the journalists and politicians are introduced to. These two factors vary according to the organizing bodies. Valérie Hoffenberg, who was president of the French branch of the AJC until 2009, insisted that tours should not be limited just to Israel. She believed that if they were, the politicians and journalists would feel as though something was being kept from them.[7] For this reason some of the tours organized by the French branch of the AJC begin in Jordan, proceed to the West Bank via the Allenby Bridge and finish in Israel. Travelling via the West Bank allows the participants to meet up with Palestinian political and economic leaders who are often close to the Palestinian National Authority.

Other tours only visit Israel. The tour organized from the 4th to the 8th of February 2007 by the Organization of Jewish CEOs and Professionals in France (Union des Patrons et Professionnels Juifs de France, UPJF) for seven journalists[8] is a case in point (Focus 2007). This tour is worthy of our attention as it clearly demonstrates the wishes of the organizers to show the essentially positive aspects of Israel. The tour begins with a visit to two "kibboutzim," during which the guide refers to the history of Israel, insisting in particular on the fact that "all the land [was] bought by the Jews, often for a very high price" (UPJF 2007a). After this visit, the group is taken to Tel-Aviv University to meet a former Israeli ambassador to the United States. In the evening, a dinner is organized with Zeev Boim, the minister for integration who notably mentions the integration of Jews from Ethiopia. It should be noted that several tours—in particular the one in 2006 of Sciences Po's journalism school—include a visit to an integration center. Visits to such centers allow for the presentation of Israel as an open country which welcomes people from different cultures. The main idea is to counteract the view spread by pro-Palestinian circles according to which Israel is a racist state which practises a form of apartheid.

On the second day, the group, led by the UPJF, visits Abu Ghosh, "an Arab-Israeli village where there has never been a problem of cohabitation with the Israelis [*sic*]," to quote a phrase from the official UPJF report edited by Edward Amiach, the vice president of the association. The group visits the church of Abu Ghosh and Edward Amiach writes: "All the images of Christ were removed by the Arabs!" This comment was made without any explanation, but we can guess that it is aimed at denouncing the supposed intolerance of certain "Arabs" in particular of those who are Muslims. After Abu Ghosh, the group

---

5 For example, the AFPS paid 54,370 euros to its Palestinian partners just for the month of October 2006 (AFPS 2006a).

6 Interview with Bernard Ravenel August 11, 2006.

7 Interview with Valérie Hoffenberg October 2, 2008.

8 The names of these journalists are not cited in any of the UPJF reports, as though they have been guaranteed anonymity.

attends a series of eight meetings at the Knesset with political leaders from different parties, including Rajeb Majadleh, the first Israeli Arab to become a minister. The main aim of this meeting is, amongst other things, to demonstrate the strength of Israeli democracy, and to show that Israeli Arabs can have brilliant political careers. On the third day, the 6th of February, the group meets a highly placed Maronite dignitary, then visits "a Muslim district in Jerusalem" to exchange a few words with a Benedictine. Only one of the nun's comments is quoted in the UPJF report: "After having tried autonomy, Palestinians wish to become Israeli again." The visit of Jerusalem continues with the Old City, in particular the Holy Sepulchre and the Wailing Wall. An incident is referred to in the report: "We were able to have first hand experience of the way Arabs use the media. On that day, there was a problem with the footbridge leading to the Temple Mount which the Israelis had to repair. About ten bearded Arabs, surrounded by dozens of cameras and journalists, were screaming about scandal and blasphemy. The footbridge, which had been damaged by a snow storm, is situated outside the Temple Mount" (UPJF 2007a). The delegation then meets with a representative from the Israeli Ministry of Foreign Affairs and continues to a place which is shown to just about all groups led by pro-Israeli organizations: the Yad Vashem Memorial. Edward Amiach comments: "After these first two days, we could already notice that our guests had a different tone of voice and a different approach, that their analyses were more enquiring, more subtle and less biased than they had been on the first day"(UPJF 2007a).

The third day is devoted to an excursion to the edge of the Gaza strip. Here again the vice president of the UPJF condemns the distortions perpetrated by the media: "We and the journalists were surprised by the sight of huge, beautiful buildings … We thought Gaza was just an immense camp. In fact, these buildings, according to our guide, are inhabited by the Palestinian bourgeoisie. Television cameras have only shown us camps and misery." The delegation continues its tour towards Sderot, an Israeli town, which is constantly the target of Qassam rockets fired from the Gaza Strip. The aim of this visit is to show that the Palestinians are not the only victims of the conflict and that some Israelis also live in fear and poverty. The group stops for lunch in "an Arab village in Negev," a village where there is no unemployment and which has produced several doctors and company directors, as is pointed out in Edward Amiach's tour notebook. The object of this meal is probably to show that Israeli Arabs are not all "second class citizens," as pro-Palestinian groups claim. The group is then taken near Hebron to see the "separation wall." Once again, the purpose is to convey a positive image of Israel by underlining "the efforts made by Israelis to facilitate and fluidify access points whilst sacrificing nothing to security" (UPJF 2007a).

In the evening, at dinner, the guest is Silvan Shalom, a former Minister of Foreign Affairs who is presented by the vice president of the UPJF as "the architect of the media change of heart" thanks to whom "Israel has been able to improve its image in the rest of the world, particularly in Europe"(UPJF 2007a).

On the last day, a lunch is organized with the Israeli minister of justice, who is responsible for legal relations between Israel and the Palestinian National Authority. This meeting does nothing to improve the terrible image of the Palestinians which was put about during the first few days of the tour. The UPJF report states that: "the journalists were inundated with information which emphasized the lack of organization, the lack of willingness, the many blockages and pressures brought to bear and the widespread corruption prevalent throughout society in the Palestinian Territories" (UPJF 2007a). After a last evening in Tel Aviv the tour comes to an end. In his conclusion, the vice president of the UPJF declares:

> I am convinced that for all these journalists there will be a 'before Israel' and an 'after Israel.' Above all, they have understood that, in the Middle East conflict, it is not a question of the 'rich, strong and wicked Israeli' and the 'poor, dominated, Arab Palestinian.' There is no one person responsible for all the troubles (Israel), but the Palestinians are largely responsible for the ills which they complain about. In the future, history will show whether this sort of tour was profitable for Israel and for its image. For my part, believing as I do that the truth can only add to the cause we are defending, and being convinced that we have only been showing a reality which many people, through lack of time and ethics, refuse to take into account, I am certain that the journalists will regard Israel with esteem and respect. (UPJF 2007a)

After a journey such as this, it is legitimate to ask oneself whether the seven journalists who took part had the impression they had been shown an edited version of the Israeli–Palestinian conflict. The visitors' book which they signed at the end of the tour does not convey that impression. The chief editor of a national radio station states: "This was my first trip to Israel … of course I set off with certain pre-conceived ideas. I am very grateful to [you] for letting me come close to reality in a country which is bubbling with life and hopes of peace. Thank you to all those who accepted to devote their time to us and to speak to us frankly. And a special thanks to those who took us under their wing so that we could share their love of Israel with them" (UPJF 2007b). In fact, amongst all the comments left in the visitors' book, only one expresses regret that they were not able to meet any Palestinian leaders (UPJF 2007b). In general, it seems as though the organizers have succeeded in getting across the message they wished to publicize. The borderline between "fact-finding" and "lobbying" is more than tenuous here.

If "fact-finding tours," which are meant to be "objective," but in fact, as a result of the organizers' directives, only show a part of the real situation, "solidarity tours," for their part, are the expression of an accepted subjectivity. Participants travel to Israel or to the Palestinian Territories in order to make a material or symbolic contribution to show their solidarity for one or other of the opposing sides.

"Solidarity Tours": Accepted Subjectivity

There are many ways to express one's solidarity. Some leaders of pro-Israeli groups claim that, in a period of crisis, when tourists stop travelling to Israel, just going there on holiday is an act of solidarity. Gil Taïeb, president of the Association for the Well-Being of the Israeli Soldier (Association pour le Bien-Etre du Soldat Israélien, ABSI), states that: "Solidarity is a big word … Just to come, to have a drink, to go shopping, to take a taxi—is already enough to show your solidarity. … Solidarity isn't a question of heroism."[9] Tourism is one of the activities which is most sensitive to reduced standards of security. In 2000, Israel had 2.4 million tourists. At that time tourism represented 3 or 4 percent of Israel's GNP (Sadan 2004). From 2001 to 2004, at the height of the Second Intifada, the annual average was about 1.2 million tourists. In 2005, French tourists represented exactly 16.4 percent of the total number of tourists visiting Israel. Aware of the importance of tourism for the Israeli economy, the Union of Jewish Students in France (Union des Etudiants Juifs de France, UEJF), which organizes "Discovery journeys" each year to Israel (and sometimes to the

9 Interview with Gil Taïeb, March 9, 2007.

Palestinian Territories), decided to maintain the 2006 edition of their summer tour despite the outbreak of war between Israel and Hezbollah. The simple fact of refusing to cancel their tour was described by the organizers as an act of solidarity.[10] In more concrete terms, their solidarity was also expressed by the distribution of toys to children in northern Israel.

A particular form of the "Solidarity Tour," called "Civil Voluntary Service," allows volunteers from all over the world to spend two or three weeks in Israeli military bases doing support work, such as general maintenance, preparing meals, folding parachutes, checking gas masks, and, sometimes, cleaning weapons. The idea of a "Civil Voluntary Service" began in 1982 and was instituted in the following year with the creation of "Sar-El," a contraction of "*sherout leisrael*" ("Service for Israel"). This Israeli organization has succeeded in attracting an average of four to five thousand volunteers each year who come from about thirty different countries. France provides a large contingent, estimated to be a quarter of the total number of volunteers. From 1983 to 2006, about 110,000 people took part in the "Civil Voluntary Service," of whom 30,000 were French (Sar-El 2007). During the war in the summer of 2006, 800 French volunteers participated in the scheme. The volunteers work about seven hours per day (except during Shabbat), wear uniforms, and are supervised by army instructors. They do not get paid for their work but they have free food and accomodation. The volunteers have to pay their own travel expenses and pay an enrolment fee of between 50 and 100 euros.[11]

The work done by the volunteers means that there is no need to call upon reservists who can thus continue their professional activities in civil life and this represents "a saving of about 1,600 euros for the state per three week period."[12] "Civil Voluntary Service" is open to everyone, Jews or non-Jews. No statistics are available for the proportion of Jewish and non-Jewish participants but pro-Israeli organizations are keen to emphasize the number of non-Jews taking part in the program (UPJF 2004). An organizer from Sar-El at Ashdod stated that the last group she had led in 2008 was made up of 18 people of whom 6 were non-Jewish.[13] According to her this proportion of non-Jewish participants was higher than in the majority of groups. However, she also stated that sometimes whole groups of non-Jewish participants, in particular groups of Protestants, took part in the "Civil Voluntary Service" program.

In times of war, other more specific acts of solidarity can take place. The ABSI, which collects large sums of money in France (several hundred thousand euros per year), use these sums to build rest and recreation centers for the military or for enroling former conscripts in university courses. During the conflict in the summer of 2006, for example, they collected 250,000 euros. From the very first days of the war, members of the association travelled to the north of Israel to distribute parcels to soldiers returning from the frontline.[14]

So far, we have dealt mainly with tours organized by pro-Israeli groups, but pro-Palestinian groups are also very active in organizing "solidarity tours." At the time of the Second Intifada, the number of these tours increased greatly, especially with the creation of the International Civil Campaign for the Protection of the People of Palestine (Campagne Civile Internationale pour la Protection du Peuple Palestinien, CCIPPP). This organization was founded at the beginning of 2001 and became a reality in June of the same year

---

10 Interview with Deborah Halimi, August 31, 2006.

11 See www.upisrael.org/combien_coute.php, accessed May 28, 2007.

12 See www.upisrael.org/lesplus.php, accessed on May 28, 2007.

13 Telephone conversation with a Sar-El representative in Ashod, March 25, 2008.

14 Interview with Gil Taïeb March 9, 2007.

when the first mission was sent to the Palestinian Territories. The main idea was to send an "international" contingent into the Palestinian Territories who, by their mere presence, would prevent the Israeli army from acting as brutally as they would have done had there been no witnesses from a foreign country present. During the mission, the participants were encouraged to act as witnesses, "by sending information, as frequently as possible, in the form of articles, reports or photos, because the international participants are often the only witnesses of extremely urgent situations which the ordinary media ignore" (CCIPPP 2001). At the beginning, the founders of the CCIPPP intended to send a mission every three or four months. However, the high demand soon encouraged the organizers to increase the frequency of the missions. Between June 2001 and March 2002, 11 missions were organized.[15] At the outbreak of war in June 2006, 119 missions had been planned, with a total of between 1,000 and 1,200 participants.[16]

There are different ways to "act as witnesses" and to express "solidarity," depending on the mission. "Bearing witness" can take many forms, from the creation of a blog to the organization of photographic exhibitions, or even writing a book (Bové 2002; Alcouloumbré and Baudoin 2003). "Solidarity" can also be shown in many ways. In the first missions, it was demonstrated in spectacular acts of intervention or forced entry. The most well-known was the forced entry into Muqataa—which was surrounded by the Israeli army—by an international group led by José Bové. The group, who had a camera, managed to photograph Yasser Arafat having to use candlelight because the electricity had been cut off.[17] José Bové wrote: "At least we have won the image war" (Bové 2002, p. 37).

Some missions can be dangerous. Several foreigners have been killed, the most well-known being Rachel Corrie. No French activists have as yet suffered the same fate, but some have been wounded (Léostic 2006). However, in most missions, "solidarity" takes the form of less dangerous actions, olive picking being one of the most widespread activities. The international participants help to pick the olives while at the same time "protecting" the Palestinian farmers from the Israeli soldiers and the settlers who try to impede the harvest. Harvesting the olives is never seen as a simple gesture of logistic support towards the farmers, but as a real act of political solidarity, an act of "resistance" to colonization (AFPS 2006b).

Some associations organize tours combining solidarity and exploration of the area. "Generation Palestine" is aimed at young people in particular. It began following the organization of two tours in 2005 and 2006 by the General Union of Palestine Students (GUPS), as part of a project named "Bridges over the Wall." As underlined by the former secretary general of the Paris office of GUPS, Racha el-Herfi, some of the participants "know absolutely nothing" about the Israeli–Palestinian conflict.[18] It is therefore necessary to teach them things. The tour is almost entirely in the Palestinian Territories, in contact with people living, for instance, in refugee camps. In Israel, the participants meet almost exclusively "refuzniks" and anti-settlement activists, such as members of the Alternative Information Centre (AIC).[19] Incidentally, it should be noted that a visit to this center is often included in the tours organized by pro-Palestinians. Michel Warschawski, one of the

15 Interview with Nahla Chahal, July 3, 2007.
16 Interview with Youssef Boussoumah, July 13, 2006.
17 This mission is featured in the film *Le Siège*, directed by Samir Abdallah.
18 Interview with Racha el-Herfi, July 5, 2006.
19 Ibid.

heads of the AIC (and who comes originally from France), is regarded as a key figure by the leaders of pro-Palestinian associations.

Finally, another way for French people to travel to the Palestinian Territories is through twinning. French collectivities which are twinned with Palestinian towns are grouped together in an organization called the Decentralised Cooperation Network for Palestine (Réseau de Coopération Décentralisée pour la Palestine, RCDP), chaired by Claude Nicolet, deputy mayor of Dunkerque, a city in the north of France, twinned with Gaza. Elected representatives travel regularly to the Palestinian Territories to monitor the progress of various projects. Some of these call for townspeople to go to the Palestinian Territories themselves. For example, as part of the partnership between Seine-Saint-Denis and the town of Jenin, Greco-Roman wrestling coaches from Bagnolet—a town in the suburbs of Paris—went to the West Bank to teach their sport to young Palestinians (Cités Unies France 2005). One specific organization, the Association for the Promotion of Twinning between Palestinian Refugee Camps and French towns, (Association pour la promotion des jumelages entre camps de réfugiés palestiniens et villes françaises, AJPF), looks after the actual twinning with refugee camps. The first partnership of this kind took place in 1989 between the town of Montataire (Picardie) and the camp of Deisheh. Today, just about all the Palestinian refugee camps in the West Bank and the Gaza strip are twinned with at least one French town. Since 2002–2003, partnerships of this kind have spread to Palestinian refugee camps in Lebanon, Syria and Jordan. Bagnolet, for example, has been twinned with Chatila since 2003. The co-president of the AJPF, Fernand Tuil, regularly accompanies French groups to the Palestinian Territories. Often they are young people from the suburbs as a large percentage of French towns twinned with Palestinian cities and refugee camps are suburban towns. For Fernand Tuil, taking young people on the spot helps to explain the complexity of the conflict to them—he calls it "an educational act"[20]—in order to avoid importing the conflict into France. It is about demonstrating that the Israeli–Palestinian conflict is primarily territorial—and not religious—so it should not be transposed to France in the shape of confrontations between Jews and Muslims.

Tourism of this kind, however, does worry the French security services because the participants may be brought into contact with members of groups who figure on the European Union's list of terrorist organizations. For some pro-Palestinian activists it is, however, unfair to stick a "terrorist" label on groups like Hamas who, according to them, seek, above all, to "resist" an occupation. Youssef Boussoumah, of the CCIPPP, describes in positive terms, a meeting that took place in 2001:

> This is the second mission and there are a lot of young French North African girls but also "Franco-French" young women and friends from the UJFP, (the Franco-Jewish Union for Peace, Union Juive Française pour la Paix), Michèle Sibony and the whole team. And we have a meeting organized with the coordination of the Palestinian resistance which includes the thirteen Palestinian factions. We are meeting them at the Agricultural Hall in Ramallah. It is an initial contact meeting to get to know one another. We arrive and they are all there, even the Islamic Jihad. It is really impressive, they are sitting in a semi circle. Very friendly, very welcoming, just waiting for us like that, calmly. Their spokesman is a Hamas representative. He speaks very good French and is very welcoming, etc. And he has been chosen to speak by all the others. It's quite extraordinary. There was the Palestinian Communist party, the PFLP, the DFLP, and of course the Fatah. It was really

20 Interview with Fernand Tuil, May 11, 2007.

> quite impressive. And he spoke on behalf of everyone and he was really very convincing. People wanted to hear more, including the comrades from the UJFP. And what's more, no questions were avoided, like suicide bombings, etc. They were very frank.[21]

These examples tend to show how the different types of journeys overlap. "Solidarity," "discovery," and "fact-finding" cannot easily be isolated from one another, and, in each case, a particular view of the Israeli–Palestinian conflict is presented to the participants. The third category, "spiritual tours," also tends to overlap with the other kinds of journey and to create a specific field of knowledge, this time marked by a religious approach.

"Political-Spiritual" Tours: Understanding and Compassion

Jerusalem is sometimes called the "Thrice Holy City" because of the importance of the Wailing Wall for Jews, the al-Aqsa Mosque for Muslims and the Basilica of the Holy Sepulchre for Christians. Pilgrimages, in the purely religious sense, are not of interest here. When religious journeys take on a political dimension, however, it is quite different.

Although the phenomenon of Evangelical Christians supporting Israel is often mentioned (Fath 2005), it is much more widespread in the United States than in France. One could even say that the movement is in its infancy in France. The Israeli authorities are counting on its development. For example, from the 25th to the 29th of September 2006, a tour was organized for representatives of the French Evangelical Church. The tour was financed by, amongst others, the Israeli Ministry of Tourism and El-Al Airlines.[22] Traditionally, however, the most visible activists among the French Christian groups have not been supporting Israel but rather the Palestinians.

In 1970, the first Christian conference in support of Palestine was organized in Beirut. About four hundred Christians from 37 countries took part in it.[23] Georges Montaron, director of the weekly *Témoignage Chrétien* ("Christian Monitor"), played a major role in setting up this conference. From the 1970s until the death of Georges Montaron in 1997, *Témoignage Chrétien* was seen to repeat the terms used by Noël Bouttier, former editor of the magazine, as "an almost ferociously and unconditionally pro-Palestinian publication."[24] From the end of the 1990s onwards, the position of the Christian publication changed. As Noël Bouttier explains:

> We didn't want to abandon the struggles which *Témoignage Chrétien* had made its own, but we wanted to move towards a finer understanding of what Israel really was, … the internal conflicts, to understand the Israeli democratic setting which is very real. This doesn't mean that there aren't Arabs who are more or less excluded from politics … What we wanted was to understand the complexity of the region and to try to get closer to all the peace-makers in Israel and the Palestinian Territories.[25]

---

21 Interview with Youssef Boussamah, July 13, 2006.

22 See the account of this visit by Henri Viaud-Murat on: www.blogdei.com/index.php/2006/10/06/36-voyage-en-israel-rapport-de-notre-frere-henri-viaud-murat-tres-interessant, accessed October 24, 2008.

23 "L'appel aux chrétiens du monde entier," May 10, 1970, Beirut. See http://leprisonnier.levillage.org/lesogres2004/article.php3?id_article=1263, uploaded 30/12/2005, accessed May 19, 2014.

24 Interview with Noël Bouttier, December 13, 2007.

25 Ibid.

At the height of the Second Intifada, in April 2002, when rumours of a massacre committed by the Israeli army in Jenin were growing, *Témoignage Chrétien*'s editorial board were wondering what editorial line to take, hesitating between full support of the Palestinians and a more balanced attitude. Finally, on April 11, 2002, the magazine appeared with the front page headline: "In the name of God, Cease Fire!" This issue carried an article entitled: "In the name of the God of Moses, of Jesus and of Mohammed, the fighting must stop," which was "an inter-religious call for peace in the Near East." The text was re-published in subsequent issues, with the addition of many signatures. By April 26, it had been signed by a hundred personalities including several Christian, Jewish, and Muslim dignitaries (*Témoignage Chrétien* 2002). The call was also open to the general public. Within a few months, more than 10,000 signatures were collected (Fédération Protestante de France 2002).

In the wake of this mobilisation, *Témoignage Chrétien* organized a journey which was not presented as a "pilgrimage to the Holy Land" but as "a journey to Israel and Palestine." This short visit, which took place in February 2003, had a dual purpose, both spiritual and political.[26] The two hundred participants were accompanied by a bishop, a pastor, a Muslim theologian, and a rabbi. They went to Jerusalem where they visited the ecumenical prayer garden on the Mount of Olives, the Old City, the tomb of Yitzhak Rabin, and Yad Vashem. They also went to Bethlehem and Nazareth, which is "obligatory" for most Christian groups.

A second journey was organized from the 7th to the 11th of November 2003, with about two hundred participants. Once again, the idea was to go both to Israel and to the Palestinian Territories, to meet people, talk to them, comfort them, exchange views, and pray. Once again "spiritual" and "political" were the key words of the tour. The program included inter-religious exchanges at Beersheba, meditation in the Negev desert, reflection in the churches of Taibe and Abu Gosh, a visit to the Old City of Jerusalem, prayers at Yad Vashem, discussions with the inhabitants of a refugee camp on the West Bank, a meeting with a leader of the United Nations Relief and Works Agency for Palestine Refugees in the Near East (UNRWA), discussions with a civil servant from the Israeli Ministry of Foreign Affairs, a meeting with a Franco-Israeli student who was victim of a terrorist attack, and the highlight of the tour, a visit to Yasser Arafat at the Muqataa. In short, it involved trying to understand the conflict, to sympathize with the victims of both camps and to demonstrate that a religious confrontation is not inevitable. When they were taking leave of Yasser Arafat, the latter embraced the Rabbi Philippe Haddad, a move which unleashed a huge polemical debate within the Jewish community in France (Haddad 2004).

A third trip was planned in 2004. This time, about one hundred and seventy people travelled to the Near East, among them about fifteen Catholic, Protestant, and Muslim scouts. One of the highlights of the trip was a ceremony in memory of Yasser Arafat in Bethlehem. Another memorable event was the meeting with Avraham Burg. Two more journeys followed. The first took place between the 20th and the 28th of March 2006 and the second from the 1st to the 5th of December 2007. Etienne Pinte, a UMP[27] Member of Parliament and mayor of Versailles, was one of the 25 people who took part in this trip.[28]

All the tours presented so far involved crossing borders. Of course, someone travelling with a pro-Palestinian group will not perceive borders in the same way as someone

26 See www.temoignagechretien.fr/journal.php?ref=paix, accessed October 30, 2008.

27 The UMP (Union pour un Mouvement Populaire) is the main French right-wing party.

28 See www.terreentiere.com/REPORTAGES/tc-israel-palestine-2007.aspx, accessed October 29, 2008.

travelling with a pro-Israeli group. For pro-Israeli groups, entering Israel is just a formality. The issue of borders, therefore, is not often mentioned in their travel reports, especially if they stay west of the Green Line. It is quite different for pro-Palestinians for whom crossing the border is a particularly difficult experience.

Travellers Facing Borders

The state of Israel controls its own borders very strictly. It also controls the borders of the Palestinian Territories. Attempts to gain direct access to these Territories are quite rare but very widely publicized. This was the case in May 2010 when a flotilla of ships carrying humanitarian aid and activists from 40 different countries, including many from Turkey, tried to force an entry through the Israeli embargo and was prevented from reaching the port of Gaza by navy commandos. During the raid, which took place in international waters, nine passengers were killed. In the great majority of cases, pro-Palestinian activists choose to enter the Palestinian territories using more indirect and less risky methods, either via Ben Gurion Airport (near Tel Aviv) or by crossing the border with Jordan (which is less common).

It is only after they have been checked by Israeli border guards that they can continue their journey to the Palestinian Territories—almost always to the West Bank, as the Gaza Strip has become more or less inaccessible because of the isolation policies imposed by the Israeli governments since Hamas came to power.

As far as the "border" between Israel and the West Bank is concerned, most of the pro-Palestinian associations in France accept the 1967 border, with the exception of a few minority groups who believe that the state of Israel, as an expression of Zionism, should cease to exist, which would, in effect, solve the whole problem of defining the borders of the state.[29] A return to the 1967 border is not seen as impossible by the activists who are in favor of this solution. Thus, they refuse to use the term "border" to describe the unilateral line which Israel is in the process of drawing. As Bernard Ravanel stated in public, when he was president of the group of French NGOs for Palestine (Plateforme des ONG françaises pour la Palestine), in the interests of peace it is neccessary to have well-defined borders between two states but "a unilateral definition or modification of the borders is not an act of peace; on the contrary, it is a declaration of war."[30] Rather than using the term "border," pro-Palestinian activists talk about the "Wall" and "checkpoints." They denounce the fact that some of these "checkpoints" look like "border posts," when they are not actually situated on a legal border according to international law.

It is thus necessary to differentiate carefully between two moments: setting foot on Israeli soil and then going into Palestinian territory. The former is felt to be a difficult moment by the members of pro-Palestinian associations, the Israeli border guards being usually suspicious, or even openly hostile towards those who take part in "solidarity tours" to help the Palestinians. Although the first mission of the CCIPPP did not have too many problems getting into Israeli territory, things got a lot more difficult for the second mission. Youssef Boussoumah, who took part in it, relates that he was questioned for five hours and that other activists had to wait even longer. It was only after the intervention of Michel Warschawski, who alerted the Israeli press and the French authorities, that the

29 Interview with Walid Attalah, August 16, 2006.

30 Communication from Bernard Ravenel during the conference organized by Génération Palestine, November 1, 2008, in Paris.

Israelis let the activists enter their territory. Since then, several missions have been turned back. Youssef Boussoumah believes that he cannot return to the occupied territories until "Palestine is freed." If he has not tried to go back, he explains, it is because he does not want to give the Israelis "the pleasure of turning [him] back at Ben Gurion."[31] Fernand Tuil also experienced some difficulties at Ben Gurion Airport.[32] He recounts that he was stuck there with 52 young people for more than 10 hours. Yet again, it was thanks to Michel Warschawski's contacts that the situation was finally resolved, but the group was not allowed to enter Palestinian territory.[33]

Since then, Fernand Tuil has been taking special precautions. When he organizes tours with French Members of parliament who have a twinning arrangement with a Palestinian refugee camp, he informs the French Ministry of Foreign Affairs and provides them with the list of participants.[34]

Certain pro-Israeli groups also send a list of participants before they travel. For example, Valérie Hoffenberg, who was then president of the French branch of the American Jewish Committee, organized a tour for personalities from the French Muslim community at the end of 2008 so that they could discover Israel. People whose names sound Arabic are usually kept longer when they arrive on Israeli soil.[35] In order to avoid problems, Valérie Hoffenberg contacted the Israeli authorities in advance, gave them the participants' names and their passport numbers and requested that a "VIP reception" should be organized for them at the airport.[36] During a previous trip, some French dignitaries were given preferential treatment when they crossed the Allenby Bridge across the Jordan River. They were late for a meeting in the Palestinian Territories and could not afford to be held up at the Israeli border post. Not only had Valérie Hoffenberg given the Israeli authorities a list of the group members, she had also obtained the telephone number of the person in charge of the border post. As the group was running late, she called to warn him that the coach was about to arrive at the border post, and as a result it only took them a few minutes to cross the border. The dignitaries involved certainly realized that they had not been treated like other tourists, but they could not believe that crossing the border could be done so quickly.[37]

For pro-Palestinian groups, providing a list beforehand can work if there are officials in the group, since refusing entry to dignitaries could cause a diplomatic incident. When the participants are "ordinary citizens," other techniques are used to avoid being refused entry into Israeli territory. For example, one method is to arrive in no particular order to avoid being recognized as a group of activists. After the José Bové incident at the Muqataa, all the members of the next mission were turned back. Subsequently, however, members of other missions succeeded in gaining entry to Israel. Nahla Chahal, one of the organizers of CCIPPP, explained that they had to be quite crafty to achieve this: "They were very well disguised: one as a doctor, another as an engineer, etc. ... We realized that they had decided to turn everyone back, and they stopped the biggest groups."[38] Later on, good

31 Interview with Youssef Boussoumah July 13, 2006.
32 Ibid.
33 Interview with Fernand Tuil, May 11, 2007.
34 Ibid.
35 Interview with Nahla Chahal, July 3, 2007.
36 Interview with Valérie Hoffenberg, October 2, 2008.
37 Ibid.
38 Interview with Nahla Chahal, July 3, 2007.

contacts were made with "Israeli anti-settlement activists" who provided "alibis"[39] for the CCIPPP activists.

As Nahla Chahal explained:

> [The Israeli anti-settlement activists] provided lists of their friends … We contacted them and asked them: "Are you willing to receive such and such a person who is a student and to say that he is coming to your house to ask you questions about his thesis?" Sometimes, they would even go to the airport to meet them. There were hundreds of people, some of whom were quite elderly. I managed to get one ninety year old man, a famous historian, to come from a kibbutz. Sometimes, the hosts only spoke Hebrew and we had to find people to translate. We had to reassure them about the fact that we were not trying to get terrorists into the country. But they trusted us. It is important to mention this as, for a whole year, they made it possible for us to get round the decision to refuse our entry.[40]

Apart from getting into Israel, another key moment is entering Palestinian territory. "Checkpoints" are an essential part of the activist experience. Pro-Palestinian activists do not distinguish between "checkpoints" between Israeli territory and the West Bank, and those which are within the West Bank. "Checkpoints" are generally seen as a symbol of occupation and of the daily annoyance Palestinians have to live with. Travellers make a point of crossing checkpoints on foot, like the Palestinians, even though, as French passport holders, at some "checkpoints" at least, they would be allowed to remain in their vehicles. "Checkpoints" are frequently mentioned in their reports. They emphasize the time spent waiting and the rough treatment of Palestinians by the soldiers. For example, a member of the French Communist Party recounts that he saw Israeli soldiers violently pointing their M-16 rifles at some Palestinians going through the Qalqilya checkpoint. When he held out his own passport, the Israeli soldiers said: "Welcome to Israel," whereas he was about to enter Palestinian territory. This Communist Party official was obviously marked by this experience as he comments: "You only have to go through the 'checkpoint' at Qalqilya twice to understand why people become terrorists."[41]

Several groups organize "training courses" to prepare travellers before their departure for the psychological violence of certain situations. One of the CCIPPP organizers explains that these courses involve "being put into a situation of conflict by role playing."[42] In fact, the participants are put into stressful situations with actors playing the role of Israeli soldiers. The aim is to avoid the tense situations which earlier missions were exposed to, at "checkpoints" and elsewhere. Youssef Boussoumah recalls an incident which he witnessed during the second CCIPPP mission:

> At that time, we were present at the "checkpoints" to try and help people get through. We put a certain amount of pressure on the soldiers to try to force them to let people through, particularly the elderly. Once, I remember, we were at the Qalandia "checkpoint" going from Jerusalem to Ramallah. There is a Palestinian camp just near there. We were shouting slogans against the occupation at the soldiers in English. We saw a group of young people arriving. We spoke directly

39 This was the word used by Nahla Chahal in the above interview.

40 Ibid.

41 Interview with a Communist Party official, August 22, 2006.

42 Interview with Youssef Boussoumah, July 13, 2006.

> to the soldiers saying: "What are you doing here? You know it's against the law, this checkpoint is not legal, we demand to see your commanding officer," etc.
>
> [Bishop Jacques] Gaillot was there. Suddenly, the young women who were with us started shouting slogans in Arabic and the group of young people who were there joined in out of sympathy and started shouting too. Perhaps it was encouraging for them. But please don't go writing that the civil missions were involved in provocation! It was the second mission and we learnt a lot from it. We changed our tactics. It's true that at the time, things almost got nasty.We managed to get out of it but at one point things were very sticky. The soldiers started to get angry – the kids even angrier. One of them started to throw a stone. Things could have degenerated very quickly. So, we preferred to leave and stopped shouting slogans in Arabic.[43]

The "checkpoint" theme is also very present in the activities organized by pro-Palestinian activists in France. For example, the Coordinating Council for a Just Peace Settlement in the Middle East (the Coordination des Appels pour une Paix Juste au Proche-Orient, CAPJPO) staged some street theater with the help of volunteer actors. The play, which lasted about twelve minutes, was supposed to recreate the atmosphere at a "checkpoint." It portrayed Palestinians—young and old people, pregnant women, etc.—lined up and being mistreated by some Israeli soldiers.[44] The play was a good illustration of the way many pro-Palestinian activists see the "checkpoints," not as places marking borders nor as points which the Israelis can use to prevent terrorist attacks, but rather as places where all sorts of arbitrary acts of humiliation are regularly carried out.

Conclusion

However different they may seem, the visits to the Near East organized by "pro-Palestinian" and "pro-Israeli" groups from France have some similarities. As opposed to tourism, which tends to decline in times of crisis, the number of "activist-tours" tends to increase when the situation becomes less stable. They are not strictly speaking "humanitarian" either, since humanitarian organizations generally claim to be neutral. On the contrary, most of the "activist-travellers" are firmly committed to one or other of the opposing sides. Even activities which seem apparently neutral, like helping to pick olives, take on a strong political meaning in the context of the Israeli–Palestinian conflict. A third point which these tours have in common is the fact that they only partially show the reality of the situation to the participants. This is partly due to the short length of the tours. But another more important reason is that the organizers only choose to show certain places and only arrange meetings with certain people. Although the places visited vary according to the group, the Yad Vashem memorial is almost always visited by pro-Israeli groups. As for the pro-Palestinians, taking part in a demonstration against the Wall in Bil'in has become almost a "ritual," in the sense used by Erving Goffman (1974).[45]

---

43 Ibid.

44 Interview with Olivia Zemor on, August 1, 2006.

45 The demonstration at Bil'in is repeated every week with the same succession of sequences. First sequence: confrontation of the demonstrators with the Israeli army; second sequence: the demonstrators advance, tear gas, and then rubber bullets are fired, and sometimes, as a last resort, live

The tours we have discussed differ in the way they are organized and in their aims. The images of the Israeli-Palestinian area which are given to the participants are very different, depending on the group involved. We have tried to define the different types of tours: "fact-finding tours," "solidarity tours," and "political-spiritual tours." The division into these categories is, however, far from perfect and some common features exist between the different types of tour. This labelling must thus be considered as a fairly loose description rather than a rigid portrayal of reality.

Finally, we wish to expand on the title of the present chapter. The expression "activists without borders" ("militants sans frontières" in the original version, in French), is, of course, a reference to the NGO Médecins sans Frontières ("Doctors without Borders"), and, more widely, to the "right to intervene." The case studied shows that a powerful state such as Israel is in a position to exercise strong resistance to this "right." Many pro-Palestinian activists have experienced this first hand by being turned back at Ben Gurion Airport. The question of passing between Israel and the West Bank is quite different because the current position of the Wall—or the "security fence," to use the Israeli term—does not correspond to any internationally-recognised border. The building of the Wall is seen by pro-Palestinian actvists as a unilateral attempt by Israel to define its own borders and annex further territories. For this reason, even the use of the word "border" has been rejected. Specific actions are being carried out against the Wall, such as the "Stop the Wall" campaign.[46] As long as the line which separates the state of Israel from the future state of Palestine continues to be a border which is not accepted by the two sides and which is not internationally recognised, it will never be—if I may so put it—a "border without activists."

---

ammunition; third sequence: the wounded are evacuated and the demonstration disperses. This could make an interesting study in terms of "interaction ritual." A fourth sequence could be added: an e-mail is sent by the association "Friends of Freedom and Justice—Bil'in," with a report including the number of wounded in the demonstration.

46 See http://www.plateforme-palestine.org/article.php3?id_article=210 and http://www.stopthe wall.org/, accessed November 5, 2008.

Chapter 13
Israel to Palestine, and Back: Meeting with Post-2000 Israeli Activists against the Occupation

Karine Lamarche

Introduction

The international press and various observers of the situation in the Middle East tend to refer to those Israelis who oppose their government's politics and advocate for a dialogue with Palestinians as "pacifists."[1] Now, among the few thousands of Israelis who have been active in that field in the past years, most do not identify with this label anymore; rather they call themselves "activists against the occupation," "left-wing activists," or even "human rights activists." Besides the limited number of "anti-occupation movement" supporters (which succeeded the "peace movement" supporters) and their being subject to strong feelings of animosity in Israel, what most characterizes the various protest groups and their members is the variety in their ideological positions and in their types of actions.[2] In addition to the traditional demonstrations held in Israel and near Jewish settlements in the West Bank, which were already organized by "peace" activists in the 1980s and the 1990s, we currently witness many demonstrations—against the wall,[3] in particular—and many non-violent direct actions,[4] as well as actions of solidarity (helping in agricultural work, rebuilding houses that were destroyed by the army, etc.), and acts of civil disobedience.[5] These activities take place in the occupied territories; Israelis, Palestinians, and international volunteers take part in them together. Israeli military forces often suppress these activities.

Until recently, most Israelis who crossed the Green Line in order to participate in non-violent protests in the occupied territories remained near areas under army control and did not go into the Palestinian villages. However, over time, some activists seem to have developed a new relation to space and to its boundaries, and they have gotten used to crossing the border—where border is understood "not only as a boundary that should be crossed or circumvented but as a place that defines, establishes, or puts at stake the terms

---

1 This is also the chosen term in the literature on the topic, whether it dates back to the time of the Oslo Accords (Bar-On 1996; Kaminer 1996) or to the period preceding the onset of the Second Intifada (Hermann 2009; Pouzol 2008).

2 Some activists are in favor of a binational state, with equal rights for the Palestinians and the Jews. Others support the two-state solution with the Green Line as a border between the states. The activists also differ in terms of their position vis-à-vis Zionism: some openly oppose Zionism, while others "simply" distance themselves from official Zionistic views.

3 I shall use the word "wall" when the construction separating Israel from the Palestinian territories is made from concrete slabs (in urban areas, for example), and to designate the structure as a whole. The rest of the time, I shall use the word "barrier."

4 Such as blocking roads, moving the concrete slabs meant to restrain Palestinians' freedom of movement, or opening the barbed-wire fence with pliers.

5 Nathalie Tenenbaum defines civil disobedience as "the deliberate accomplishment of an action prohibited by a law or by an enforced regulation" (2005, p. 6). See also Hayes and Ollitrault (2012).

of one's crossing and of his or her relation with the other"[6]—and to blending in with the Palestinian environment as much as possible. Entry into Palestinian autonomous areas is forbidden to Israelis and they therefore need to take evasive action or practice dissimulation.[7] These "border-crossing activists," as we shall call them, have thus developed new forms of cooperation with their Palestinian partners.

The first part of this chapter addresses the development of protest practices that have generated a new connection to space, in a context characterized by an increasing separation between Israel and the Palestinian territories. The second part will focus on the activists themselves: what do they do *on the Palestinian side* and how do they do it? How does this modify their perception of the situation and their image of *Palestine*, of the Palestinians, and of their own society?[8] My answers to these various questions will be based on the data that I collected over 10 months in the field, between 2006 and 2010, in preparation for my PhD thesis in social sciences dealing with the Israeli activists engaged against the occupation since the Second Intifada.[9] The data collected in that period includes a great number of observations of the activists' actions in the West Bank (in Bil'in and Na'lin, in particular, and in the South Hebron Hills) and in Israel, of approximately sixty biographical interviews of Israeli activists, and of about fifteen informative interviews.[10]

A New Type of Struggle in Israel/Palestine?

Many Israelis viewed the Palestinian popular uprising that started in September 2000 as a betrayal. The left felt equally betrayed and the "doves" were all the more disappointed with the Palestinians since their hopes for peace had been great. As a result, most Labor Party voters moved to the center and some of them even joined Kadima in 2006, while Meretz became even weaker. This period also marked the end of Shalom Achshav ("Peace Now") as a mass pacifist movement. Indeed, in that period, while the organization was kept alive by taking tentative stances and by its monitoring of the settlements, it lost the capacity to mobilize the masses that had characterized it since the end of the 1970s. Very few Shalom Achshav supporters and very few leftists in general were convinced that the struggle should continue and that it must be pursued together with the Palestinians, even when the Israeli government's separation policy reached its peak. This policy has had both physical and legal implications: on the one hand, it has further divided up the West Bank with checkpoints, with by-pass roads, and with the separation wall;[11] on the other hand,

---

6 This definition was established by the team conducting the present research ("Interactions Across Borders and Activism," a team lead by Stephanie Latte-Abdallah within the framework of the MOFIP research program http://mofip.mmsh.univ-aix.fr/Pages/Default.aspx, accessed January 7, 2015).

7 Since the Oslo Accords the West Bank has been divided in three areas. Area A includes the major Palestinian cities and the Israelis are theoretically not allowed to enter.

8 I write Palestine in italics since I refer to an ad hoc category and designate the space that the activists identify as such. This therefore does not necessarily match the understanding that other stakeholders or international authorities have of that same space.

9 "De l'autre côté du mur. Ethnographie des engagements de militants israéliens contre l'occupation pendant la seconde Intifada (2000–2010)," PhD dissertation in social sciences, under the supervision of Michel Offerlé, presented on December 7, 2011, at the EHESS.

10 In this case, the Israeli activists were not interviewed about their own life path but about the movement in which they were active; other participants in the activities—Palestinians, in particular—were interviewed together with them.

11 This division started prior to the Second Intifada but the latter catalyzed it.

its laws and regulations have made the Palestinian territories separate areas.[12] While this policy of separation did not explicitly aim to suppress the contacts between activists from both sides—when the Second Intifada broke out, such contacts were already few and far between—it did result in limiting them and in subjecting them to increased monitoring.[13] Paradoxically, it is precisely at a time when the border between Israel and *Palestine* reappeared and when a very strong ethno-national antagonism resurfaced that connecting with the other side through repeated and often illegal border crossings started to be viewed in a new light.

We can identify two phases in the development of the activism against the occupation. The first phase begins a few weeks after the outbreak of the Second Intifada, in the fall of 2000. The Palestinian territories faced a humanitarian disaster. Many villages were under curfew, were repeatedly on lockdown, and were deprived of the most basic supplies while being subjected to constant incursions on the part of the Israeli army. The Ta'ayush movement saw the light in this emergency situation. Jews and Israeli Arabs united to show solidarity with the Palestinians living in occupied territories by holding demonstrations, organizing convoys to deliver food, clothes, blankets, or medicine, helping with agriculture, rebuilding houses that were demolished by the army, etc.[14] These were humanitarian actions in that they attempted to fulfill concrete needs, yet, by confronting the Israeli military forces almost every time,[15] these activities were also political to a great extent, since they aimed to condemn the violence and the iniquity generated by the occupation. Thanks to Ta'ayush, thousands of Israelis discovered the reality of the occupation and of the military control in the territories in the early 2000s.

While the movement has not completely died, the frequency of its activities and the number of its supporters have decreased in recent years; the activists themselves acknowledge that what is left of the movement today pales in comparison with what existed in the first years of the Intifada.[16] Beyond the traditional theories on demobilization and on social movements' lifespans, according to which the activists' weariness and exhaustion, the internal divisions within groups, or the emergence of new collectives play a part in a social movement's decline (Fillieule 2005), the evolution of the Ta'ayush movement may also be ascribed to changes in the dynamics between the movement's supporters and the Israeli military forces. Initially, the army prevented the convoys from reaching their destination in the Palestinian villages, thus "forcing" the Israeli activists to force their way through the border, to find alternative passages, and to confront the soldiers at the risk of being arrested. Eventually, the army chose to allow the convoys in, therefore eliminating the combative aspect of these humanitarian acts. As a consequence, the Ta'ayush movement turned away from these type of activities and engaged in actions such as demonstrations and other forms of protest that are more explicitly political. Because of this shift in activities, the movement also lost a great number of its active supporters.

---

12 Despite the argument that the Judea and Samaria area (i.e. the West Bank) is an integral part of the State of Israel.

13 Interview with Michael Sfard, an Israeli lawyer who defends activists and organizations engaged in the struggle against the occupation, April 2007, Tel-Aviv.

14 Interview with Gadi Algazy, co-founder of the Ta'ayush movement, November 2007, Tel-Aviv. For an overview of the activities of Ta'ayush in the early 2000s, refer to the testimonial book by David Shulman (2006).

15 In this vein, we can recall Gene Sharp's "political Jujitsu" to designate the use of the adversary's force in order to destabilize them (Sharp 1973, p. 657).

16 Jerusalem activists, who are active in the South Hebron Hills for the most part, retain the name "Ta'ayush."

The second phase in the development of activism against the occupation in the post-Oslo period started around 2003–2004. While the Ta'ayush movement predominated for several years, a number of micro-movements targeting very precise geographical areas and/or causes began to emerge. They were active in the villages organizing protests against the separation barrier, the South Hebron Hills, the city of Hebron itself, and even in East Jerusalem. The Ta'ayush movement's shrewd combining of politics and humanitarianism opened breaches into *Palestine*, through which the few hundreds of Israelis engaged against the occupation then sneaked, at a time when the "peace camp" in existence before Oslo disappeared for good. We should note that when the Ta'ayush supporters crossed over to the Palestinian side, they intentionally did so in plain sight, without trying to sneak in or circumvent the barrier. Entire buses full of activists would drive to the location of the activities, which had necessitated weeks of preparation and discussions with the local political authorities. 2003–2004 therefore saw the beginning of a new way of relating to Palestinian space and Palestinian individuals on the part of the Israeli activists.

It should be said that the humanitarian emergency that characterized the First Intifada years somehow gave way to a "routine" situation during the 2000s, which may account for the change in the engaged groups' repertoires and perspectives. Indeed, although the situation in the occupied territories was far from being tolerable, the lockdowns and curfews became less frequent, and so did the military incursions in the Palestinian towns and villages. Additionally, another crisis emerged in the meantime with the building of the separation wall, beginning in 2003. While a major part of the Israeli population came out in favor of this physical separation meant to reduce the number of suicide bombing attacks that had increased significantly over the years, only a minority of activists could perceive the disruption it was to cause.[17]

The movement of the Anarchists Against the Wall was formed in 2003 as a protest against the wall being built at the time. Some of its members were former Ta'ayush activists. Others came from the alternative protest scene in Tel-Aviv, which includes punks, LGBT activists, antispecists defending animal rights, etc. At the time, the village of Mash'a, whose land was annexed to the Israeli side of the barrier, became the meeting point for Palestinians, Israelis, and internationals searching for ways to fight the occupation together.[18] The first joint demonstrations, organized that same year, were repressed in bloody crackdowns, and the media began to call the Israelis participating in the joint protests the "Anarchists Against the Wall."[19] Following Mash'a, other villages like Budrus, Biddu, Beit Likkya, Bil'in, and Na'lin, organized protests against the barrier. Every week, half a dozen to a dozen "Anarchists" would join these demonstrations.[20] They also took part in direct actions in Palestinian areas (blocking roads, dismantling checkpoints, breaching openings in the barrier, etc.) and in Israel proper.

---

17 Interview with Gadi Algazy, November 2007, Tel-Aviv.

18 Those internationals belong to several groups of solidarity with the Palestinians, in particular the International Solidarity Movement (IMS), which trains volunteers coming mainly from Europe and Northern America in non-violent resistance.

19 In 2004, Gil Naamati, an Israeli who was taking part in a protest against the wall, was gravely hurt in Mash'a. The army had opened fire with real bullets. Later, in the presence of Israelis, the army most often used non-lethal munitions. In the following years, however, over twenty Palestinians were killed in other villages and many Israeli and international demonstrators were wounded, sometimes gravely, by the rubber bullets and the tear gas grenades used against them.

20 Not all the Israelis who take part in the group's activities identify with the anarchist philosophy.

If the Anarchists Against the Wall movement is without a doubt the ideal type of the new kind of engagement described in this chapter, we can identify other groups trying to create bridges with *the other side* by inconspicuously or even clandestinely crossing the Green Line and the different points separating Jews and Palestinians, by establishing collaborations, etc. We can thus mention the Combatants for Peace movement, created in 2005 by former Israeli soldiers and Palestinian freedom fighters, or the Bnei Abraham ("The Sons of Abraham"). The latter is a small group comprised mostly of people who come from religious Jerusalemite families, who organized acts of solidarity and protested together with Palestinians living in Hebron and in the South Hebron Hills.

These various groups all share the same wish to break away from the "paternalism" that long marked initiatives aimed at helping the Palestinians. Even though many Israeli activists tend to identify with one movement more than with another, they rarely belong to the movements as members.[21] By avoiding partisanship, activists retain the freedom to attend the political events that speak to them and for which they make themselves available, rather than only participating in their own movement's political activities. Finally, these groups favor "grassroots" activities and their repertoire of actions revolves around civil disobedience and non-violent direct actions. These types of protests do not necessitate a great number of participants and can be organized in a minimum amount of time with very few means. This may explain why they are favored by people whose potential resources that can be mobilized have decreased over the years.

Although it unfortunately exceeds the scope of this chapter, we should mention the connection that exists between the activists in question and the proponents of alter-globalization, who were particularly active in the first decade of the twenty-first century. For example, several members of the Anarchists Against the Wall movement were engaged in the social and political protests in Europe and North America, from which they imported specific protest capital when they returned to Israel/Palestine (Matonti and Poupeau 2004, p. 8). Similarly, when they come to take part in the struggle against occupation, the international volunteers who had joined the IMS or other groups bring with them the knowledge and skills that they have acquired in their previous experience of political socialization.

Protest Practices and Representations

In an interview conducted in 2008, Michel Warschawski, president of the Alternative Information Centre (AIC) and former Matzpen activist, points out that some Israeli activists have established a remarkably new relationship with the Palestinians and with the borders of the space in which they live. Here is an excerpt of the interview:

> Have I told you about the story of little Yossi?[22] I often tell this anecdote because it is very symptomatic … We're on our way back from a meeting; we're at the very edge of Bil'in. I am driving back from a meeting in Tel-Aviv with Yossi and, all of a sudden, Yossi tells me:
>
> "*Would you mind dropping me off at Bil'in*?"

21 Only a few groups offer a membership program.

22 An employee at the AIC and a former member of the Anarchists Against the Wall movement.

"*Where is Bil'in?*" He tells me where it is because at that time, I didn't even know where it was, and I couldn't believe it: "*Are you crazy?! In the West Bank in the middle of the night? No way!*"

"*OK, no problem, then just drop me off on the road.*"

"*You're even crazier than I thought! You want me to drop you off like that on the road in the middle of the West Bank at midnight?!*"

"*Why, what's the problem?*"

"*This is not Manhattan!*"

"*Ok, then you know what, just make a little detour through Na'lin*—that's when I started to learn these names, Na'lin, Bil'in—*and drop me off at the grocery store.*"

"*The grocery store in Na'lin at midnight? But there's a curfew starting at 5:00 p.m. and everybody goes to sleep!*"

"*No, no, my friends are expecting me.*"

And then I can hear him speak on the phone in the car, half in Arabic, half in Hebrew, half in English, with Mohammad K.—I learned his name later. That was a kind of relationship that was completely unknown to me. Completely cool, completely easy. So I tell him:

"*OK, OK, I'll drop you off at the grocery store. But this Mohammad, is he a communist? FPLP? Fatah? Hamas?*"

"*Pfff ... I have no idea!*"

"*What do you mean, you have no idea? On what basis do you cooperate?*"

"*Well, we're against the wall!*"

"*But on what basis? You have a document, right?*"

And so he looks at me like it's all Greek to him and I tell myself: "What's that stuff?" Of course, I was very curious so I followed the whole thing closely and I spared no time going to Bil'in. But I discovered a new basis. In my generation, if we didn't hold twenty meetings on a document, no Palestinian would agree to work with an Israeli and take part in political action. We had to be clean ... On the right of return, and on this, and on that ... If you didn't write "withdrawal from all occupied territories, including East Jerusalem, and the right of return" they got killed by their own community, by the ones who were in political competition with them. They were lying in wait to trip them up on that point. "Oh, so you collaborate with Jews?"—"No, but they're for the right of return, they're for this and they're for that." They needed proofs and they needed to be able to show them. Whereas in this case, everything happened like that! That's when I understood that there was ... I think that this phenomenon is not just local; it happens in cooperation, in

> action rather than in political positioning. The positioning is expressed through action. "If these people are ready to get in fights with the army and get their face smashed, they're good people. We can work with them." Period. That's a major difference with the time before the Second Intifada. For me, that's very telling about the political shift, and also about what it entails in the Israeli-Palestinian relations. In order to cooperate, the few hundred Yossis and the few hundred Mohammads don't need to spend most of their time in collective preparatory work ...
>
> What I understood in the car with Yossi was that, for him, going to Na'lin or Bil'in was not very different from going to Nazareth, which was not very different from going to north Tel-Aviv. Whereas in our case, we had a very distinct feeling, which wasn't only a feeling but was actually a reality, of crossing a border: a security border, a spatial border, a political border.

From this excerpt, we shall highlight three elements that are relevant to our study. The first one is the idea suggested by Michel Warschawski that pragmatic action—meant to express in and by itself the positioning of the movement's members—takes precedence over ideological discourse. The second one is the fact that some Israelis are ready to expose themselves to the physical and legal risks of military repression, a fact which matters significantly when they establish new ways of collaborating with the Palestinians. While the Israeli activists do not run the same risks as the Palestinians, the former pay dearly for participating in protests. In so doing, they distance themselves from the position that long characterized the Israeli leftists, who were quite enthusiastic about speaking of coexistence in Tel Aviv or in Jerusalem, yet who were much less prompt in participating physically in the opposition together with the Palestinians. This in turn raised the Palestinians' suspicion toward what they call "normalization" initiatives.[23] Finally, the third element underlined by Michel Warschawski is the ease with which today's activists—the "Anarchists," in particular—go to the Palestinian villages, which is indicative of their uninhibited relation to borders and territorial boundaries. During my numerous field observations, it became quite clear that these activists were familiar with the territory in which they travelled; they did not hesitate to travel on Palestinian roads and to use Palestinian transportation in order to reach distant villages or even to go to cities located in Area A, into which they are theoretically forbidden entry.

These activists also distinguish themselves in the way they mentally partition the territory and in the way this partitioning affects their perceptions. The very act of crossing over into Palestine and of being there have become challenges in and of themselves, while the type of relations the activists establish with the Palestinian population and the Palestinian space is at the core of their engagement. Indeed, while the "peace" activists sometimes went *into the occupied territories* but never *into Palestine*, the activists described in this chapter travel to areas that are far from Israeli military control. To reach these areas, they travel on Palestinians roads, at times using their public transportation, and build relationships with Palestinians; sometimes these blossom into genuine friendships. The activists learn to play

---

23 Normalization (*tadbiyeh* in Arabic) is understood here as cooperating with Israeli institutions, businesses, organizations, or individuals, thus creating an impression of normality despite the Israeli occupation.

cat and mouse not only with their own country's military forces but also with the Palestinian police, which does not necessarily approve of their presence in Palestinian towns.[24]

This particular type of engagement requires the activists to acquire a specific kind of knowledge, a certain set of skills, and an array of meanings and significations in order to create a specific protest habitus. While referring to the distinction made by Anthony Giddens, Karel Yon explains that "the notion of protest habitus advantageously includes what constitutes the … activists' practical consciousness, what is not conceived in their behaviors, their automatisms in their socialization into the institution, and their discursive consciousness understood as their ability to justify and to express their acts and their objectives thanks to a partisan culture" (Yon 2005, p. 142). This notion helps us take into account the fact that these activists cannot be defined only by "a series of social coordinates" but also by the variety of their engagement in specific and heterogeneous social contexts (Lahire 1998).

Some of these activists were politically socialized into the Ta'ayush movement and started going over to the Palestinian side during the large-scale activities that the movement organized. It is only later, after having established contacts with certain Palestinians living in the territories and after being invited to come back alone to visit them, that these activists experienced their crossing over to the other side in a new way, as they learned to find their way on a map, to circumvent the checkpoints, to use local transports, etc. Others started to come to Palestinian villages when they joined the protests against the wall. During certain periods, private cars and minivans could drop them off on site, but at other times, these activists had to walk several miles, take Palestinian taxis, or even get to the site of the demonstration the day before and spend the night there because the army blocked access to the site. The activists often recount these experiences when they describe the process by which they became more and more comfortable *in Palestine*. They also often mention the experiences that formed their strong feeling of antagonism towards the settlers, thus putting into perspective the feeling of antagonism they had experienced towards the Palestinians at one time. Alon, an Israeli in his thirties, who is active in the South Hebron Hills, thus told me:

> In the beginning, it was something very spiritual to meet Palestinians because you know, before, I was afraid of them. I had met Palestinians only in the army. I had never been in Palestine. And so I had to work on my ingrained fear of the Arab people … It was a long process and when I finally traveled to Palestine for the first time, everything went well. I was like: "Whoa, it's so beautiful" and also talking with the people: "Salam Aleikoum—Aleikoum Salam" … it was another culture … there was the feeling of being abroad, which is nice. I like that because it was really a foreign country for me. [Q: And now, do you still feel like you're in a foreign country?] Today? Less so in the villages, because it was already like … I feel at home. … You know, the second or third time I went to the South Hebron Hills, I told my girlfriend: "If I was dropped off here in the middle of the night and I had to find refuge somewhere, I would never go to a settlement. I would go to a Palestinian village, even if I don't know anyone there."

The moment when activists find themselves in what they identify as *Palestine* often plays a major role. Demonstrating in Tel-Aviv or monitoring a checkpoint at the entrance of Jerusalem is one thing, but finding oneself in a Palestinian village that was taken over by

---

24 Israeli activists, which I once accompanied to Ramallah following a demonstration against the barrier, were thus arrested by the Palestinian police and taken to the police station, where they underwent a short interrogation and were admonished at length before being taken back to Area B.

the army, and running away to avoid the soldiers' bullets is another thing. At that moment, the activists feel that whatever is left of the binary opposition between *us* (Israelis) and *them* (Palestinians) crumbles. The identification process does not follow the criteria that existed beforehand. Danger changes sides. The menace is not the Palestinian anymore but the soldier who blocks the road, who sprays tear gas on the crowd and arrests demonstrators; or it is the settler with his gun over the shoulder. Through these experiences, the Israeli activists thus undergo a transformation: there is often a *before* and an *after* in their representations and interpretations of themselves and of the world around them.

Finally, it is not unusual for the Israeli activists to establish ties with the Palestinians of the villages where they most often visit, ties that go beyond simple protest cooperation. They often come for a visit, regularly share their meals after the activities, frequently call them on the phone, and sometimes take care of finding them food, medicine, or various appliances that the Palestinians do not have access to or cannot purchase because they are too expensive; the activists then become sorts of "solidarity smugglers." Every Israeli who wants to establish contacts with the other side sees learning Arabic—even if he or she rarely learns more than the basics—as an essential milestone, not so much because they need the language to communicate with the Palestinians (indeed, many of them speak Hebrew perfectly) but because they wish to break away from the relationship that is considered "colonialist" between the Israelis and the people whose land they occupy. Adopting the Palestinian political culture—through symbols such as Handala,[25] the Palestinian flag, slogans in Arabic, or references to the Nakbah—can also be seen as another milestone in the process by which some activists blend more and more in the Palestinian landscape, both physically and cognitively.

One of the most striking consequences of this process is the activists' socialization to a feeling of adversity, which often means that they distance themselves from the other Israelis and even confront them, while continuing to see the Palestinians as belonging to another world. Thus, even if the Israeli activists become physically and mentally closer to the Palestinians, the former are well aware of the distance between themselves and the latter, and they know that such distance cannot be shortened by their common opposition to the occupation. There are indeed many social and ideological differences that separate the Israeli activists and their partners in the struggle. Beyond the classic religious difference between Muslims and Jews, the Palestinians with whom the Israelis cooperate today are for the most part villagers attached to rural traditions and to a strong gender specific division of labor.[26] The activists have held many debates about the Palestinians' approach, which is often viewed as sexist. Indeed, the Palestinians sometimes tried to prevent women from participating in certain activities;[27] and they have succeeded in getting the women to come dressed in modest clothes.[28] The treatment of animals is another point of contention

25 This young cartoon character created by Palestinian cartoonist Naji al-Ali has become an iconic symbol of Palestinian resistance. Anti-occupation activists often wear Handala figurines as pendants.

26 In the past, the few instances of protest cooperation involved Palestinians of the urban upper class. As for the Israeli activists, they often belong—like their peace-supporting predecessors—to the non-religious Ashkenazi elite; they pursued academic studies and lead relatively privileged lives.

27 Indeed, in certain villages, the Palestinians asked the Israelis "not to bring women" to demonstrations or other forms of protest. The Israelis refused to abide and declared that if the women could not come, the men would not come either.

28 Today, the participants in the demonstrations receive e-mail messages inviting them to "take into account the cultural differences in terms of clothes." The women in particular are expected not to

between Israeli activists and the Palestinians of the territories. While many Israeli activists are vegetarians or vegan and proponents of anti-speciesism, the consumption of meat and "animal exploitation" are part of the Palestinians' daily lives.[29] In the villages where demonstrations are held, beef heads and carcasses suspended from hooks are not a rare sight. Elsewhere, in the most distant camps in South Hebron, donkeys and camels are treated in a way that is likely to offend the supporters of animal liberation.

The cultural differences between Palestinians and Israelis do not affect social relations alone. One side's behavior or ways of relating to others are sometimes misunderstood by the other side. The term *chutzpa* in Hebrew designates a certain roughness combined with impudence that is considered by many as a typical Israeli trait. *Chutzpa* is in sharp contrast with the Palestinian codes of courtesy. Dafna was a Ta'ayush activist for many years and she has maintained a privileged relationship with the Palestinians living in the small town of Nu'aman, where she often goes alone. She explains:

> There is also something I both like and don't like: it's the fact that the Palestinians never say "No," even if that's what they think. You always have to think about what you've just been told because it doesn't always correspond to what was said, grammatically speaking … And being forward is precisely an Israeli characteristic: what we say is what we think! Now that's a real gap … Sometimes, I was ashamed of myself because I said things too directly and that made the Palestinians feel uncomfortable. There are questions that you should never ask, for instance.

Finally, although they are against the occupation and are in favor of the creation of a Palestinian state or of a binational state, the Israeli activists do not necessarily see Palestinian nationalism in more positive terms than Zionism.[30] Indeed, while international volunteers readily chant "Palestine will live, Palestine will win" or "Free Free Palestine" during protests, and while they wave the Pan-Arab colored flag and wear the traditional kaffiyeh or other visible symbols of their support to their hosts, few Israeli activists take on such signs and behaviors and rather adopt the ones marking their opposition to the occupation, to the settlers, and to the army. Doron is not an anarchist but he regularly takes part in the protests in Bil'in. He makes that last point clear:

> I never wave the Palestinian flag. I never chant Palestinian nationalist slogans. I am always very careful about that. I sometimes say: "Leave Palestine" or stuff like that but I don't like it so much and I always make sure I don't chant things that would turn me into a Palestinian nationalist because I don't like nationalism, period.

These words reflect the position of the majority of the activists interviewed: however supportive of the struggle for Palestinian independence they are, they do not intend to turn into supporters of any kind of nationalism. For that matter, while some anti-occupation Israelis publically call themselves "Palestinians of Jewish origin," "Palestinian Jews," or even "residents of occupied Palestine," thus showing their wish to question the official

---

wear tops exposing their shoulders or their upper chest.

29 Even though very few of them actually eat meat on a daily basis, meat dishes being served on special occasions.

30 Some of them, who are anarchists, would rather be living in a Middle East without states, but that does not prevent them from supporting the Palestinian claim to statehood.

labels that symbolically designate their place in the Israeli-Palestinian space, they usually do so when they are abroad, not when they are *in Palestine*, where this inversion would seem ludicrous. The Israeli activists in question, far from blending in the Palestinian landscape and taking on its codes and its values, are generally acutely aware of their estrangement from that world.

At the same time, the more these activists grow closer to Palestinians, the more they feel that they belong to a minority in their own country, a feeling that hampers their interaction with their friends and family who do not share their opinion, or even with people they run into on the street. Tal, a woman in her sixties, grew up in a very nationalistic family. It is only at the end of the 1990s that she "opened her eyes" to the political situation, as she says, and that she joined various protest groups: Profil Hadash,[31] at first, then Ta'ayush, and Machsom Watch.[32] She explains the difficulty she had in keeping her past social relations:

> Today, I feel very uncomfortable outside of the [activist] environment. I can't bear anymore talking with people who don't think like us. It hurts me. And I don't have the patience to listen to what they have to say on the situation. ... Before, I was sometimes invited at friends on Friday evenings, to talk over a good meal, but we always ended up talking about politics ... And then, all their racism resurfaced, and I had to bite my tongue the whole evening to keep from saying what I deeply thought. And that's something I did all my life. I can't take it anymore. I'm not going to change people. If I say what I really think, it will just make them angry ... And I don't go to people to make them angry, so there is no more room for me there.

In the case of Matan, who had received a bullet in the eye a few weeks before our interview and who still did not know whether he would recover his eyesight, the feeling of estrangement from the other Israelis is such that he sees in each of them his potential aggressor:

> When I'm walking on the street and I see young people in their twenties, I tell myself: "Maybe that's the soldier who shot me!" ... So either you're strong enough to say: "OK, you shot me an hour and a half ago but now, it's OK, I can live with you," either you somehow separate from society.

These words seem revealing of one of the most significant characteristics of the engagement of the activists in question: such engagement is not strictly political; it rubs off on all social activities, on all relationships with others, on all representations, on all ways of being and thinking. This engagement thus contributes to the creation of a habitus whose specificity partly lies in the fact that it falls in between two social worlds that are foreign to each other. On the one side, there is the Israeli far-left, with activists who have many cultural, economic, and social resources; on the other side, there is rural Palestine, where traditions prevail, and where the ones who take part in the protests have fewer resources.

Each visit to the Palestinian side leaves marks that are interiorized and that, over time, generate dispositions that prompt the activists to act, speak, feel, and think differently, such

31 "New profile" in Hebrew. This movement condemns the militarization of Israeli society and supports conscientious objectors.

32 *Machsom* means checkpoint in Hebrew. Members of this movement, who all are women, have been monitoring checkpoints in the West Bank and in East Jerusalem since the start of the Second Intifada.

dispositions differing from the ones they had initially acquired during their socialization process. They are confronted with situations—such as circumventing military roadblocks, using Palestinian transportation, feeling safe in areas that are far from military control, or sharing moments in the lives of families living under occupation—that define and reinforce tangible perspectives and mental boundaries, which, in turn, guide the activists' understanding of the world. In this kind of engagement, the passage to the other side in and of itself becomes the statement of an ideological position while the process by which spaces and situations that were considered dangerous become familiar usually comes with radically calling into question the various narratives through which the feeling of belonging to the national Israeli-Jewish community is expressed.

Conclusion

The scope of this chapter prevented me from analyzing in more depth the various life stories that prompted Israelis to become "border-crossing" activists; at any rate, we have seen that the situations experienced by the activists in the process contributed in defining and/or reinforcing their view of the world.[33] As they get to know *the other side* and those that were long presented as the enemies, they draw new frontiers and gradually move away from their own tribe: the Israeli Ashkenazi Jews. Indeed, they are pushed to call into question their former loyalty to Zionism and their understanding of the occupation; often they feel that they no longer belong to the society in which they grew up. This is one of the fundamental differences between the engagement of the activists described in this chapter and the engagement of the "pacifists" who, in the 1980s and 1990s, demonstrated in Tel-Aviv and Jerusalem to put an end to the war in Lebanon and to the settlement initiatives in the West Bank. While being both a left-wing activist and a committed Zionist was then possible (and even the norm), today these two labels seem to have become almost contradictory. It is true that many Israelis engaged in the struggle against the occupation claim a strong attachment to their country, thus demonstrating a kind of patriotism that some would call Zionism. Yet, their attachment moves away from the Israeli ethos, characterized by a Jewish-centered and militaristic Zionism. There are thus few Israeli activists today, who stand up for the idea that their State be Jewish and democratic, that the army be a moral occupation force, and that the separation be a necessary solution to "divorce" the Palestinians. For that matter, while a great majority of the "pacifists" would loyally serve in the army and in reserves, even if it meant that they would "shoot and cry," most of the Israeli activists against the occupation today find a way to evade military service or become conscientious objectors.[34]

Moreover, while yesterday's pacifists enjoyed a certain audience in Israeli society, a very clear split now divides the activists and the rest of the Jewish population, who accuses the former of putting Israel's security in jeopardy and of stirring up anti-Semitism by spreading antipatriotic messages abroad. Avigdor Lieberman, founder of the extreme-right party Israel Beiteinu ("Israel, our home"), has left his mark on people's mind with his "no loyalty, no citizenship" motto. Several laws proposed by Israel Beiteinu and the Likud were passed recently; laws that condemn foreign countries' calls to boycott Israel or products

33 For further reading on this topic, see Lamarche (2013).

34 This phrase was used to refer to the "Peace Now" supporters who would loyally serve in reserves and would then demonstrate against the occupation.

from the occupied territories, and laws that sanction those who promote conscientious objection and those using the term "Nakba."

Thus the process by which activists cross over to *the other side* is fueled at the same time by their passages to and from *Palestine* and their discovery of "another world", but also by the reactions prompted at the institutional and social levels by their getting away from the Israeli predominant practices and views on the separation. By being branded *outsiders* by their fellow-citizens and by bearing the burden of transgression, these activists are all the more inclined to adopt the alternative narrative, discursive, and cognitive schemas they have access to through their engagement.

Chapter 14
Bodily Relief: Some Observations on Martyrdom Operations in Palestine

Esmail Nashif

Introduction

Eventually everyone will die, but the manner of death is not usually optional, and depends largely on the sociohistorical configuration that determines the individual's way of life. Moreover, death and the ways in which it occurs are basic elements in the (re)production apparatuses of any given society (Kellehear 2007; Durkheim 2002 [1897]). Viewing it from this perspective, we may say that exploring the possible manners of dying in a certain society can reveal some of the major aspects of that society's ways of making life possible. In this regard, the aspects that might be revealed by such an exploration would vary from one society to another, and in the same society at different historical junctions. Here I want to address a specific manner of dying in a specific society at a certain moment of its history. The manner is that of martyrdom operations, and the society is the Palestinian one at its late colonial junction.[1]

The modern history of the manners of dying in Palestine is intrinsically related to its situated (de)colonization processes. While the Palestinian society as a whole is subordinated to the colonial regime, each of its segments interacts differently with this regime.[2] Our main focus here is the segment that resides in the West Bank and the Gaza Strip, and the transformative processes it has been experiencing since the Oslo agreements, especially during the heydays of the Second Intifada, 2000–2007.[3] During that period it appears that martyrdom operations were among the main experiences of death in this segment. The centrality of the experience was not determined by how many Palestinians did, in fact, carry out a martyrdom operation. Rather, the deep impact of this manner of dying has to do, I will argue, with the redefinition and reorganizational practices and processes that it entailed, and probably still does. In the last two decades, the Palestinians living in these areas and their transformative experiences have been studied extensively. But the manners of their dying, in spite of being a salient feature of their daily life experiences, have not received much attention.[4]

---

1 Probably more than any other phenomenon related to Palestine, the martyrdom operation has generated a corpus of confused and contradictory academic literature. For examples of different positions and research styles on the topic of martyrdom operation, see: Abufarha (2009), Asad (2007), Pedahzur (2006), Bloom (2005), Gambetta (2005), Pape (2005), Reuter (2004), and Hage (2003).

2 For an overview of the different segments of the Palestinian society see Farsoun and Aruri (2006).

3 Mbembe (2003) provides a good introduction to the era, topic, and segment of the Palestinian society on which this chapter is focused.

4 One of the few attempts to conceptualize death in Palestine is Ghanim's (2008).

One could chart these manners of dying and trace the points at which they emerged. For example, the shattering events of the War of 1948 instituted a certain manner of dying, namely, massacres, which seems to be still prevalent although it is repeatedly dislocated to other geographies, temporalities, and segments of Palestine and Palestinian society (Pappe 2007). Seen from the historical perspective, the martyrdom operation is relatively recent in the chronicles of Palestinian society, particularly that segment of it that resides in the West Bank and will be the focus of our discussion here. In the literature on the subject, its emergence is usually dated to 1994 (Asad 2007, pp. 39–64; Ricolfi 2005, pp. 77–129). This date was chosen because in that year the actual performance of the martyrdom operation was observed by various researchers, politicians and the media. Needless to say, such a complex and multi-staged operation could not have been carried out instantaneously, and this manner of dating is, to say the least, problematic. Martyrdom operations could have begun only as part of larger transformative processes that engulfed the colonized populations in Palestine in general, and the West Bank and Gaza in particular. Hence, any attempt at understanding them must start by re-situating them in these sociohistorical processes of transformation and the different ways through which they were localized and practiced by different real living social Palestinian subjects.[5] The overwhelming and total experience of martyrdom operation experience for the involved Palestinian subjects and communities have direct implications on how we could approach it analytically. Palestinians, collectively as well as individually, are acutely aware of their history, and in that sense they experience the martyrdom operation as a continuation and as a break in their history of armed resistance. The manner of localizing and experiencing these martyrdom operations as a continuation and as a break with history could vary largely according to specific contexts and individual Palestinians. Notwithstanding these localizations and individual experiences, there are formal processes of colonization that are immanent in the colonial regime, and which determine the rhythms of and intensities of resistance, martyrdom operations included. These interrelated threads compose and stage martyrdom operations as we experience it. For these reason, I will suggest three vantage points for looking at these transformations and localizations that generate the martyrdom operations: historical, ethnographic, and structural.

These three vantage points will enable us to extricate the horizontal, individual, and vertical layers of the martyrdom operations from their composite orders. Horizontally, the martyrdom operations will be positioned as a form of armed struggle that both emerges from and contains previous ones, namely Palestinian forms of armed struggle. Previously, this type of armed resistance was mainly practiced by PLO factions against Israel (Sayigh 1997). The focus will be on the transition from using external tools of resistance to using the individual body itself as a tool and an arena of resistance. Individually, in contrast with the previous type of the political affiliation of the PLO's armed struggler, the martyrdom operator is mainly affiliated with local tightly-knit social and familial communities (Abufarha 2009; Ophir et al. 2009). Vertically, we will approach it simultaneously as an event in itself and as a symptom indicating deeper sociohistorical structuration processes. These are two interrelated levels of analysis and interpretation. As an event the martyrdom operation could be seen as a structural variation on the body-space relations. As a symptom

5 On the interrelation between the general sociohistorical processes and the ways in which they are localized and practiced, see, for example, Mignolo (2000) and Williams (1977).

it is an expression of the impossibility of re-formulating the interrelations of the Palestinian national bodies with the Palestinian national spaces of the late colonial context.[6]

These three sites of exploration, i.e., the historical, the ethnographic, and the structural, could be seen as different constitutive moments of martyrdom operations. The conducting of such an operation, though, integrates these moments into redefining and reorganizational practices and processes for the Palestinian subjects and communities. Hence, one of the main arguments that I will raise later is the notion that the martyrdom operation is, among other things, a strategy of forming anew the Palestinian subjectivities and communal relations beyond the separate localities of neighborhoods, villages, and segments—an option rendered almost impossible in the current colonial design of things.

In the current moment of the colonial regime in Palestine it is impossible to (re) build the independent Palestinian national community. This impossibility is actualized through the systematic practices of the Israeli regime. These practices are coupled with the positionalities, perceptions and alliances practiced by the Palestinian traditional national elites. This coupling and alliance between the Israeli regime and the Palestinian national one resulted in the Oslo Accords signed in 1993. Theses accords did not resolve the colonial contradictions that regulate Palestinian daily realities,[7] but rather displaced the colonial conditions to unprecedented scales of (un)systematic violence. These displacements brought major shifts in the ways in which Palestinians perceive their national collectivities. Mainly, these shifts brought back into focus the centrality of the body and the space interrelations in the conflict between the Palestinians and the Israeli regime (Bishara 2002, p. 117). The argument that we will explore here is that the implications of reordering the body-space relations in the West Bank and Gaza created a moment of opening in the previous order that regulated bodies, spaces and their mutual networks of movement. Palestinian subjects did use this opening in various ways in order to relocate their positions in the colonial regime. Still these ways of relocations were conditions on the contexts, histories of localities, and structural relations of dominations and counter-domination in and through which those subjects operated. One of these relocations is the redefinition and reorganization of the resistance codes of behavior, its subjects, and its tools. It seems, as I will try to show later on, that the body-space axis of communal-subjective relations is the major site through which the "new order" has settled. The context of martyrdom operations, the operators, and the communal infrastructures for conducting it, are part of the "new order" of the colonial setting in Palestine. Moreover, I will try to fathom how martyrdom operation as a manner of dying was not only a choice among other manners of dying but it came to predominate, momentarily at least, at this specific juncture of reordering.

The martyrdom operation is seen here, then, as a way of forming anew the Palestinian national collectivities. The main research question for us in this chapter becomes to explicate the real social processes that served this manner of formation. The history of the Palestinian armed struggle, the ethnographic moment of a Palestinian city, and structural analyses are the stations used in this chapter to cope with the research question. The insights achieved by reconnecting these stations can shed light on how the manners of dying in Palestine are indicative of the ways Palestinians live their colonial predicament at this juncture of their tragic history.

---

6 On the reordering of the spatiotemporalities of colonial Palestine since Oslo, see Weizman (2007) and Ophir et al. (2009).

7 See Said (2001).

Historical Aspects of the Re-collecting of the Body

The reemergence of the Palestinian armed struggle in the 1960s took a specific collectively organized form of intervention in the realities of the colonial condition. The formative processes were not restricted to a certain layer, segment, or domain of Palestinian society. They started at different focal points, only to be geared up to the whole of that society. Clearly, not everyone was engaged in the same manner and/or directly in armed struggle, but the main trope and subject of discourse centered on "armed struggle" and "resistance" (Sayigh 1997). What interests us here is the uncovering of that formative moment and its main characteristics.[8] These lasting formations, I will contend, became one of the major layers that made the martyrdom operation, as a particular form of resistance, an unarticulated possibility of action among the new Palestinian national collectivities.

Since the late 1960s, then, the reorganization of this national collectivity meant, among other things, the building of a new "individual" as a historical subject. The new Palestinian rejected the existing order of defeat, exile, and of being a refugee and endeavored actively to achieve the ultimate aim of liberating Palestine.[9] The main strategy for reaching this ultimate aim was armed struggle. While the political and institutional dimensions of armed struggle have been partly researched, the socioeconomic and cultural dimensions have hardly been addressed (Farsoun and Aruri 2006). The argument that a major segment of Palestinian society started to reproduce itself mainly via "resistance," with the armed struggle at its core, can be fully understood only if we closely inspect the interaction between the socioeconomic and the cultural spheres. Here I will try to sketch the basic characteristics of this vital axis of the then new Palestinian collectivity.

The interfaces between the socioeconomic and the cultural spheres in the Palestinian context are determined by three main poles and their histories: the body, the tools of liberation, namely political organizations, and a certain way of perceiving historical realities. If the new order revolved mainly around instituting the old-new sociopolitical Palestinian goals—liberation, return, and statehood—then these poles were reorganized in a specific manner. The two previous decades of the 1950s and 1960s, as a transition period, were characterized by fragmentation and dispersion. In contrast, the emerging order was instituted through the apparatuses of centralization of the PLO, such as print media education, and military training systems. These concrete apparatuses relinked the body and the tools via the perception that the national collectivity was the historical subject. Let us look closely at these formative moments.

The centrality of the body as a fundamental arena in Palestine's colonial condition stems from the ways in which Palestinians perceive and experience the colonial settler regime of the Zionist movement (Peteet 1994, Shalhoub-Kevorkian 2009). The major outcome of the War of 1948, was the dismantling of the sociomaterial infrastructure of Palestinian society. The different, but acute repetition, of the tragic war of 1967, resulted in the further destruction of the sociomaterial infrastructures in the West Bank and the Gaza Strip (Roy 1995). After losing these infrastructures, the majority of the Palestinians were left, simply with their individual and social bodies as almost the sole means of (re)production owned by them. The sociopolitical reorganization of these means of (re)production as the basis for the reemerging Palestinian national collectivities was the main galvanizing and

8 See Sayigh (1979) for detailed descriptions of personal narratives of subjects involved in these processes.

9 See, for example, Sayigh (1979) and Kanafani (1969).

contested site between the different Palestinian, Arab, and international political agents. As can be observed in the various expressive media of the Palestinian national culture(s), in painting, literature, and poetry for example, the body is presented time and again as both a site of contradictions and a horizon of resolutions for the colonial conflict (Nashif 2005). As the last arena owned by the Palestinians, the individual and the social bodies carried their collectivity. At the same time these same bodies carried the impossibility of realizing such a collectivity, e.g., refugeehood, assassinations, martyrdom, oppression. Despite these contradictory positions, and maybe because of them, the body became instrumental in the process of trying to merge the Palestinians into a national homogeneous collectivity. It constituted the material as well as the symbolic bases of the national tools of liberation, namely, political organization with armed struggle as its main course of action. For this the body in the Palestinian context became a sort of over-determined, à la Althusser, arena that regenerates itself at different contexts of this colonial situation. Hence, it is not a coincidence that the body still functions as the last refuge in the recurrent moments of acute collective crises (Kanaaneh 2002).

The history of the Palestinian national tools of liberation in the form of political organizations can be dated back to the second decade of the twentieth century (al Sharif 1995). Since the mid-1960s, the crucial and unique aspect in the Palestinian case has been the moment of rebuilding these tools of organized political activities as processes of institutionalizing the Palestinian body/bodies. To put it simply, this means that the differentiated spheres of the social collectivity were gathered under the umbrella of the political sphere due to the specific nature of the colonial condition in Palestine (Nashif 2008a, p. 196). In a sense, the massive loss of the land's body was replaced by the political body as a substitute carrying the diverse possible Palestinian bodies. The interrelations between the political substitute and the social and individual bodies are mainly characterized as being a nurturing-disciplining type of dynamic structuration for both sides. Being nested in such a bundle of interrelations generates a certain mode of understanding and intervening in history.

It seems that the totality of experiences of the wars of 1948 and 1967 resulted in homogenizing effects on the diverse Palestinian ways of perceiving history. The Palestinian as a historical subject/agent was not a new invention that appeared in the wake of the war of 1967. Rather, this turning point created a historical momentum into which different subjectivities, familial, religious, regional, professional, and others, were forced to merge into one over-arching manner of perceiving historical realities, the national one (Kimmerling and Migdal 1994, pp. 159–275).This perception has grown and found expression in collective codes of "how can we liberate Palestine." These discursive formations still reign today with minor changes (Khalili 2009, p. 113).

The interrelations between the individual and the social bodies, the tools of liberation e.g. political organizations, and the national historical agency generated what we might call a Palestinian political economy of liberation. In many respects, the PLO exemplified this political economy of liberation, but was not restricted to it. The political economy of liberation is the recurrent pattern in which Palestinian subjects mobilize the tools of liberation in order to organize and use the bodies in a systematized manner, as implied by and derived from the dominant national historical perception. Seemingly, this principle operates as follows: the amount of the body that you lend/release to the tool/organization depends on the quality of your national consciousness, as an individual subject. But the amount of the body invested in and through the tool/organization is historically determined, which brings us back to the dialectics of body-mind and their bridging via institutionalization. The Palestinian variety of this could be formulated as follows: The greater the loss of the body at its different levels, the better the quality of the mind for commitment to liberation. As for the bridge, namely the institution, although it still functions as a nest/container for the body-mind dyad, it simultaneously acts as the arena of contestation, negotiation, inclusions/exclusion, and merging for individual as well as collective agents. What is at stake here is the owning of the bodies as the means of production, and the material and symbolic surplus values that are gained by putting these bodies to work. The capital—material as well as symbolic—thus accumulated is controlled by the leaders and their circles in the upper echelons of the political organizations; and in many respects it is personified as standing on the verge of private ownership. Thus, the centrality of the body in reproducing the collectivities of the Palestinians is restored by being part of these larger processes of the triad of body-tool-agency. This triad is a chain of production, which is the skeleton of the new Palestinian collectivity that matured and came to dominate most of the segments, each with its own variety of the triad, of Palestinian society after 1967. On first sight, it seems as if these processes were slowed down due to the direct occupation of the West Bank and the Gaza Strip. But in spite of, and at times because of, the practices of occupation, the processes of instituting the nationally-based triad of body-tool-agency never stopped. As is well documented by now, the First Intifada in 1987 was the culmination of these processes (Nassar and Heacock 1990).

In the context of the Palestinian armed struggle we see that the body of the armed struggler is the carrier of the tools of liberation. At the same time the body of the martyr is sublimated as a manifestation of the ultimate realization of the triad of the political economy of liberation (Pitcher 1998). That is to say, the total investment of the body via the organization is the restaging of the Palestinian subject in history. Hence, the main argument in our context is that since the mid 1960s this triad of political economy of liberation has constituted the pre-dispositional structuration mode for martyrdom operations. Three decades later, it became a concrete historical praxis and a political intervention in the late colonial condition in Palestine.

Distracted Ethnography

One of the functions of a nation-state apparatus is to work on the maintenance of the daily taken-for-granted spatio-temporal continuities and activities, and above all to maintain a deep trust in them on the part of the masses (Anderson 1983, Althusser 1971). Colonial apparatuses, by extension and contrast, are concerned mainly with repeatedly disrupting the realities upon which the colonized stand in order to generate a collective sense of deep mistrust in continuities and activities (Young 1995). Martyrdom operations collapse this dichotomy, and in many respects render the two poles of coloniality equal. Most of the mainstream academic literature describes the colonizers' horror when a martyrdom operation is carried out and ignores the horror of the colonized which is generated by the same act (Asad 2007, p. 65). It was the experience of the deep shattering mistrust in the social realities that brought me to delve into the nature of martyrdom operations.

Since their staging in 1994, I have witnessed and experienced martyrdom operations in colonial Palestine at many levels and in different forms of engagement. The descriptions cited here refer to the period between 2000 and 2007, during which I lived and worked in Ramallah. My work consisted basically of conducting ethnographic practices, research, and teaching at Bir Zeit University. The focus of my research and ethnographic activities did not start with military operations and their spatial dimensions, but was on the phenomenon of political imprisonment. Any military operation, Palestinian, or Israeli, would redefine the real and imagined ethnographic spatiality and temporality. The Palestinian martyrdom operations and the Israeli targeted assassinations with massive air-to-earth missiles were probably the two extremes beyond which the ideas and practices of expected/acceptable spatio-temporal grids would literally melt into thin air. Thus, without any premeditated intention, I started to rethink the basic premises underlying my research on political prisoners and their dynamic communal processes.

There are many similarities between the collective act of the hunger strike in prison and the martyrdom operation. Both practices use the body, the individual and the collective ones, as their main instrument of intervention. Both of them work by blinding the dialectic between colonizer and colonized bondage relations, eliminating the colonizer's ability to control the colonized.[10] This is achieved by destroying the possessions (the body) of the colonized which he controls, and hence blinding the colonizer. Moreover, both of these practices occur in the context of intense confrontation resulting from the collapse of the previous equilibrium in the colonial condition. These shared characteristics can lead us to rethink the basic premises of how to build a community on national grounds, among other kinds of collectivities. But the differences between these two resistive practices and their collective implications are no less telling. Martyrdom operations leave no room for maneuver; if they succeed they lead to an end, at least literally for the martyr. In a hunger strike there is a wide margin for negotiation, managing the conflictive wills and achieving some aims. But, in fact, these two metaphors/strategies of community building and shapes of subjectivity, more than they contradict each other, imply one another in their dialectic movement.[11]

In this way, interweaving the practical and the analytical, I became engaged in investigating the martyrdom operations. The gradual process of engagement led to the growth of a sensibility of horror on my side as a researcher. There are certain skills and tactics of ethnographic work that one probably acquires only by conducting ethnography.[12] And such is the context of ethnography in times of war, in this case the Palestinian Second Intifada.[13] The Palestinians have a commonsense wisdom on how to conduct a confrontation with the occupation forces and representatives. A common sense is a collective process of encoding/decoding directives of behavior and feelings that is carried out individually and collectively during engagements with the occupiers. These processes that generate the codes of conduct break down in times of intense crises, only to be restored with new codes of conduct accumulated during the latest crisis.[14] The period between two such events, namely the break and the restoration of commonsense wisdom of confrontation, is what interests us here for two reasons. First, the martyrdom operation is currently the ultimate break of common sense, which means that most of the current codes of conduct on a daily basis become irrelevant after such an operation. Second, my ethnographic research was

---

10 For a comparison of these practices with the Irish political imprisonment experiences, see Feldman (1991).

11 To compare with the case of the Irish hunger strike, see Beresford (1987). And for comparison with the militarization of the body in rebuilding communities see the Japanese case, Ohnuki-Tierney (2002).

12 Compare this experience with what C. Nordstrom calls "ethnography of a warzone." See Nordstrom (1997).

13 For comparison with other researchers doing ethnographies during the Second Palestinian Intifada, see Allen (2006, 2008) and Hammami (2005).

14 For examples of such dynamics see Rosenfeld (2004).

conducted during such a transitory period, in which the Palestinian codes of conduct were crashed time and again for longer periods of time, roughly between late 2000 and early 2007.

During the transitory period I learned certain techniques of how not to be horrified by the growing expressions of military violence toward me as an individual Palestinian and as part of the targeted collective of Palestinians. The exceptional context of martyrdom operation caused most Palestinians to act differently: alert, suspicious, mistrusting, and above all horrified. The horror was not based on the rational calculation of how the colonizers would react. I argue that it has to do with the processes of re-building the Palestinian collectivities themselves on national grounds. Moreover, the horror is related to the acute collective awareness of the impossibility of these rebuilding processes in the current colonial power regime.

Suspended Bodies

A deeply involved ethnographer, and this regardless of the context of involvement, would face many difficulties in trying to generate a linear story of her ethnography. An ethnographer working in the Palestinian context could not be but deeply involved in the realities of his/her fieldwork. This deep involvement could take many shapes and be expressed in different narrative styles.[15] In the context of this research, the ethnographic involvement was centered mainly on the experiences of the body in the colonial scheme of oppression. This centralization around and on the body was a major characteristic of the Palestinian context itself during the Second Intifada. In this ethnography, the ethnographic body—that is, data generated through ethnographic practices, and the body of the ethnographer (bodily experiences of the ethnographer during fieldwork)—were intermixing, merging, and re-shaping each other's domain constantly. The choice of narrating these experiences in a fragmentary manner, partly, at least, represents the moments of the intermixing between these two bodily aspects of this ethnography. The underlying argument is that this dynamic of intermixing is the determining feature of the time of the break of commonsense wisdom of confrontation.

Several months after the beginning of the Second Intifada in late September 2000, a certain question grew out of my daily fieldwork experiences. "When should I feel fear, and behave accordingly?" I asked myself each time when a new, higher level of violence was practiced on me directly and/or on me as part of the Palestinian collective, meaning the measures taken by the occupation forces and referred to as collective punishment.

A soldier, that anonymous violent agent-machine, is pointing a gun at my chest. He is shouting in broken Arabic with a heavy Hebrew accent, "Go home." Three speeding jeeps, full of invisible soldiers, encircle us repeatedly. We, a group of lecturers and students, were stuck at Surda checkpoint just when the curfew began.[16] Some of us tried to avoid the checkpoint by taking a side road, but it did not work and we hid in an abandoned house on the outskirts of Surda village. The soldiers noticed us; within two minutes we

15 The first recorded controversy in anthropology regarding the manner of involvement in doing ethnography in Palestine with the Palestinians was around the work of Ted Swedenburg. See Swedenburg (1989, 1992) and Shokeid (1992). See also Furani and Rabinowitz (2011) for a preliminary general review of the anthropological literature on Palestine.

16 Surda Checkpoint was located between Ramallah and Bir Zeit University during the time when I was conducting the ethnography.

were surrounded by three jeeps. It was early evening at Qalandya checkpoint.[17] Hundreds of Palestinians, if not more, crowded at the distance created by the soldiers from the checkpoint itself. A group tried to step forward in the direction of the soldiers; the soldiers shot bullets and sound bombs, then gas bombs. After several hours of waiting, some people took the western bypass via Rafat village, while others took the eastern one via Qalandya refugee camp and stone quarries. I joined one of the groups that took the eastern option. After several hundred meters I became aware of a sharp noise like the buzzing of a large bee flying around. Someone shouted, "They are shooting at us." It was around two in the morning when my partner woke me up. The noise of vehicles, tanks, helicopters, and jet fighters was so overpowering that it left no place to run to. Israel was staging an invasion of Palestinian cities in the West Bank. It was called by the Israeli army and media the Protective Shield.

These scenes were almost daily experiences and practices. With the passage of time they became more violent and pervasive. In a parallel and interconnected manner, my sensibilities to the "raw," naked type of violence became numb. The daily mundane talk between Palestinians on this subject followed the pattern: "Were you hit by a tank? That's nothing, wait until a jetfighter attacks you. Oh, then you will have a real story!" By this manner of reframing the violent events one could survive the daily flow and be prepared for the worst. Still, in all such interrelations that generated violence there was a thin line by which one could guess, anticipate, or even measure the other side's reaction and the quantity and quality of violence that would be used. This thin line was crucial during the period between the break and the restoration of common sense, literally in order to survive. This dynamic economy of violence was interrupted and ceased to be relevant only in one context, the martyrdom operations. What was it in this type of punctuation of spatio-temporalities that could halt other grids of (dis)continuities?

The socio-visual character of the Palestinian city during the Second Intifada was changing rapidly. Each day everyone was updated by his\her social network on the changes that would affect his daily life. The updates also included information on how to cope with these changes\punctuations. Whatever the type and measure of the break, you reconnected. Reconnection means forming new ways of daily practices to restore the spatio-temporal grids, or what might be called your daily infrastructure. The bakery on the street corner, for example, was closed last night, so you know that in the morning you need to look for a new one, and you do. The same applies to the route you take to your school or work. The curfew also brings temporality to the forefront of the city's space. More often than not, in periods of curfew you excavate your own temporal infrastructure. Palestinian collectivities are small. Almost everyone knows somebody who could be an acquaintance of yours. The face-to-face presence becomes more and more important for the maintenance of the new emerging localities, because the "cuttings" of the grids were aimed at remodeling the imagined Palestinian national collectivity as a number of local disconnected ones, such as families, clans, neighborhoods, and villages.[18] The martyrdom operations, in one sense, are shapes of the resistance that emerges from these localities.

In being a member of a locality, as a type of collectivity, you are either part of the one social whole or there is no one for you. Put differently, you are either totally an insider or a total outsider. There is no such thing as a community that is not imagined by its members.

17 For a detailed account of the Qalandya camp and checkpoint see Abourahme (2011).

18 For comparison of how this strategy was used on a different segment of the Palestinian society, see A. Cohen (1965), and T. Asad's critique of it (1975).

Every collectivity, whatever its shape or form, includes an imaginary layer as a constituent part. In this situation, the imagined platform of the Palestinian locality is a metaphor for a continuum between nothingness and oneness. If you passed through a Palestinian city at the moment of breaking news of a martyrdom operation, you would experience the following scene that rapidly oscillates between nothingness and oneness.

The city holds its breath, as if time was frozen. The cold stillness of all the bodies of humans and objects brings them into oneness. The breeze coming from the west seems so alien with its joyful motions, fresh smells of trees, and empty land. The moment extends as if time stands still. Some people step out of a grocery to see the end of their street. Their stepping outside is timeless, their bodies are held in mid-air. The gas station, the butcher's shop, and the grocery are unusually empty. A heavy smell of unexpected fear predominates. It paints the fronts of the buildings. A mass of people stands still in the city square, as if they have all drunk from the same well or been sprinkled with that magic dust of the fairies who postpone the coming of the real. Nobody can look up or down. The horizon is the only refuge, if any. In the last fragment of this frozen moment the mass of people looks like a surge of water from a broken dam, but rather than rushing forward it pours in every possible and impossible direction. The stiff and convoluted movement of the mass as one block is suddenly shattered into invisible atoms running in what the socio-visual observer could see as a chaotic non-patterned flow. The guard of the bank, in his impeccable uniform, is the last official holding on to his bureaucratic mask. The gold shop owners run in frenzy, closing whatever they can reach, closing and locking it. Several cars are left in the middle with their fronts slightly turned toward the sidewalk. An old peddler selling juice is smiling with the last call, "Drink it! Have the last cup … Rush over here and take it!" With his call the last three security personnel of the Palestinian National Authority run away and disappear, melting away like a handful of salt, as the Arabic idiom goes. The peddler drinks the juice and starts to run too, with his smile hanging on. The square, the streets, and the alleys show spots of their faces, becoming emptier with the passing seconds. Less people, less temporal fragments, and more stretches of the spatial presence reemerge at the moving heels of the (un)familiar faces. The line of buildings that starts at the downtown square and extends to the eastern side of the city regains its familiar contours: three elongated triangular fronts emerge, and then there appears a wide rusty green gate with a lion-shaped knocker. It is familiar, clear, and deserted. The quiet, almost smooth, sonic view is probably the most deeply disquieting aspect, for it drives you to look for the noise, any noise, even that of the Israeli jet fighter or missile.[19] It builds a thirst in you for the presence of the code, any social code so that you can practice anticipation. Then the outburst of loud TV sets, radios, cars honking, and human voices shatter the thirst, but you will never be satiated with water again. As you run to somewhere a glimpse of the made-up face of the newsreader from the cable network news broadcast appears with breaking news in black on a deep red background, crossing your overstrained perceptions: when was the last time I saw him/her while I was relaxing?

Eyes are looking for contact to become individuals, to come out of the heavy feelings of oneness/nothingness. They grab the smallest, the vaguest hint of the familiar, but to no avail. No individual could reemerge standing on the nothing side of it. To reproduce individuals one first needs a collective. Hence, the eyes run wildly, seeking contact to resume sociality, order, a code. It starts the moment the bodies clash, bump into each other. Again, it is pain

19 For a comprehensive overview of the literature on socially constitutive sonic layers, see Bull and Back (2003) as well as Attali (1985).

that defines limits and sociality. Then rationality steps in, details of the event, details of bodies, details to coordinate the one and the individual.

The break of the common sense codes of conduct is expressed first and foremost in the ways the community's textures and tissues are punctuated. These basic textures and tissues are made of eye contact and sonic views through which individual Palestinians practice their collectivities. For the individual body and the collective one in the face to face presence in the new Palestinian localities are interwoven mainly via the sonic view and socio-visual practices. Due to a long colonial history, but more acutely so in the Second Intifada, these socio-bodily practices of technologically unmediated hearing and seeing are the inter-subjective matter of the Palestinian communities. The moment of break, then, is almost directly expressed via silence and the loss of eye contact (Nashif 2008b).

This is the moment of the event of a martyrdom operation as it is experienced in a Palestinian city. The brutally compressed spatio-temporality changed the practices and the shapes of the individual as well as the national bodies and collectivities and their real and imagined interrelations. These processes, among others, gave rise to the historical possibility of reprocessing the predisposition for martyrdom to form new practices. Although the new formations are organized around the political economy of liberation as we described it in the previous section, still they were not ready-made types of resistance transplanted into a receptive social and historical context. Rather, they accumulated in the wake of the various colonial modalities, and acquired their final shape and timing through Palestinian individual subjects and local communities who could carry them out.

Relocating the Structure(s)

The structural analysis of the martyrdom operation is the focus of this section. It is not taking for granted that such an approach is relevant to the phenomenon under discussion. Rather, on the one hand, the structural analysis is derivative from the nature of the colonial regime in Palestine. And, on the other hand, the martyrdom operation reprocesses this exact "nature" as a counter-dominance formal relations of resistance. Hence, the underlying assumption is that structural analysis is mandatory in order to comprehend the formal features of martyrdom operations in the Palestinian context.[20]

The colonial regime in Palestine is built on the relational binary of the colonizer and the colonized. This does not mean that it does not change on the level of practices and processes of colonization vis-à-vis the Palestinians. The moment of the Second Intifada is a structural one that is mainly manifested in the measures taken by the Israeli armed forces towards Palestinian society in the West Bank and the Gaza Strip. These measures treat the Palestinians as one undifferentiated collective body, and their inner logic of operation is one of totality that decodes the whole of the Palestinian body, be it an individual or a collective one. For example, if there is a checkpoint, every Palestinian must pass through it. Regardless of the type of resistance practiced by the Palestinians, collective punishments are the regular Israeli response. The same is true with the practice of ensuring the death of Palestinian strugglers, which became common during this period (Azoulay and Ophir 2008). These are formal, structural, relations that put the colonized and the colonizer at bipolar interrelations. This moment of the colonial regime, in a sense, demands a structural approach analysis to

20 The argument that each phenomenon carries in it its methods of analysis is based on the intervention of Horkheimer (2002).

trace its working order. The martyrdom operation is inherently constructed in a relational manner with this colonial moment.

The massive use of power during the Second Intifada by the Israeli armed forces is only one indicative practice of how such a structural nature is conducted. The interlocked Palestinian bodies in these paths of the colonial regime could hardly escape the grip of these formal oppressive and massive enforcements of the regime. Hence, at least partly, the martyrdom operation is a structural mirroring and refraction of the latest moment of the colonial regime in Palestine. The historical and the ethnographic analyses of the previous sections tell the structural mirroring and refraction as practiced and experienced by real individuals and communities. Here, I will turn to the structurally interlocked bodies to explicate their formal acts of mirroring as resistive ones. By doing these analytical reconstructions, I will try to critically and openly engage with these kinds of strategies of formation of Palestinian subjectivities and communities (Pitcher 1998, p. 27). Building on Hannah Arendt's arguments regarding constituting national communities, Asad (2007) argues that "suicide bombings" are part of a moment of violence that constitutes Islamic communities. Here, I agree with Asad on the aspect of formation entailed in martyrdom operations. But, at least in the Palestinian context, it does not seem to constitute a temporal point of origin.

There are three main characteristics of the martyrdom operation that differentiate it from other types of Palestinian armed resistance. First, while in the classical manner of armed resistance the body is the carrier of the tool of resistance, here the body is the tool itself. The operator does not carry a tool; his/her body is the main explosive tool. Second, there is an irony in this type of resistance that is not found in other types of dialectical relations in the Palestinian colonial condition. The Israeli practices are directed at eliminating the Palestinian body at its different levels of existence and manifestation. The operator eliminates his/her body, as the Israeli regime aims, but he takes the Israeli one with him too. This is a new manner of resolution that is entailed in the colonial contradictions. The third characteristic of the martyrdom operation is the movement of the Palestinian body in forbidden spaces and times. The two real and imagined enclaves of the Palestinian and Israeli sociopolitical bodies are trespassed and transmuted by the path created by martyrdom operation. They are trespassed and transcoded in a manner that deeply confuses and redefines the borders of each well-entrenched national. The operator leaves the Palestinian national body only to merge with the Israeli body, and in this way to redefine the power relations between them, as if his movement declared the end of both national bodies and the search for a third through a certain mixture of both.[21]

In order to reweave these characteristics into their relational whole we still need to draw the schematic path of a martyrdom operator in order to rebuild their structure. Here I am using a variation of the interpretive model developed by Jameson (1981). Schematically, the model works as follows: in the first analytical move we explore the inner structure of the martyrdom operation, then we position it in the current social formation of Palestinian society, which we approach by tracing the path of the operator, and finally we try to relocate it in history generally perceived, and this is the topic of the final section of the chapter.

In a specific locality with an infrastructure, meaning accumulated forms of resistance to the occupation, there are processes of preparation and a moment of readiness to conduct

21 For a detailed study of the Israeli body(ies) see Weiss (2002). For a different interpretation of the mixture created by the conflictive and merging dynamics of the colonial condition in Palestine, see Falk (2004).

a martyrdom operation. In the context of the Second Intifada, localities in the West Bank are disconnected from nearby communities of resistance, but reimagined as a continuum of disconnections.[22] The path, then, is recreated as practicing the reimagined compressed spatio-temporalities, with a no less transformative counter-compression agency (Makdisi 2010). The operator has to pass and trespass all the Palestinian and Israeli barriers, the first of which is the mundane Palestinian social life. In order to begin walking the path, he or she has to adopt a double but interconnected sociality. In the context under discussion, the confined locality and the objective condition of survival enforce a certain form of secrecy which is less dependent on organizational structure, as in the previous PLO-dominated period of resistance, than on familial and neighborhood social networks. Hence, walking the path reshapes the locality as well as being shaped by it, by merging anew different forms of collectivity with the national ones (Taraki 2006).[23]

There are many intermediate layers of spatio-temporalities between the Palestinian and the Israeli communities. These are mainly filled with filtering devices and no-man's-land time patches. Leaving the Palestinian enclaves in order to enter the Israeli ones, the operator must pass through these in-between areas of space and time.[24] Regardless of his choice, the in-betweenness seems to act as an opening for "cleaning" identities, that is, for redefining and entrenching certain self-perceptions concerning "who am I?". It has to recycle certain aspects so as to intensify the Palestinian-Israeli dichotomy, into a four-pole matrix. I am a Palestinian, acting Israeli, I look like part of the collective, but I am totally its hidden negation. The moment he or she passes through this transformative passage and enters the Israeli enclave, the operator actualizes the symbolic transformation, which collapses the existing colonial order of hierarchies and separations. The second moment into the Israeli enclave is that of the explosion itself. Now it materializes the literal and metaphorical unification of both bodies as a total negation of the colonial order of things, bodies, and agencies.

The above descriptions and analyses locate four basic relational axes that cohere to generate the inner structure of the martyrdom operation. These are: the individual body of the operator, the two national/communal bodies, the intermixing of the bodies, and the body of space/time. We could claim that in a certain context, with a certain bundle of historical layers, the individual body that is the object of the social relations of oppression returns to negate itself as a locale of oppression by destroying materially and symbolically the conditions that made it a site of oppression. The conditions that the individual oppressed body seeks to abolish are: first, two separated colonized communities; second, the fragmentation of the colonized communities into disconnected localities; and third, the colonizer's unification as

---

22 For more on these developments, see, for example, Bornstein (2003), Rosenfeld (2004), Hammami (2005), Taraki (2006), Allen (2006, 2008), Nashif (2008a), and Makdisi (2010).

23 The different chapters in the volume edited by Lisa Taraki (2006) clearly indicate the transformative nature of the Second Intifada. The forms of hybrid social realities that reigned during the "national" era are no longer relevant to the new challenges of the current one. Hence, Palestinians reconstruct familial and other social structures in order to cope with and survive the new phase of occupation.

24 Bornstein (2003) documents and analyses the crossings and trespassing of the Palestinian workers, who leave their hometowns in the West Bank to work in Israel. This is one example of the history of the borderline culture that has been vexed since the War of 1948. One of the first monographs dealing with this history and its implications for the Palestinian social and political realities is Cohen (1965). This is to say, the spatial practices of the operator are one of the latest manifestations of this borderline culture in Palestine.

one whole sealed locality. The body intermixes both by re-gathering the colonized and by inserting itself into the whole sealed locality of the colonizer.

Certain phenomena are bound to their historical context by reprocessing its possibilities of changing. Such is the martyrdom operation. It springs from the sociohistorical context of the colonial condition in Palestine only to articulate its total negation. We will revisit this sociohistorical context in the light of the insights we gained in the previous sections, where we focused on the moment of the martyrdom operation experientially and structurally. After looking at the operation as a manifestation of historical agency aimed at transforming the lived realities, it is time to re-position it in the larger processes of the (de)colonization of Palestine.

Conclusion

Martyrdom operation, as a manner of dying, is one strategy, among others, of (re)producing a (de)colonized social order in Palestine. Tracing and locating the main junctures of constructing the martyrdom operation was challenged repeatedly by the basic contradictory nature of looking for life through an active manner of ending it. The argument is that this main contradictory feature is inherent in the nature of the colonial order, which is aimed at and acts to dismantle the Palestinian national collectivity. The multi-layered exploration presented here show that the martyrdom operation carries this contradiction, looking for life by actively ending it, to its historical conclusion.

The three main interconnected layers that were explicated in this chapter are the historical, the ethnographic, and the structural ones. The main insight that comes out from these layers is that the martyrdom operations are a form of intervention in exactly the same colonial realities that gave rise to these operations in the first place. The first layer is the accumulation of colonial experiences that culminated in a certain political economy of liberation in the late 1960s. In this political economy that determines the conduct of the Palestinians the body is assigned a central position. The argument is that this position of the body is a kind of proto-form that predisposes it for martyrdom operations. The second layer is the ethnographic research method that permitted a certain manner of narrating the Palestinians' experiences of martyrdom operations as native's data. In this ethnographic site, the academic distinction between the ethnographic body and the body of the ethnographer is questioned time and again. The questioning processes are refracted in a fragmentary mode of narrating the event of the martyrdom operation. The third layer is the structural analysis, which outlines the specific but historical manner of carrying out the martyrdom operation and hence also enables us to recognize it as such. We are able to recognize it, for its formal features re-position it relationally as a mirroring structure of resistance inside the colonial system in Palestine. In this sense, these operations are a derivative strategy of re-forming new Palestinian subjectivities and communities. But their derivativeness is actualized precisely at their moment of impossibility due to the nature of the colonial regime. In order to further our understanding of the martyrdom operation, we will re-focus our attention on the three layers as expressions of one whole.

The three analytical layers reveal different temporal modalities that resolve their contradictions via the body spatially, i.e., reshuffle it so as to subordinate it to temporal logic. The martyrdom operations are manifestations of a certain moment/site in the processes of the (de)colonization in Palestine. They have taken place on the one hand as part of the temporality of colonization, and on the other hand inside Palestine as the real and imagined spatial grids. Moreover, and probably more important, the operations are a collective means of transforming the conditions of possibility of being a Palestinian in the here and now of the decolonization processes. The event of actualizing martyrdom operations is the only arena in which its different possible pre-conditions act as a whole. Hence, the genealogical conception of history that is based on the idea of tracing the scattered power practices is the most relevant frame of analysis to our case.

The genealogical method is applied as we trace the consecutive forms of armed resistance and bodily practices through their diverse moments and sites (Foucault 1977). Forms of resistance and their bodily practices do not emerge sequentially from the previous ones; rather, the colonial condition structures what it already entails in potential or predispositional forms. The potential and predispositional forms, however, are not determined or elaborated by reconnecting the past with the future via the present. They simply sketch the impossible without naming it, and in such a way determine the particular impossibilities, such as the Palestinian national collectivity in the late colonial order (Mbembe 2003). To repeat, the colonial regime totally annexes the national time and space of the Palestinians in order to annul them. The counter practices, the martyrdom operation by totally ending life, bring time back to a zero state, i.e., the new beginning of the Palestinian national community.

It seems that there is a certain way of dying that operates during the constitution of new social orders. Unlike other manners of dying, which derive from the way in which society is organized, this category is located at the interface between two or more social orders. The transitional dynamics enable these manners of dying to contain both the contradictions of the previous order and the possible orders to come. Martyrdom operations bear these transitional features in a certain configuration. The over-emphasis on the interrelation of body-space, inherent in this manner of dying, raises many questions regarding the virtual neutralization of the temporal aspects. If any manner of dying is a composition of spatio-temporal relations that reorder the individual and social bodies, martyrdom operations in the Palestinian context seem to halt the passage of time in order to start it at a new beginning. The Israeli occupation practices during this period redefine the Palestinian national collectivity as separate and desperate localities, with no nationally imagined Palestinian time. These same structural relations of oppression are used in order to counter these resistance practices. In countering them, though, historical layers interact with the current patterns and construct the martyrdom operation as a manner of dying relevant to this historical moment in colonial Palestine.

Bibliography

ABOURAHME, Nasser, 2011. "Spatial Collisions and Discordant Temporalities: Everyday Life between Camp and Checkpoint." *International Journal of Urban and Regional Research*, 35(2), 453–61.

———, 2009. "The Bantustan Sublime: Reframing the Colonial in Ramallah." *City*, 13(4), 499–509.

ABUFARHA, Nasser, 2009. *The Making of a Human Bomb: An Ethnography of Palestinian resistance*. Durham, NC: Duke University Press.

ABU MOAILEK, Yasser, 2005. "Ex-Car Thief Aims to Revive Business in Gaza." *The Electronic Intifada*, November 8 [Online]. Available at: http://electronicintifada.net/v2/article4286.shtml, accessed January 6, 2014.

ABU MUKH, Leena, 2006. *Movements to and from the Palestinian Territories under Israeli Occupation after Oslo (1993–2006)*. Florence: European University Institute.

ABUNIMAH, Ali, 2006. *One Country: A Bold Proposal to End the Israeli-Palestinian Impasse*. New York: Metropolitan Books; Henry Holt and Company.

ABU ZAHRA, Nadia, 2007. *Legal Geographies in Palestine: Identity Documentation, Dispossession, Repression and Resistance*. PhD. Supervised by Tony Lemon and Dawn Chatty. University of Oxford

AGAMBEN, Giorgio, 1997. *Homo sacer, le pouvoir souverain et la vie nue*. Paris: Seuil.

AGBARIEH, Asma, 2004. "Illegal West Bank Workers in Um al-Fahm: Wrong Side of the Wall." *Challenge*, 88 [Online]. Available at: www.workersadvicecenter.org/Chall88-Asma-Wall.htm, accessed November 2009.

AIDA (association of International Development Agencies), 2011. "Restricting Aid: The Challenge of Delivering Assistance in the Occupied Palestinian Territory." Research Report, June.

AL-AHRAM, 2013. "Abbas Issues Anti-Hamas Statement on Egypt's Sinai." *Al-Ahram*, November 13 [Online]. Available at: http://english.ahram.org.e.g./NewsContent/1/64/86325/Egypt/Politics-/Abbas-issues-antiHamas-statement-on-Egypts-Sinai-.aspx, accessed May 19, May 19, 2014.

ALCOULOUMBRE, Catherine and BAUDOIN, Laurent, 2003. *Choses vues en Palestine. Campagne internationale pour la protection du peuple Palestinien*. Paris: Le Temps des cerises.

ALI, Nur, 2009. "Shhab: Gaza Tunnels Constitute a Threat for Egypt." *Al-Youm al-Saba*, December 29 [Online]. Available at: http://www.youm7.com/News.asp?NewsID=171788, accessed May 19, 2014. (In Arabic).

AL-HUSSEINI, Jalal and BOCCO, Riccardo, 2008. "Les négociations israélo-palestiniennes de juillet 2000 A Camp David: Reflets du processus d'Oslo." *Relations Internationales*, 136, 51–72.

ALLEN, Lori, 2006. "The Polyvalent Politics of Martyr Commemoration in the Palestinian Second Intifada." *History and Memory*, 18(2), 107–38.

———, 2008. "Getting by the Occupation: How Violence Became Normal During the Second Palestinian Intifada?" *Cultural Anthropology*, 23(3), 453–87.

ALMOG, Oz, 2000. *The Sabra*. Berkeley: University of California Press.
AL-MUGHRABI, NIDAL, 2008. "Palestinians Blow Up Border Wall, Flood Into Egypt." *Reuters* [Online]. Available at: http://www.reuters.com/article/idUSL23892772008 0123, accessed May 23, 2014.
AL SHARIF, Mohamad, 1995. *The Search for Entity: A Study in Palestinian Political Thought, 1908–1993*. Nicosia: The Center for Socialist Studies and Research in the Arab World.
Al-SURANI, Ghazi, 2008. "A document on the economic, social and political effects of the tunnels in Rafah" (in Arabic) *Al-Hiwar Al-Mutamadin*, 2495, December 14 [Online]. Available at: http://www.ahewar.org/debat/show.art.asp?aid=156356, accessed May 19, May 19, 2014.
ALTHUSSER, Louis, 1971. *Lenin and Philosophy, and Other Essays*. New York: Monthly Review Press.
———, 1996. *For Marx*. London: Verso.
ALTUNIŞIK, Meliha Benli, 2008. "The Possibilities and Limits of Turkey's Soft Power in the Middle East." *Insight Turkey*, 10(2), 41–54.
AMARA, Mohammad Hasan, 2000. "Israeli Palestinians and the Palestinian Authority." *MERIA*, 4(1), 38–44.
AMIDROR, Major General (reserves) Yaakov, 2007. *Counterinsurgency War: The Israeli Experience*. Jerusalem: Jerusalem Center for Public Affairs.
AMIR, Merav, 2011. "On the Border of Indeterminacy: The Separation Wall in East Jerusalem." *Geopolitics*, 16(4), 768–92.
———, 2013. "The Making of a Void Sovereignty: Political Implications of the Military Checkpoints in the West Bank." *Environment and Planning D: Society and Space*, 31, 227–44.
AMIREH, Amal, 2003. "Between Complicity and Subversion: Body Politics in Palestinian National Narrative." *The South Atlantic Quarterly*, 102(4), 747–72.
AMIRY, Suad, 2010. *Nothing to Lose but Your Life*. Doha: Bloomsbury Qatar Foundation.
ANDERSON, Benedict, 1983. *Imagined Communities*. London: Verso.
ANDREAS, Peter, 2001. *Border Games: Policing the US-Mexico Divide*. Ithaca: Cornell University Press.
ARETXAGA, Begonia, 2001. "Terror as a Thrill: First Thought on the War on Terror." *Anthropological Quarterly*, 75(1), 139–50.
ARIELI, Shaul and SFARD, Michael, 2008. *The Wall of Folly*. Tel Aviv: Yediot Aharonot. (In Hebrew).
ARNON, Arie, 2007. "Israeli Policy towards the Occupied Palestinian Territories: The Economic Dimension, 1967–2007." *Middle East Journal*, 61(4), 573–95.
ARNON, Arie and BAMYA, Said (eds), 2007. *Economic Dimensions of a Two-States Agreement between Israel and Palestine*. Aix-en-Provence: Aix Group.
ARNON, Arie, LUSKI, Israel, SPIVAK, Avia and WEINBLATT, Jimmy, 1997. *The Palestinian Economy*. Leiden: E.J. Brill.
ARNON, Arie and WEINBLATT, Jimmy, 2001. "Sovereignty and Economic Development: The Case of Israel and Palestine." *The Economic Journal*, 111, 291–308.
ASAD, Talal, 1975. "Anthropological Texts and Ideological Problems: An Analysis of Cohen on Arab Villages in Israel." *Economy and Society*, 4: 251–82.
———, 2007. *On Suicide Bombing*. New York: Columbia University Press.

ASSOCIATED PRESS, 2010. "Hamas: Egypt gasses 4 smugglers." April 29 [Online]. Available at: http://www.jpost.com/Middle-East/Hamas-Egypt-gasses-4-smugglers, accessed May 19, May 19, 2014.

ATTALI, Jacques, 1985. *Noise: The Political Economy of Music*. Minneapolis: University of Minnesota Press.

AZOULAY, Ariella and OPHIR, Adi, 2008. *This Regime Which is Not One: Occupation and Democracy between the Sea and the River (1967–)*. Tel Aviv: Resling. (In Hebrew).

AZOULAY, Caroll, 2008. "L'alyah de France se place au-dessus de la mêlée." *Guysen Israel News*, January 8.

BACKMAN, René, 2006. *Un mur en Palestine*. Paris: Fayard.

BALZACQ, Thierry 2005. "The Three Faces of Securitization: Political Agency, Audience and Context." *European Journal of International Relations*, 11(2), 171–201.

BAR, Neta and BEN-ARI, Eyal, 2005. "Israeli Snipers in the Al-Aqsa Intifada: Killing, Humanity and Lived Experience." *Third World Quarterly*, 26(1), 133–52.

BARAK Oren and Gabriel SHEFFER, 2007. "The Study of Civil-Military Relations in Israel: A New Perspective." *Israel Studies*, 12(1), 1–27.

BARDA, Yael, 2012. *The Bureaucracy of the Occupation*. Tel Aviv: Hakibutz Hameuchad; Jerusalem: Van-Leer Institute. (In Hebrew).

BAR-ON, Mordechai, 1996. *In Pursuit of Peace: A History of the Israeli Peace Movement*. Washington: United States Institute of Peace Press.

BARTH, Frederik, 1995. "Les groupes ethniques et leurs frontières." In: POUTIGNAT, Philippe and SREIFF-FENART, Jocelyne (eds). *Théorie de l'Ethnicité*. Paris: PUF, 203–49.

BARTHE, Benjamin, 2007. "Les voleurs de Nablus." *Le Monde*, March 14.

BARZAK, Ibrahim, 2010. "Easing of Gaza blockade leaves many smugglers idle." *The Washington Post*, November 2 [Online]. Available at: http://www.washingtonpost.com/wp-dyn/content/article/2010/11/02/AR2010110204600.html, accessed May 19, 2014.

BATAILLE, Georges, 1985. *Visions of Excess: Selected writings, 1927–1939*. Minneapolis: University of Minnesota Press.

BAYART, Jean François, 2004. *Le Gouvernement du monde: Une critique politique de la globalization*. Paris: Fayard.

BAYAT, Asef, 2000. "From 'Dangerous Classes' to 'Quiet Rebels': Politics of the Urban Subaltern in the Global South." *International Sociology*, 15(3), 533–57.

BBC, 2010a. "Gazans Cut Through Egypt's Border Barrier." BBC Online, May 6 [Online]. Available at: http://news.bbc.co.uk/2/hi/8664316.stm, accessed May 19, 2014.

———, 2010b. "Egypt Police Raid Gaza Car Smuggling Tunnel." BBC Online, July 7 [Online]. Available at: http://www.bbc.co.uk/news/10545626, accessed May 19, 2014.

BEAUJOUAN, Nicolas, 2010. "Interview de Nisreen, Mazzawi Militante LGBT Palestinienne." Le Blog de la Commission LGBTI du NPA, May 10 [Online]. Available at: http://npa-lgbti.overblog.com, accessed October 6, 2010.

BECK, Ulrich, 1987. "The Anthropological Shock: Chernobyl and the Contours of the Risk Society." *Berkeley Journal of Sociology*, 32, 153–65.

———, 2003. "The Silence of Words: On terror and war." *Security Dialogue*, 34(3), 255–67.

BEN AMI, Shlomo, 2006. *Scars of War, Wounds of Peace: The Israeli-Arab Tragedy*. Oxford University Press.

BEN-ARI, Eyal, 2001. "Tests of Soldierhood, Trials of Manhood: Military Service and Male Ideals in Israel." In: MAMAN, Daniel, BEN-ARI, Eyal and ROSENHEK, Zeev (eds). *Military, State and Society in Israel*. New Brunswick and London: Transaction Publishers, 239–67.

BEN-ARI, Eyal and BILU, Yoram, 1997a. "Introduction." In: BEN-ARI, Eyal and BILU, Yoram (eds). *Grasping Land: Space and Place in Contemporary Israeli Discourse and Experience*. Albany: Suny Press, 1–24.

———, 1997b. "Saints' Sanctuaries in Israeli Development Towns: On a Mechanism of Urban Transformation." In: BEN-ARI, Eyal and BILU, Yoram (eds). *Grasping Land: Space and Place in Contemporary Israel Discourse and Experience*. Albany: Suny Press, 61–83.

BEN-ARI, Eyal and LOMSKI-FEDER, Edna, 1999. "Introduction." In: BEN-ARI, Eyal and LOMSKI-FEDER, Edna (eds). *The Military and Militarism in Israeli Society*. Albany: State University of New York Press, 1–34.

BEN-ARI, Eyal, MAYMON, Meirav, GAZIT, Nir and SHATZBERG, Ron, 2004. *From Checkpoints to Flowpoints: Sites of Friction between the Israel Defense Forces and Palestinians*. Final report submitted to the Friedrich Ebert Foundation, Israel, December.

BEN-ARI, Eyal, ROSENHEK, Zeev and MAMAN, Daniel, 2001. "Introduction. Military, State and Society in Israel: An Introductory Essay." In: MAMAN, Daniel, BEN-ARI, Eyal and ROSENHEK, Zeev (eds). *Military, State and Society in Israel*. New Brunswick and London: Transaction Publishers, 1–39.

BEN-DOR, Anat, KAGAN, Michael and CAHANA, Aelad, 2008. *Nowhere to Run: Gay Palestinian Asylum-Seekers in Israel*. Tel Aviv: Tel Aviv University, Buchmann Faculty of Law.

BENNAFLA, Karine and PERALDI, Michel (eds), 2008. "Frontières et logiques de passage: L'ordinaire des Transgressions." *Cultures & Conflits*, 72, 7–12.

BEN-NAFTALI, Orna, GROSS, Aeyal M. and MICHAELI, Keren, 2009. "The Illegality of the Occupation Regime: The Fabric of Law in the Occupied Palestinian Territory." In: OPHIR, Adi, GIVONI, Michal, and HANAFI, Sari (eds). *The Power of Inclusive Exclusion: Anatomy of Israeli Rule in the Occupied Palestinian Territories*. New York: Zone Books, 31–88.

BENNET, James, 2005. "The Day After Peace: Designing Palestine." *New York Times*, May 15.

BENVENISTI, Meron, 1988. *The Sling and the Club*. Tel Aviv: Keter. (In Hebrew).

BENZIMAN, Uzi, 2004. "We have an Alibi." *Haaretz*, October 14.

BEN-ZVI, Yael, 1998. "Zionist Lesbianism and Transsexual Transgression. Two Representations of Queer Israel, Power and Sexuality." *Middle East Report*, 28, Spring, 26–9.

BERESFORD, David, 1987. *Ten Men Dead: The Story of the 1981 Irish Hunger Strike*. New York: Atlantic Monthly Press.

BERTRAND, Claire and BERTRAND, Olivier, 2006. *Itinéraires et Combats d'une Avocate Israélienne*. Paris: L'Harmattan.

BIALE, David, 1997. *Éros Juif*. Hébraïca Series. Arles: Actes Sud.

BICHLER, Shimshon, and Jonathan NITZAN, 2001. *From War Profits to Peace Dividends, Jerusalem*. Jerusalem: Karmel. (In Hebrew).

BISHARA, Azmi, 2006. *The Checkpoint*. Tel Aviv: Babel.

BISHARA, Marwan, 2002. *Palestine/Israel: Peace or Apartheid*. London: Zed Books.

BLIGH Alexander, 1999. "The Intifada and the New Political Role of the Israeli Arab Leadership." *Middle Eastern Studies*, 35(1), 134–164.

BLOOM, Mia, 2005. *Dying to Kill: The Allure of Suicide Terror*. New York: Columbia University Press.

BOCCO, Riccardo, BRUNNER, Matthias, DANEELS, Isabelle, HUSSEINI, Jalal, LAPEYRE, Fredéric and RABAH, Jamil, 2003. *Palestinian Public Perceptions on Their Living Conditions. Report VI*. October 2003. Graduate Institute of Development Studies (IUED), University of Geneva.

BOCCO, Riccardo, BRUNNER Matthias, DANEELS, Isabelle, LAPEYRE, Fredéric and RABAH, Jamil, 2002a. *An Overview of Palestinian Public Assessment of Its Needs and Conditions Following The Recent Israeli Military Operations in The West Bank, Report IV*. April 2002.

———, 2002b. *Palestinian Public Perceptions on Their Living Conditions. Report V*. December 2002. Graduate Institute of Development Studies (IUED), University of Geneva.

BOCCO, Riccardo and MANSOURI, Wassila, 2008. "Aide internationale et processus de paix: le cas palestinien, 1994–2006." *A Contrario*, 5(2), 6–22.

BOCCO Riccardo and MEIER, Daniel, 2005. "Penser la notion de frontières au Moyen-Orient." In: Bocco, RICCARDO and MEIER, Daniel (eds). *Frontières au Moyen-Orient* (special issue), *A Contrario*, 3(2), 3–10.

BOLE-RICHARD, Michel, 2013. *Israël: le nouvel Apartheid*. Mayenne: Les liens qui libèrent.

BONTEMPS, Véronique, 2009. "Tu dois prendre le risque, tu n'as pas le choix: Entre expérience collective et précarisation extrême, récits d'ouvriers palestiniens travaillant en Israël." November 12 [Online]. Available at: http://mediamed.mmsh.univ-aix.fr/chaines/mofip/limites-frontieres2/Pages/ouvriers-palestiniens.aspx, accessed May 25, 2014.

BORNSTEIN, Avram, 2002. *Crossing the Green Line Between the West bank and Israel*. Philadelphia: University of Pennsylvania Press.

———, 2008. "Military Occupation as Carceral Society. Prison, Checkpoints and Walls in the Israeli-Palestinian Struggle." *Social Analysis*, 52(2), Summer, 106–30.

———, 2010. "Palestinian Prison Ontologies." *Dialect Anthropol*, 34, 459–72.

BOTIVEAU, Bernard, CONTE, Edouard and SIGNOLES, Aude (eds). *Palestine, Etudes rurales*. Paris: Éditions de l'École des hautes études en sciences sociales, 173–4.

BOTIVEAU, Bernard and SIGNOLES, Aude (eds), 2004. *D'une Intifada l'autre. Les Quotidiens en Palestine, Revue Egypte Monde Arabe*. Cairo: Complexe/Cedej, No. 6.

BOUILLON, Markus E., 2006. *The Peace Business: Money and Power in the Palestine-Israel Conflict*. London: I.B. Tauris.

BOUSSARD, Valérie, MERCIER, Delphine and TRIPIER, Pierre, 2004. *L'aveuglement organisationnel*. Paris: CNRS.

BOVÉ, José, 2002. "Les membres de la 11ème mission et José Bové." In: BOVÉ, José and BRAUMAN, Rony. *Retour de Palestine*. Paris: Mille et une nuits.

BRAVERMAN, Irus, 2008. "'The Tree is the Enemy Soldier': A Sociolegal Making of War Landscapes in the Occupied West Bank." *Law and Society Review*, 42(3), 449–82.

———, 2011a. "Civilized Borders: A Study of Israel's New Crossing Administration." *Antipode*, 43(2), 264–95.

———, 2011b. "Hidden in Plain View: Legal Geography from a Visual Perspective." *Law, Culture and the Humanities*, 7(2), 173–86.

BREINER, Yehoshua, 2009. "Down and Out in the Seam Zone." *Walla News*, April 29. (In Hebrew).

BREUER, Theresa, 2013. "Unrest in Egypt: Closed Tunnels Could Ruin Hamas." *Spiegel Online*, July 30 [Online]. Available at: http://www.spiegel.de/international/world/egypt-and-gaza-closed-smuggling-sinai-tunnels-could-ruin-hamas-a-913927.html, accessed May 19, 2014.

BULL, Michael and BACK, Les (eds), 2003. *The Auditory Culture Reader*. Oxford: Berg.

BUZAN, Barry and WAEVER, Ole, 1997. "Slippery? Contradictory? Sociologically Untenable? The Copenhagen School Replies." *Review of International Studies*, 23, 241–50.

CARDOSO, Fernando H. and FALETTO, Enzo, 1979. *Dependency and Development in Latin America*. Berkeley: California University Press.

CAVANAUGH, Kathleen, 2007. "The Israeli Military Court System in the West Bank and Gaza." *Journal of Conflict and Security Law*, 12(2), 197–222.

CHANDOUL, Mustapha, BOUBAKRI, Hassan, GILDAS, Simon, and COSTA-LACOUX, Jacqueline, 1991. "Migrations clandestines et contrebande à la frontière tuniso-libyenne." *Revue de Migrations Internationales*, 7(2), 155–62.

CLOT, Ziyad, 2010. *Il n'y aura pas d'Etat Palestinien—Journal d'un Négociateur en Palestine*. Paris: Max Milo.

COBHAM, David, 2001. "Economic Aspects of the Israeli-Palestinian Conflict: Introduction." *The Economic Journal*, 111(472), 249–53.

COHEN, Abner, 1965. *Arab Border Villages in Israel: A study In Continuity and Change in Social Organization*. Manchester: Manchester University Press.

COHEN, Gili, 2014. "Top Israeli Think Tank: If Talks Fail, Israel Should Withdraw from 85% of West Bank." *Haaretz*, January 28 [Online]. Available at: http://www.haaretz.com/news/diplomacy-defense/.premium-1.570945, accessed February 4, 2014.

COHEN, Samy, 2009. *Tsahal à l'épreuve du terrorisme*. Paris: Le Seuil.

COLLINS, John, 2008. "Dromocratic Palestine." *Middle East Report*, 248, Fall, 8–13.

———, 2011. *Global Palestine*. London: Hurst and Company.

CONTE Edouard, 2005. "L'autre Mur: Mariages bannis et citoyennetés fragmentées en Israël-Palestine." *Etudes rurales*, 173–4, 127–52.

COOK, Catherine, HANIEH, Adam and KAY, Adah, 2004. *Stolen Youth. The Politics of Israel's Detention of Palestinian Children*. London and Sterling, VA: Pluto Press.

COSTA-LACOUX, Jacqueline, 1991. "Migrations clandestines et contrebande à la frontière tuniso-libyenne." *Revue des Migrations internationales*, 7(2), 155–62.

CRENSHAW, Kimberle, 1991. "Mapping the margins: Intersectionality, Identity Politics, and Violence against Women of Color." *Stanford Law Review*, 43(6), 1,241–99.

CURMI, Brigitte, 2002. "Les enjeux de l'après-Oslo. Le mouvement associatif dans les Territoires palestiniens." In: NÉFISSA, Sarah Ben and HANAFI, Sari. *Pouvoir et associations dans le monde arabe*. Paris: CNRS, 95–123.

CYPEL, Sylvain, 2005. *Les emmurés. La société israélienne dans l'impasse*. Paris: La Découverte.

DACCA, Walid, 2009. *Consciousness Molded. Or: Of the Re-Identification of Torture*. Israel: Jalbu'a Prison. (Most of this testimony has subsequently been published in: BAKER, Abeer and MATAR, Anat (eds), 2011. *Threat. Palestinian Political Prisoners in Israel*. London: Pluto Press).

DARDOT, Pierre and LAVALE, Christian, 2009. *La nouvelle raison du Monde*. Paris: La Découverte.

DARY, Joss and SIEFFERT, Denis, 2002. *La guerre israélienne d'information. Désinformation et fausses symétries dans le conflit israélo-palestinien*, Paris: La Découverte.

DAYAN, Hilla, 2009. "Regimes of Separation: Israel/Palestine and the Shadow of Apartheid." In: OPHIR, Adi, GIVONI, Michal and HANAFI, Sari (eds). *The Power of Inclusive Exclusion: Anatomy of Israeli Rule in the Occupied Palestinian Territories*. New York: Zone Books, 281–322.

DEBIE, Franck and FOUET, Sylvie, 2001. *La paix en miettes. Israël et Palestine 1993–2000*. Paris: PUF.

DE CERTEAU, Michel, 1984. *The Practice of Everyday Life*. Berkeley: University of California Press.

DE CROUSAZ, Pascal, 2005. "Le facteur démographique dans la détermination par Israël de ses frontières avec les Palestiniens." *A Contrario*, 3(2), 66–98.

DEMING, Edwards W., 1986. *Out of the Crisis*. Cambridge, MA: MIT Press.

DEREK, Gregory, 2004. *The Colonial Present: Afghanistan, Palestine, Iraq*. Malden: Blackwell.

DIECKHOFF, Alain, 2003. "Israël: Le retour du complexe de Massada." In: DIECKHOFF, A. and LEVEAU, R. (eds). *Israéliens et Palestiniens. La Guerre en Partage*. Paris: Balland.

——— (ed.), 2008. *L'Etat d'Israël*. Paris: Fayard.

DURKHEIM, Emile, 2002. *Suicide: A study in Sociology*. London: Routledge.

EFRAT, Elisha, 2006. *The West Bank and Gaza Strip: A Geography of Occupation and Disengagement*. London and New York: Routledge.

ELAD, Moshe, 2008. "Palestinian headed to civil war." *Ynetnews*, September 22 [Online]. Available at: http://www.ynet.co.il/english/articles/0,7340,L-3600099,00.html, accessed September 23, 2012.

ELYACHAR, Julia, 2005. *Markets of Dispossession: NGOs, Economic Development, and the State in Cairo*. Durham: Duke University Press.

EPHRON, Dan, 2010. "Easing the Import Blockade Won't Help Gaza." *Newsweek*, June 23.

ERICSON Richard, 1994. "The Division of Expert Knowledge in Policing and Security." *British Journal of Sociology*, 45(2), 149–75.

EUROMID OBSERVER, 2013. "The Gaza Siege after the Egyptian Shutdown of The Tunnels." July [Online]. Available at: http://www.euromid.org/report/SlowDeath.pdf, accessed May 19, 2014.

FALK, Avner, 2004. *Fratricide in the Holy Land: A psychoanalytic view of the Arab Israeli conflict*. Madison: University of Wisconsin Press.

FARSAKH, Leila, 2002. "Palestinian Labor Flows to the Israeli Economy: A Finished Story?" *Journal of Palestine Studies*, 32(1), 14–5.

———, 2005. *Palestinian Labour Migration to Israel*. London and New York: Routledge.

FARSOUN, Samih and ARURI, Naseer, 2006. *Palestine and the Palestinians: A Social and Political History*. Boulder, CO: Westview Press.

FATH, Sébastien, 2005. "Le poids géopolitique des évangéliques américains: Le Cas d'Israël." *Hérodote*, 119, 25–40.

FELDMAN, Allen, 1991. *Formations of Violence: The Narrative of the Body and Political Terror in Northern Ireland*. Chicago: Chicago University Press.

FILC, Dani and RAM, Uri (eds), 2004. *The Power of Property: Israel in the Globalization Age*. Jerusalem: Van Leer Institute. (In Hebrew).

FILLIEULE, Olivier, 2005. *Le désengagement militant*. Paris: Belin.

FINKELSTEIN, Norman, 2005. *Beyond Chutzpah: On the Misuse of Anti-Semitism and the Abuse of History*. Berkeley: University of California Press.

FISCHER, Michael M.J., 2006. "Changing Palestine-Israel Ecologies: Narratives of Water, Land, Conflict, and Political Economy, Then and Now, and Life to Come." *Cultural Politics: An International Journal*, 2(2), 159–91.

FISCHER, Stanley, ALONSO-GAMO, Patricia, VON ALLMEN, Ulrick Erickson, 2001. "Economic Developments in the West Bank and Gaza since Oslo." *The Economic Journal*, 111, 254–75.

FISCHER, Stanley, HAUSMAN, Leonard J., KARASIK, Anna D. and SCHELLING, Thomas C., 1994. *Securing Peace in the Middle East: Project on Economic Transition*. Cambridge, MA: Harvard University.

FISCHER, Stanley, RODRIK, Dani and TUMA, Elias (eds), 1993. *The Economics of Middle East Peace: Views from the Region*. Cambridge, MA: MIT Press.

FLEISCHMANN, Ellen, 2003. *The Nation and Its New Women: The Palestinian Women's Movement, 1920–1948*. Berkeley: University of California Press.

FMEP (Foundation for Middle East Peace), 2008. *Settlement Population Outside the Separation Barrier* [Online]. Available at: http://www.fmep.org/settlement_info/settlement-info-and-tables/stats-data/settlement-population-outside-the-separation-barrier, accessed May 23, 2014.

FORTE, Tania, 2001. "Shopping in Jenin: Women, Homes and Political Persons in the Galilee." *City and Society*, 12(2), 211–43.

FOUCAULT, Michel, 1969. *L'archéologie du savoir*. Paris: Gallimard.

———, 1977. "Nietzsche, Genealogy, History." In: FOUCAULT, Michel. *Language, Counter-Memory, Practice: Selected Essays and Interviews by Michel Foucault*. New York: Cornell University Press.

———, 2004a, *Sécurité, territoire, population*. Paris: Gallimard.

———, 2004b. *Naissance de la biopolitique*. Paris: Gallimard.

FREEDMAN, Marcia, 1990. *Exile in the Promised Land*. Firebrand: Ithaca Press.

FRISCH Hillel, 1996. "The PLO and the Palestinians in Israel: Politicization or Radicalization?" *Nationalism and Ethnic Politics*, 2(3), 446–64.

FURANI, Khaled, and RABINOWITZ, Dan, 2011. "The Ethnographic Arriving of Palestine." *Annual Review of Anthropology*, 40, 475–91.

GAMBETTA, Diego, 2005. *Making Sense of Suicide Missions*. Oxford: Oxford University Press.

GARB, Yaakov, 2008. *A Survey of Freight Traffic between Israel and the Southern West Bank and the Tarqumia Crossing in Particular* [Online]. A report funded by USAID and submitted to the Economic Cooperation Foundation. Available at: www.ygarb.com/publication, accessed January 6, 2014.

GAZIT, Nir, 2009. "Social Agency, Spatial Practices, and Power: The Micro-foundations of Fragmented Sovereignty in the Occupied Territories." *International Journal of Politics, Culture and Society*, 22(1), 83–103.

GAZIT, Shlomo, 1985. *The Carrot and the Stick, The Israeli Administration of Judea and Samaria*. Tel Aviv: Zmora-Bitan. (In Hebrew).

GHANEM, Asad and OZACKI-LAZAR, Sarah, 2003. "The Status of the Palestinians in Israel in an Era of Peace: Part of the Problem but not Part of the Solution." *Israel Affairs*, 9(1–2), 263–89.

GHANIM, Honeida, 2008. "Thanatopolitics: The case of the Colonial Occupation in Palestine." In: LENTIN, R. (ed.). *Thinking Palestine*. London: Zed Books, 65–81.

GIDDENS, Anthony, 1991. *The Consequences of Modernity*. Cambridge: Polity Press.

GOFFMAN, Erving, 1974. *Les Rites d'Interaction*. Paris: Editions de Minuit.

GORDON, Neve, 2002. "Outsourcing Violations: The Israeli Case." *Journal of Human Rights*, 1(3), 322–37.

———, 2006. "Human Rights." In: RAM, Uri and BERKOVITCH, Nitza (eds). *In-Equality*. Jerusalem: Ben Gurion University of the Negev Press. (In Hebrew).

———, 2007. "Of Dowries and Brides: A Structural Analysis of Israel's Occupation." *New Political Science*, 29(4), 453–78.

———, 2008. *Israel's Occupation*. Berkeley: University of California Press.

———, 2009a. "From Colonization to Separation: Exploring the Structure if Israel's Occupation." In: OPHIR, Adi, GIVONI, Michal and HANAFI, Sari (eds). *The Power of Inclusive Exclusion: Anatomy of Israeli Rule in the Occupied Palestinian Territories*. New York: Zone Books.

———, 2009b. "The Political Economy of Israel's Homeland Security Industry." Working Paper for The New Transparency Project, Queen's University, April 28 [Online]. Available at: http://www.sscqueens.org/sites/default/files/The%20Political%20 Economy%20of%20Israel%E2%80%99s%20Homeland%20Security.pdf, accessed May 13, 2014

GRAYMAN, Ambre, 2007. "J'aime Israel et je le prouve." *Guysen Israel News*, February 21.

GREENBERG, Hanan, 2006. "Civilians Man Border Crossing." *Ynetnews*, January 16 [Online]. Available at: http://www.ynet.co.il/articles/0,7340,L-3200772,00. html, accessed May 13, 2014.

GREILSAMMER, Ilan, 1978. *Les communistes israéliens*. Paris: Presses de la Fondation Nationale des Sciences Politiques.

GRINBERG, Lev, 1991. *Split Corporatism in Israel*. Albany: State University of New York Press.

———, 1993a. *The Histadrut Above All*. Jerusalem: Nevo. (In Hebrew).

———, 1993b. "The Crisis of Statehood: A Weak State and Strong Political Institutions." *Journal of Theoretical Politics*, 5, 89–107.

———, 1996. *Introduction to Political Economy*. Tel Aviv: Security Publisher. (In Hebrew).

———, 2004. "Génocide Symbolique." *La Libre Belgique*, March 29 [Online]. Available at: http://www.lalibre.be/debats/opinions/article/160137/genocide-symbolique.html, accessed May 5, 2011.

———, 2007a. "The Economic Strangling Ring: Three Turning Points in 40 Years of Economic-Military Domination." *Theory and Critic*, 31, Winter. (In Hebrew).

———, 2007b. *Imagined Peace: The Dynamics of Political Spaces in Israel/Palestine*. Ramallah: Mu'assasat al-ayâm, Madar. (In Arabic).

———, 2010. *Politics and Violence in Israel/Palestine. Democracy versus Military Rule*. London and New York: Routledge.

GUPTA, Akhil and FERGUSON, James, 1997. "Beyond 'Culture': Space, Identity and the Politics of Difference." In GUPTA, Akhil and FERGUSON, James (eds). *Culture, Power, Place: Explorations in Critical Anthropology*. Durham, NC: Duke University Press, 33–51.

GUTTMAN, Nathan and KEINON, Herb, 2006. "Wolfensohn Resigns as Quartet Envoy." *Jerusalem Post*, April 5 [Online]. Available at: http://www.jpost.com/Home/Article. aspx?id=20722, accessed May 23, 2014

HAGE, Ghassan, 2003. "Comes a Time We all are Enthusiasm: Understanding Palestinian Suicide Bombing in Times of Exighophobia." *Public Culture*, 15(1), 65–89.

HAJJAR, Lisa, 2005. *Courting Conflict. The Israeli Military Court System in the West Bank and Gaza*. London: University of California Press. HALEVI, Hava, SHLONSKY, Hagith and MACHSOM WATCH, 2008. *Guilty. Membership and Activity in Unlawful Associations. Military Courts 2008*. Jerusalem: MachsomWatch.

HALPER Jeff, 2003. "Why is Israel Really Building the Wall?" In: PENGON (ed.). *The Wall in Palestine. Facts, Testimonies, Analysis and Call to Action*. Jerusalem: Palestinian Environmental NGOs Network, 172–4.

HAMIDI, Camille, 2010. "Catégorisations éthniques ordinaires et rapport au politique : Éléments sur le rapport au politique des jeunes des quartiers populaires." *Revue française de science politique*, 60(4), 719–43.

HAMMAMI, Rema, 2004. "On the Importance of Thugs: The Moral Economy of a Checkpoint." *Middle East Report*, 231–2, 26–34.

———, 2005 'On the Importance of Thugs: The Moral Economy of a Checkpoint', *Jerusalem Quarterly*, 22–23, 16–28.

HANDEL, Ariel, 2009a. "Where, Where to and When in the Occupied Territories? An Introduction to Geography of Disaster." In: GIVONI, Michal, HANAFI, Sari. and OPHIR, Adi. (eds). *The Power of Exclusive Inclusion*. New York: Zone Books, 179–222.

———, 2009b. "Chronology of the Occupation Regime, 1967–2007." In: GIVONI, Michal, HANAFI, Sari and OPHIR, Adi. (eds). *The Power of Exclusive Inclusion*. New York: Zone Books, 603–34.

———, 2013. "Gated/Gating Community: The Settlements' Complex in the West Bank." *Transactions of the Institute of British Geographers*, 39(4), 504–17.

HANIEH, Adam, 2006. "The Politics of Curfew in the Occupied Territories." In: BEININ, Joel and STEIN, Rebecca L. (eds). *The Struggle for Sovereignty: Palestine and Israel 1993–2004*. Stanford, CA: Stanford University Press, 324–37.

HARKER, Christopher, 2010. "New Geographies of Palestine/Palestinians." *The World Arab Geographer*, 13(3–4), 199–216.

HARLE, Vilho, 1994. "On the Concepts of the 'Other' and the 'Enemy'." *History of European Ideas*, 19(1–3), 27–34.

HARLEY, John B., 1989. "Deconstructing the Map." *Cartographica*, 26(2), 1–20.

HARLOW, Barbara, 1996. *After Lives: Legacies of Revolutionary Writing*. London: Verso.

HARTSHORNE, Richard. 1939. *The Nature of Geography*. Lancaster, Penn.: Association of American Geographers.

HASS, Amira, 2002. "Israel's Closure Policy: An Ineffective Strategy of Containment and Repression." *Journal of Palestine Studies*, 31(3), 5–20.

———, 2009. "Tul Karem Checkpoint: The Workers Are Allowed to Enter No More than 5 Pitah." *Haaretz*, June 29. (In Hebrew).

HASSON, Shlomo, 1998. "From Frontier to Periphery in Israel: Cultural Representation in Narratives and Counter Narratives." In: YIFTACHEL, O. and MEIR, A. (eds). *Ethnic Frontiers and Peripheries: Landscapes of Development and Inequality in Israel*. Boulder, CO: Westview Press, 115–40.

HAVKIN, Shira, 2008. *La privatisation des checkpoints: Esquisse d'une transformation d'un dispositif de pouvoir*. Master's thesis, supervised by Jocelyne Duboist and Elsa Dorlin, University of Paris 1 [Online]. Available at: www.fasopo.org/reasopo/jr/mem_havkin.pdf, accessed April 24, 2014.

———, 2011. "The Reform of Israeli Checkpoints: Outsourcing, Commodification and Redeployment of the State." *Les Études du CERI*, 174bis, May [Online]. Available at: http://www.sciencespo.fr/ceri/sites/sciencespo.fr.ceri/files/Etude174_english.pdf, accessed April 13, 2014.

HAYES, Graeme and OLLITRAULT, Sylvie, 2012. *La Désobéissance Civile*. Paris: Presses de Sciences Po.

HAZAN, Haim, 1994, *Old Age: Constructions and Deconstructions*, Cambridge: Cambridge University Press.

HEIDEGGER, Martin, 1996. *Being and Time*. Albany: State University of New York Press.

HERB, Guntram H., 2008. "The Politics of Political Geography." *The SAGE Handbook of Political Geography*, 21–40.

HERMANN, Tamar, 2009. *The Israeli Peace Movement. A Shattered Dream*. New York: Cambridge University Press.

HERTZOG, Sergio, 2005. "The Relationship between Economic Hardship and Crime: The Case of Israel and the Palestinians." *Sociological Perspectives*, 48(2), 189–211.

HEVER, Shir, 2010. *The Political Economy of Israel's Occupation*. Pluto Press: New York.

HIBOU, Béatrice (ed.), 1999. *La Privatisation des Etats*. Paris: Karthala.

HIER, Sean P., 2003. "Risk and Panic in Late Modernity: Implications of the Converging Sites of Social Anxiety", *British Journal of Sociology*, 54(1), 3–20.

HILAL, Jamil (ed.), 2007. *Where Now for Palestine? The Demise of the Two State Solution*. London: Zed Books.

HOCHBERG, Gil, 2010. "Israelis, Palestinians, Queers: Points of Departure." *GLQ: A Journal of Lesbian and Gay Studies*, 16(4), 493–516.

HOCHBERG, Gil, 2010. "'Check Me Out.' Queer Encounters in Sharif Waked's Chic Point: Fashion for Israeli Checkpoints." *GLQ: A Journal of Lesbian and Gay Studies*, 16(4), 577–97.

HOFMAN, Gil and JERUSALEMPOST.COM STAFF, 2008. Israel Beitenu to Quit Coalition, *Jerusalempost.com*, uploaded January 15, 2008, available at: http://www.jpost.com/Israel/Israeli-Beiteinu-to-quit-coalition, accessed January 7, 2015.

HORKHEIMER, Max, 2002 [1968]. "Traditional and Critical Theory." In: HORKHEIMER, Max (ed.). *Critical Theory: Selected Essays*. New York: Continuum.

HOUELLEBECQ, Michel, 1997. *Rester Vivant et Autres Textes*. Paris: J'ai Lu.

HUSSEINI-Al, Jalal and SIGNOLES, Aude (eds), 2011. *Les Palestiniens entre Etat et Diaspora*. Paris: Karthala/IISMM.

INTELLIGENCE AND TERRORISM INFORMATION CENTER, 2008. *Has the Palestinian Authority Fulfilled its Security Commitments, and to what Extent, in Accordance with the Road Map? Interim Report*. January 9 [Online] Available at: http://www.terrorism-info.org.il/en/article/19109, accessed May 1, 2011.

INTERNATIONAL CRISIS GROUP, 2008. *Ruling Palestine II: the West Bank Model?* Middle East Report No. 79, July 17.

ISSACHAROFF, Avi, 2007. "New Fatah Gaza Force Training for Future Clashes with Hamas." *Haaretz*, April 1 [Online]. Available at: http://www.haaretz.com/news/new-fatah-gaza-force-training-for-future-clashes-with-hamas-1.217164, accessed October 8, 2012

ISSACHAROFF, Avi and AZOULAY, Yuval, 2008. "PA to Deploy Nearly 500 Armed Police Around Jenin." *Haaretz*, May 4 [Online]. Available at: http://www.haaretz.com/print-edition/news/pa-to-deploy-nearly-500-armed-police-around-jenin-1.245202, accessed May 4, 2008

JAMESON, Fredric, 1981. *The Political Unconscious: Narrative as a Socially Symbolic Act*. Ithaca, NY: Cornell University Press.

JAYYUSI, May, 2004. "Subjectivity and Public Witness: An Analysis of Islamic Militancy in Palestine." Unpublished Paper, presented at the SSRC Conference on the Public Sphere in the Middle East, Beirut.

KAMINER, Reuven, 1996. *The Politics of Protest. The Israeli Peace Movement and the Palestinian Intifada*. Brighton: Sussex Academic Press.

KANAANEH, Rhoda, 2002. *Birthing the Nation: Strategies of Palestinian women in Israel*. Berkeley: University of California Press.

KANAFANI, Ghassan, 1969. *Um Saad*. Beirut: The Arab Institute for Research.

KATZ, Yaakov and JPOST STAFF, 2006. "Car parts smuggled from Israel to Iraq." *Jerusalem Post*, December 3 [Online]. Available at: http://www.jpost.com/Israel/Article.aspx?id=15791, accessed February 28, 2010.

KELLEHEAR, Allan, 2007. *A Social History of Dying*. Cambridge: Cambridge University Press.

KELLY, Tobias, 2006. *Law, Violence and Sovereignty among West Bank Palestinians*. Cambridge: Cambridge University Press.

KEMP, Adriana and RAIJMAN, Rebecca, 2008. *Migrants and Workers: The Political Economy of Labor Migration in Israel*. Jerusalem: Van Leer Jerusalem Institute and Hakibutz Hameuchad. (In Hebrew).

KERSHNER, Isabel, 2005. *Barrier: The Seam of the Israeli-Palestinian Conflict*. New York: Palgrave Macmillan.

KHALIDI, Rashid, 2006. *The Iron Cage: The Story of the Palestinian Struggle for Statehood*. Boston: Beacon Press.

KHALILI, Laleh, 2009. *Heroes and Martyrs of Palestine: The politics of National Commemoration*. Cambridge: Cambridge University Press.

KHALILI, Laleh and SCHWEDLER, Jillian (eds), 2010. *Policing and Prisons in the Middle East: Formations of Coercion*. London: Hurst and Company.

KHOURY, Bassem, 2009. "Le Hamas ne Sait Plus Quoi Faire de Son Argent." *Le Monde*, October 11 [Online]. Available at: http://www.lemonde.fr/proche-orient/article/2009/10/10/bassem-khoury-le-hamas-ne-sait-plus-quoi-faire-de-son-argent_1252157_3218.html, accessed May 19, 2014.

KIMMERLING, Baruch, 1993, "Patterns of Militarism", *European Journal of Sociology*, 2, 1–28.

KIMMERLING, Baruch and MIGDAL, Joel S., 2003. *The Palestinian People: A History*. Cambridge, MA: Harvard University Press.

KORIN LIBER, Stella, 1999. "The IDF Starts to Outsource." *Globs*, September 28. (In Hebrew)

KOTEF, Hagar and MERAV, Amir, 2007. "(En)Gendering Checkpoints: Checkpoint Watch and the Repercussions of Intervention." *Signs: Journal of Women in Culture and Society*, 32, 973–96.

———, 2011. "Between Imaginary Lines Violence and its Justifications at the Military Checkpoints in Occupied Palestine." *Theory, Culture & Society* 28, 55–80.

KOTLER, Philip, JATUSRIPITAK Somkid and MAESINCEE, Suvit (eds), 1997. *The Marketing of Nations: A Strategic Approach to Building National Wealth*. New York: The Free Press.

KRAHULIK, Karen, 2005. "Aswat (Voices): An Interview with Rauda Morcos." *Peace and Change*, 30(4), 516–17.

LABAN, Ahmad, 2005. "Unsafe Streets." *Palestine Report*, 12(1), June 22 [Online]. Available at: http://www.palestinereport.ps/article.php?article=814, accessed May 2, 2011.

LAGNADO, Lisette, 2007. "Home and Homeland: About Ahlam Shibli's Photo-Series 'Eastern LGBT'." *Nafas Art Magazine* [Online]. Available at: http://universes-iuniverse.org/eng/nafas/articles/2007/ahlam_shibli/, accessed November 27, 2013.

LAHIRE, Bernard, 1998. *L'homme Pluriel. Les Ressorts de l'Action.* Paris: Nathan.

LAMARCHE, Karine, 2009. "Political Activism and Legitimacy in Israel: Four Groups between Cooperation and Transgression." In: MARTEU, Elisabeth (ed.). *Civil Organizations and Protest Movement in Israel.* New York: Palgrave Macmillan.

———, 2010. "Du pacifisme à la dissidence histoire et enjeux des mobilisations israéliennes contre l'occupation." In: BENBASSA, Esther (ed.). *Israël Palestine: Les Enjeux d'un conflit.* Paris: CNRS, 147–61.

———, 2011. "S'engager corps et âme." *Cultures & Conflits*, 81–2, 125–50.

———, 2013. *Militer contre son camp ? Des israéliens engagés aux côtés des Palestiniens.* Paris: PUF.

LARZILLIERE, Pénélope, 2010. "Fermeture ou transformation des 'possibles protestataires'? L'évolution de l'engagement politique dans les Territoires palestiniens." In: BENBASSA, Esther (ed.). *Israël/Palestine: Etat des lieux.* Paris: CNRS, 195–207.

LATTE ABDALLAH, Stéphanie, 2006. "Notes sur quelques figures récurrentes du corps et du genre dans les guerres de Palestine." *Quasimodo*, 8–9 [Online]. Available at: http://www.revue-quasimodo.org, accessed June 10, 2009.

———, 2009. "Frontières militantes et enfermement. Les prisonniers palestiniens en Israël." In: LATTE ABDALLAH, Stéphanie and PARIZOT, Cédric (Org.). *Limites et frontières des espaces israéliens et palestiniens II*, November 17, 2009 [Online]. Available at: http://mediamed.mmsh.univ-aix.fr/chaines/mofip/limites-frontieres2/Pages/frontieres-militantes.aspx, accessed May 23, 2014.

———, 2010a. "Dedans/dehors: vivre et agir sur l'incarcération politique après Oslo." Presented at *In the Shadow of Separation: Territorial Regimes and New Socio-Economic Practices.* November 24, 2010. Al-Quds University, Abu Dis.

———, 2010b. "Incarcération des femmes palestiniennes et engagement (1967–2009)." *Le Mouvement Social*, 231, 9–27 [Online]. Available at: www.cairn.info/revue-le-mouvement-social-2010-2-page-9.htm, accessed May 26, 2014.

———, 2013. "Des féminités mobilisées et incarcérées en Palestine." *Critique Internationale*, 60, 53–69.

———, 2014a. "Entre dedans et dehors: vécus parentaux des détenus politiques palestiniens en Israël, "*Champ pénal/Penal Field*, Special Issue, XI [Online]. Available at: www.champpenal.revues.org, accessed May 19, 2014.

———, 2014b. "Washington s'engage mais Israéliens et Palestiniens sont pessimistes, entretien avec Stéphanie Latte Abdallah (propos recueillis par Sybille de Larocque)", *JOL presse*, February 10, http://www.jolpress.com/israel-palestine-jerusalem-colonies-colonisation-paix-mahmoud-abbasnetanyahu-article-824342.html, accessed May 30, 2014.

———, 2014c. "Au fil des années, Israël est de plus en plus isolé sur la scène internationale, entretien avec Stéphanie Latte Abdallah (propos recueillis par Sybille de Larocque)", *JOL presse*, February 12, http://www.jolpress.com/israel-palestine-colonies-jerusalem-refugies-frontieres-paix-john-kerryarticle-824343.html, accessed May 30, 2014.

LATTE ABDALLAH, Stéphanie and PARIZOT, Cédric (eds), 2011. *A l'ombre du Mur: Israéliens et Palestiniens entre séparation et occupation.* Arles: Actes Sud/MMSH.

LEBRUN, Céline and SALINGUE, Julien, 2013. *Israël, un Etat d'Apartheid?* Paris: L'Harmattan.

LEGRAIN, Jean-François, 1996. "La Palestine: De la terre perdue à la reconquête du territoire." *Cultures & Conflits*, 21–2, 171–221.
———, 1997. "Palestine: Les bantoustans d'Allah." In: BOCCO, Riccardo, DESTREMEAU, Blandine and HANNOYER, Jean (eds). *Palestine, Palestiniens. Territoire national, espaces communautaires*. Beyrouth, Liban: CERMOC, 85–101.
———, 2010. *Palestine: Un État ? Quel État ?* Paris: Note de l'IFRI [Online]. Available at: http://www.ifri.org/downloads/notemomjflegrain.pdf, accessed May 25, 2014
LEVON, Erez, 2010. *Language and the Politics of Sexuality. Lesbians and Gays in Israel*. New York: Palgrave Macmillan.
LEVY Yagil, 1998. "Militarizing Inequality: A Conceptual Framework." *Theory and Society*, 27(6), 873–904.
———, 2003. "Social Convertibility and Militarism: Evaluations of the Development of Military-Society Relations in Israel in the Early 2000s." *Journal of Political and Military Sociology*, 31(1), 71–96.
———, 2006. "The War of the Peripheries: A Social Mapping of IDF Casualties in the Al-Aqsa Intifada." *Social Identities: Journal for the Study of Race, Nation and Culture*, 12(3), 309–24.
———, 2007a. *From the "People's Army" to the "Army of the Peripheries"*. Tel Aviv: Carmel. (In Hebrew).
———, 2007b. "The Embedded Military: Why Did the IDF Perform Effectively in Executing the Disengagement Plan?" *Security Studies*, 16(3), 382–408.
———, 2010. *Who Governs the Military? Between Control of the Military and Control of Militarism*. Jerusalem: Eshkolot Library and Magnes University Press. (In Hebrew)
LIA, Brynjar, 2006. *A Police without a State. A History of the Palestinian Security Forces in the West Bank and Gaza*. Reading: Ithaca.
LOMSKI-FEDER, Edna, 2004. "Life Stories, Wars and Veterans: On the Social distribution of Memories", *Ethos*, 32(1), 1–28.
LOUËR, Laurence, 2000. "Qu'est-ce qu'une communauté? Les citoyens arabes dans le système communautaire israélien." *Bulletin du CRFJ*, 7, 37–46.
———, 2001a. "L'identité palestinienne en contexte: Se penser et se dire palestinien en Israël aujourd'hui." *Les Annales de l'Autre Islam*, 8: 399–417.
———, 2001b. "L'Intifada d'Al-Aqsa: Quelle place pour les citoyens arabes dans l'Etat Juif?" *Cultures & Conflits*, 41, 101–19.
———, 2002. "Un seul peuple de la Galilée à Hébron." *Critique Internationale*, 16, 36–9.
———, 2003a. *Les citoyens arabes d'Israël*. Paris: Balland.
———, 2003b. "Une Intifada des arabes d'Israël ?" In: DIECKHOFF, Alain and LEVEAU, Rémy (eds). *Israéliens et Palestiniens. La guerre en Partage*. Paris: Balland, 153–80.
LOZANO, Beverly, 1989. *The Invisible Work Force: Transforming American Business with Outside and Home-based Workers*. New York: Free Press.
LUSTICK, Ian, 1980. *Arabs in the Jewish State*. Austin: Texas University Press.
MA'AN NEWS AGENCY, 2009. "Hebron Police Destroy 100 Unlicensed Cars." *Maan News Agency*, September 28 [Online]. Available at: http://www.maannews.net/eng/ViewDetails.aspx?ID=228218, accessed May 1, 2011.
———, 2010a. "US Tunnel Detection System Left Unfinished." *Maan News Agency*, June 8 [Online]. Available at: http://www.maannews.net/eng/ViewDetails.aspx?ID=290433, accessed May 19, 2014.

———, 2010b. "Traders Say Tunnel Industry Down." *Maan News Agency*, June 26 [Online]. Available at: http://www.maannews.net/eng/ViewDetails.aspx?ID=294608, accessed May 19, 2014.

———, 2010c. "Gaza Tunnel Trade Takes Turn." *Maan News Agency*, October 7 [Online]. Available at: Online: http://www.maannews.net/eng/ViewDetails.aspx?ID=321720, accessed May 19, 2014.

MAKDISI, Saree, 2010. *Palestine Inside Out: An Everyday Occupation*. New York: Norton.

MALLEY, Robert and AGHA, Hussein, 2001. "Camp David: The Tragedy of Errors." *The New York Review of Books*, July 12 [Online]. Available at: http://www.nybooks.com/articles/archives/2001/aug/09/camp-david-the-tragedy-of-errors/?pagination=false, accessed August 12, 2009.

MALONEY, William F., 2004. "Informality Revisited." *World Development*, 32(7), 1, 159–78.

MAMAN, Daniel and ROSENHEK, Zeev, 2009. *The Bank of Israel: Political Economy in the Neo-Liberal Era*. Jerusalem: The Van-Leer Jerusalem Institute and Hakibbutz Hameuchad. (In Hebrew).

MANSBACH, Daniela, 2007. "The Power of the Duality: The Protest of the 'Checkpoint Watch' Movement." *Theory and Criticism*, 31, 77–99. (In Hebrew).

———, 2009. "Normalizing Violence: From Military Checkpoints to 'Terminals' in the Occupied Territories." *Journal of Power*, 2(2), 255–73.

MAOZ, Eilat, 2008. "The Privatization of the Checkpoints and the Late Occupation." *Coalition of Women for Peace*, October 10 [Online]. Available at: http://www.coalitionofwomen.org/?p=862, accessed May 13, 2014. (In Hebrew).

MARTEU, Elisabeth (ed.), 2009a. *Civil Organizations and Protest Movement in Israel*. New York: Palgrave Macmillan.

———, 2009b. "Relations Entre Associations Féminines Palestiniennes des Deux Côtés de la Ligne Verte." *EchoGéo*, 8.

MASSAD, Joseph, 2007. *Desiring Arabs*. Chicago: University of Chicago Press.

MATONTI, Frédérique and POUPEAU, Franck, 2004. "Le Capital Militant. Essai de Définition." *Actes de la recherche en sciences sociales*, 5(155), 4–11 [Online]. Available at: www.cairn.info/revue-actes-de-la-recherche-en-sciences-sociales-2004-5-page-4.htm, accessed May 25, 2014.

MBEMBE, Achille, 2003. "Necropolitics." *Public Culture*, 15(1), 11–40.

MERCIER, Delphine, 2009. "Affranchissement et exception au coeur des frontières profitables. Le cas des zones franches d'exportation industrielle en Amérique Centrale." *Critique* Économique, 25, 109–32.

MICHAEL, Kobi, 2007. "The Dilemma behind the Classical Dilemma of Civilian-Military Relations: The 'Discourse Space' Model and the Israeli Case during the Oslo Process." *Armed Forces & Society*, 33(4), 518–46.

MIGNOLO, Walter, 2000. *Local Histories/Global Designs: Coloniality, Subaltern Knowledge, and Border Thinking*. Princeton: Princeton University Press.

MONMONIER, Mark, 1991. *How to Lie With Maps*. Chicago: University of Chicago Press.

MORRIS, Benny, 2009. *One State, Two States: Resolving the Israel/Palestine Conflict*. New Haven: Yale University Press.

MOUSSA, Ghaïda, 2011. *Narrative (sub) Versions: How Queer Palestinian Womyn "Queer Palestinian Identity."* Unpublished thesis, Ottawa University.

NASHIF, Esmail, 2005. "Attempts at Liberation: Body Materialization and Community Building among Palestinian Political Captives." *Journal of Arab Studies*, 7–8, 46–79.

———, 2008a. *Palestinian Political Prisoners: Identity and community*, London, Routledge.

———, 2008b. "Images of the Fragmented." In: GRAY Nicola and HASHSHASH, M. Abu (eds). *The Young Palestinian Artists Award for 2006*. Ramallah and London: A.M. Qattan Foundation, 24–31.

NASSAR, Jamal and HEACOCK, Roger, 1990. *Intifada: Palestine at the Crossroads*. New York: Praeger.

NEVO, Baruch and SHUR-SHMUELI, Yael, 2004. *The Army and Society Project: The IDF and the Israeli Economy*. Jerusalem: The Israel Democracy Institute. (In Hebrew).

NEWKEY BURDEN, Chas, 2007. "Tel Aviv: The Final Gay Frontier." *Ynetnews*, August 1 [Online]. Available at: http://www.ynetnews.com/articles/0,7340,L-3349664,00.html, accessed July 27, 2008.

NORDSTROM, Carolyn, 1997. *A Different Kind of War Story*. Philadelphia: University of Pennsylvania Press.

NRG Ma'ariv, 2004. "A Vital Human Structure Should be Enabled: The Main Conclusions of the Commission for a Revolution in the Checkpoints." *NRG Ma'ariv*, August 24 [Online]. Available at: http://www.nrg.co.il/online/1/ART/772/991.html, accessed May 13, 2014. (In Hebrew).

ODEH, Aïcha, 2004. *Dreams of Freedom*. Ramallah: Muwatin. (In Arabic).

OHNUKI-TIERNEY, Emiko, 2002. *Kamikaze, Cherry Blossoms, and Nationalisms: The Militarization of Aesthetics in Japanese History*. Chicago: University of Chicago Press.

OLMSTED, Jennifer, 2009 "Locked in, Locked out of Work." *Apartheid and Beyond, Middle East Report*, 253, 28–31.

OPHIR, Adi, GIVONI, Michal and HANAFI, Sari (eds), 2009. *The Power of Inclusive Exclusion: Anatomy of Israeli Rule in the Occupied Palestinian Territories*. New York: Zone Books.

PAPE, Robert, 2005. *Dying to Win: The Strategic Logic of Suicide Terrorism*. New York: Random House.

PAPPE, Ilan, 2007. *The Ethnic Cleansing of Palestine*. Oxford: Oneworld Publications.

PARIZOT, Cédric, 2001a. *Le mois de la bienvenue. réappropriations de mécanismes électoraux et réajustements de rapports de pouvoir chez les Bédouins du Néguev, Israël*. PhD thesis. Paris: EHESS.

———, 2001b. "Gaza, Beersheba, Dhahriyya: Another Approach to the Negev Bedouins in the Israeli-Palestinian Space." *Bulletin du CRFJ*, 9 [Online]. Available at: http://bcrfj.revues.org/1742, uploaded 20 February 2008, accessed May 19, 2014.

———, 2006a. "Crossing Borders, Reinforcing Boundaries: Kin-Nections of Negev Bedouin in Gaza, West Bank and Jordan." *Cairo Papers in Social Science*, 29(1), 58–84.

———, 2006b. "En attendant le Mur: Gestions israéliennes des mobilités palestiniennes pendant la seconde intifada (2000–2006)." *Migrations Société*, 18(107), 15–39 [Online]. Available at: http://halshs.archives-ouvertes.fr/docs/00/11/39/73/PDF/en_attendant_le_mur-c.pdf, accessed May 23, 2014.

———, 2006c, "Désengagements électoraux: L'abstention arabe lors des élections législatives israéliennes du 28 Mars 2006." *La Lettre InterGéo 2*, 4–7 [Online]. Available at: http://www.prodig.cnrs.fr/IMG/pdf/Lettre2_2006.pdf, accessed May 23, 2014.

———, 2008a. "L'émergence de frontières interpalestiniennes." *Revues d'Etudes Palestiniennes*, 107.

———, 2008b. "Tightening Closure Securing Disorder." *Refugee Watch*, 31, 54–74.

———, 2009a. "Après le Mur: Les représentations israéliennes de la séparation avec les Palestiniens." *Cultures & Conflits*, 73, 53–72.

———, 2009b. "Hardening Closure, Securing Disorder: Israeli Closure Policies and the Informal Border economy between the West Bank and the Northern Negev (2000–2006)." In: BECHEV, Dimitar and NICHOLAIDIS, Kalypso (eds). *Mediterranean Frontiers: Borders, Conflict and Memory in a Transnational World.* London: Tauris Academic Studies, 177–94.

———, 2009c. "Temporalities and Perception of the Separation between Israelis and Palestinians." *Bulletin du CRFJ*, 20 [Online]. Available at: http://bcrfj.revues.org/6319, accessed May 5, 2014.

———, 2010. "Séparation et régime de mobilité entre Israël et la Cisjordanie (2000–2009): Réglementations, pratiques et acteurs." In Esther Benbassa (ed.) *Israël-Palestine: Les Enjeux d'un conflit*, Paris: CNRS, 129–46.

———, 2012. "The Israeli-Palestinian Conflict from 1993 to 2012." *Medmem: Mediterranean Memory* [Online]. Available at: http://www.medmem.eu/en/folder/40/amorce-processus-paix-impasse, accessed February 5, 2014.

———, 2014. "An Undocumented Economy of Control." In: ANTEBY, L., BABY COLLIN, V., MAZZELLA, S., MOURLANE, S., REGNARD, C. and SINTÈS, P. (eds). *Borders, Mobility and Migration.* Brussels: Peter Lang, 93–112.

PARSONS, Nigel, 2010. "Israeli Biopolitics, Palestinian Policing. Order and Resistance in the Occupied Palestinian Territories." In: KHALILI, L. and SCHWEDLER, J. (eds). *Policing and Prisons in the Middle East: Formations of Coercion.* London: Hurst and Company, 57–76.

PARSONS, Nigel and SALTER, Mark B., 2008. "Israeli Biopolitics: Closure, Territorialisation and Governmentality in the Occupied Palestinian Territories." *Geopolitics*, 13(4), 701–23.

PEDAHZUR, Ami (ed.), 2006. *Root Causes of Suicide Terrorism: The globalization of martyrdom.* London: Routledge.

PELHAM, Nicolas, 2011. "Gaza's tunnels: The Burrowing Business." August 10 [Online]. Available at: http://www.economist.com/blogs/newsbook/2011/08/gazas-tunnels, accessed May 19, 2014.

PENGON (ed.), 2003. *The Wall in Palestine. Facts, Testimonies, Analysis and Call to Action,* Jerusalem: PENGON.

PERES CENTER AND PALTRADE, 2006. *The Untapped Potential: Palestinian-Israeli Economic Relations: Policy Options and Recommendations.* December. Tel Aviv, Ramallah: Paltrade and Peres Center for Peace.

PERI Yoram, 1985. *Between Battles and Ballots: Israeli Military in Politics.* Cambridge: Cambridge University Press.

PERRY, Mark and WOODWARD, Paul, 2007. "Document details 'U.S.' Plan to sink Hamas", *Asia Times*, May 16, 2007, available at http://www.atimes.com/atimes/Middle_East/IE16Ak04.html, accessed January 16, 2015.

PETEET, Julie, 1994. "Male Gender and Rituals of Resistance in the Palestinian *Intifada*: A cultural politics of violence." *American Ethnologist*, 21(1), 31–49.

———, 2008. "Stealing Time." *Middle East Report: Waiting, the Politics of Time in Palestine*, 248, Autumn, 14–17.

———, 2009. "Beyond Compare." *Middle East Report, Apartheid and Beyond*, 253, 16–25.

PETTI, Alessandro, 2007. *Archipelaghi e Enclave: Architettura dell'ordinamento Spaziale Contemporaneo*, Milan: Mondadori Bruno.

———, 2008. "Asymmetries." In: FARQUHARSON, Alex (ed.). *The Impossible Prison.* Nottingham: Centre for Contemporary Art.

PICAUDOU, Nadine (ed.), 2006. *Territoires palestiniens de mémoire*, Paris: Karthala; Beirut: IFPO.

PICAUDOU, Nadine and RIVOAL, Isabelle (eds), 2006. *Retours en Palestine*. Paris: Karthala.

PILE, Steve and KEITH, Michael (eds), 1997. *Geographies of Resistance*. London: Routledge.

PITCHER, Linda, 1998. "The Divine Impatience: Ritual, Narrative, and Symbolization in the Practice of Martyrdom in Palestine." *Medical Anthropology Quarterly*, 12, 8–30.

PITERMAN, Sylvia and MACHSOM WATCH, 2009, *Invisible Prisoners - continued - The Obstacle Race on the Way to a Magnetic Card*, Machsom Watch, available at http://www.machsomwatch.org/files/InvisiblePrisonersTheMagneticCardEng1009.pdf, accessed January, 16, 2015.

PORTES, Alejandro, CASTELLS, Manuel and Lauren BENTON (ed.), 1989. *The Informal Economy: Studies in Advanced and Less Developed Countries*, Baltimore: Johns Hopkins University Press.

PORTES, Alejandro, and HALLER, William, 2005. "The Informal Economy." In: SMELSER, Neil J. and SWEDBERG, Richard (eds). *Handbook of Economic Sociology*. Second edition. New York: Princeton University Press, 403–25.

PORTUGALI, Juval, 1993. *Implicate Relations: Society and Space in the Israeli-Palestinian Conflict*. Dordrecht and Boston: Kluwer Academic Publishers.

POUZOL, Valérie, 2007. "L'engagement de l'ombre: Homosexualité et militantisme pour la paix dans le mouvement des Femmes en noir (Israël, 1998–2004)." In: Perreau, Bruno (ed.), *Le choix de l'homosexualité*. Paris: EPEL, 56–75.

———, 2008. *Clandestines de la paix. Israéliennes et Palestiniennes contre la guerre*. Paris: Complexe.

PUAR, Jasbir K., 2007. *Terrorist Assemblages: Homonationalism in Queer Times*. Durham: Duke University Press.

RABINOWITZ, Dan, 2003. "Borders and their Discontents Israel's Green Line, Arabness and Unilateral Separation." *European Studies*, 19, 217–31.

RADI, Lamia, 2002. "La crise de la territorialisation du nationalisme diasporique palestinien: Une gouvernance transnationale inachevée." In: DIECKHOFF, Alain and KASTORYANO, Riva (eds). *Nationalismes en mutation en Méditerranée orientale*. Paris: CNRS, 197–225.

RAM, Uri, 2008. *The Globalization of Israel: Mcworld in Tel Aviv, Jihad In Jerusalem*. New York: Routledge.

RAM, Uri and BERKOVITCH, Nitza (eds), 2007. *In-Equality, Beer-Sheva*. Beer-Sheva: The Bialik Institute, Ben-Gurion University. (In Hebrew).

RANCIÈRE, Jacques, 1995. *On the Shores of Politics*. London: Verso.

———, 2004 [1990]. *Aux bords du politique*. Paris: Gallimard.

REKHESS, Elie, 1989. "Israeli Arabs and the Arabs of the West Bank and Gaza: Political Affinity and National Solidarity." *Asian and African Studies*, 23(2–3), 119–54.

———, 2002. "The Arabs in Israel after Oslo: Localization of the National Struggle." *Israel Studies*, 7(3), 1–44.

REUTER, Christoph, 2004. *My Life is a Weapon: A Modern History of Suicide Bombing*. Princeton: Princeton University Press.

RICOLFI, Luca, 2005. "Palestinians 1981-2003", In D. Gambetta (ed.) *Making Sense of Suicide Missions*. Oxford University Press, 77–129.

RITCHIE, Jason, 2010. "How do you say 'Come out of the closet' in Arabic? Queer Activism and the Politics of Visibility in Israel-Palestine." *GLQ: A Journal of Lesbian and Gay Studies*, 16(4), 557–75.

ROSENFELD, Maya, 2004. *Confronting the Occupation: Work, Education, and Political Activism of Palestinian Families in a Refugee Camp*. Stanford: Stanford University Press.

ROUHANA, Nadim, 1997. *Palestinian Citizens in an Ethnic Jewish State*. New Haven: Yale University Press.

ROSE, Gillian, 2003. "On the Need to Ask How, Exactly, is Geography 'Visual'?" *Antipode*, 35(3), 212–21.

ROSENHEK, Zeev, 1999. "Migrant Workers in the Israeli Welfare State: Trends of Exclusion and Inclusion." *Social Security*, 56, 97–112. (In Hebrew).

ROY, Ananya, 2009. "Why India Cannot Plan its Cities: Informality, Insurgence and the Idiom of Urbanization." *Planning Theory*, 8(1), 76–87.

ROY, Sarah, 1995. *The Gaza Strip: The Political Economy of De-development*. Washington, DC: The Institute for Palestine Studies.

———, 2001. "Decline and Disfigurement: The Palestinian Economy after Oslo." In: CAREY, Roane (ed.). *The New Intifada, Resisting Israel's Apartheid*. London: Verso.

SADAN, Ezra, 2004. "How is Israel's Economy Affected by the Security Situation?" *Jerusalem Issue Brief*, 4(11), December 19.

SAID, Edward, 2001. *The End of the Peace Process: Oslo and After*. New York: Vintage.

SALIM, Abd al-Halim, 2010 "Tons of Explosives and Missiles found in Sinai before being Smuggled to Gaza." *Al-Youm al-Saba*, May 12 [Online]. Available at: http://www.youm7.com/News.asp?NewsID=226325, accessed May 19, 2014. (In Arabic).

SARFATI, Philippe, 2006. "Enorme solidarité en faveur d'Israël, mais … reste la reconstruction." *Guysen Israel News*, September 11.

SASSON-LEVY, Orna, 2002, "Constructing Identities at the Margins: Masculinities and Citizenship in the Israeli Army", *Sociological Quarterly*, 43(3), 357–83.

———, 2008, "Individual Bodies, Collective State Interests: The Case of Israeli Combat Soldiers", *Men and Masculinities*, 10(3), 296–321.

SAYIGH, Rosemary, 1979. *The Palestinians: From Peasants to Revolutionaries*. London: Zed Books.

SAYIGH, Yezid, 1997. *The Armed Struggle and the Search for State: The Palestinian National Movement 1949–1993*. Oxford: Clarendon Press.

SCHNEIDER, Friedrich and ENSTE, Dominik, 2000. "Shadow Economies: Size, Causes, and Consequences." *Journal of Economic Literature*, 38(1), 77–114.

SHAFIR, Gershon and PELED, Yoav, 2000. *The New Israel: Peacemaking and Liberalization*. Boulder, CO: Westview Press.

———, 2002. *Being Israeli: The Dynamics of Multiple Citizenship*. New York: Cambridge University Press.

SHALHOUB-KEVORKIAN, Nadera, 2009. *Militarization and Violence against Women in Conflict Zones in the Middle East: A Palestinian Case-study*. Cambridge: Cambridge University Press.

SHARP, Gene, 1973. *The Politics of Nonviolent Action*. Boston: Porter Sargent.

SHEHADEH, Raja, 1982. *The Third Way: A Journal of Life in the West Bank*. London: Quartet Books.

———, 2008. *Palestinian Walks: Notes on a Vanishing Landscape*. London: Profile Books.

SHELAH, Ofer, 2003. *The Israeli Army: A Radical Proposal*. Tel Aviv: Zmora-Bitan. (In Hebrew).

———, 2005. "The Dark Side of Disengagement." *NRG Ma'ariv*, July 1.
SHENHAV, Yehuda, 1995. *The Organization Machine: A Critical Inquiry into the Foundations of Management Theory*. Tel Aviv: Shoken. (In Hebrew).
SHIKLER, Omer, 2005. "The IDF Acts to Improve the Passages and the Checkpoints." *NFC*, July 22. (In Hebrew).
SHOKEID, Moshe, 1992. Commitment and Contextual Study in Anthropology. *Cultural Anthropology*, 7(4), 464–77.
SHULMAN, David, 2006. *Ta'ayush. Journal d'un combat pour la paix. Israël Palestine 2002–2005*. Paris: Seuil.
SCHULMAN, Sarah, 2011. "A Documentary Guide to 'Brand Israel' and the Art of Pinkwashing." *Mondoweiss. The War of Ideas in the Middle-East* [Online]. Availale at: http://mondoweiss.net/2011/11/a-documentary-guide-to-brand-israel-and-the-art-of-pinkwashing.html, accessed October 10, 2013.
———, 2012. *Israel/Palestine and the Queer International*. Durham, NC: Duke University Press.
SMOOHA, Sammy, 1984. *The Orientation and Politicization of the Arab Minority in Israel*. Haifa: The Jewish-Arab Center.
———, 1989. "The Arab minority in Israel: Radicalization or Politization?" *Studies in Contemporary Jewry*, 5, 59–88.
SMITH, Charles, 2007. *Palestine and the Arab-Israeli Conflict*. Bedford: San Martin's.
SMITH, Rafi and ARIAN, Reli, 2007, *Public Attitudes towards Israeli Police 2006*, State of Israel, Ministry of Public Security, Bureau of the Chief Scientist. (in Hebrew).
SNEGAROFF Claire and Michael BLUM, 2005. *Qui sont les colons? Une enquête de Gaza à la Cisjordanie*. Paris: Flammarion.
SOLOMON, Alisa, 2003. "Viva la Diva Citizenship: Post Zionism and Gay Rights." In: BOYARIN, Daniel, ITKOWITZ, Daniel and PELLEGRINI, Ann (eds). *Queer Theory and the Jewish Question*. New York: Columbia University Press, 149–65.
SORKIN Michael (ed.), 2002. *The Next Jerusalem: Sharing the Divided City*. New York: Monacelli Press.
SORKIN, Michael, 2005. *Against the Wall: Israel's Barrier to Peace*. New York: New Press.
STEIN, Rebecca, 2010. "Explosive. Scenes from Israel's Gay Occupation." *GLQ: A Journal of Lesbian and Gay Studies*, 16(4), 517–36.
STRANGE, Susan, 1996. *The Retreat of the State: The Diffusion of Power in the World Economy*. Cambridge: Cambridge University Press.
SWANN, Dennis, 1988. *The Retreat of the State: Deregulation and Privatisation in the UK and US*. New York: Harvester.
SWEDENBURG, Ted, 1989. "Occupational Hazards: Palestine Ethnography." *Cultural Anthropology*, 4(3), 265–72.
———, 1992. "Occupational Hazards Revisited: Reply to Moshe Shokeid." *Cultural Anthropology*, 7(4), 478–95.
SWIRSKI, Shlomo, 2005. *The Price of Occupation 1967–2005*. Tel Aviv: Mapa. (In Hebrew).
RAZOUX, Pierre, 2006. *Tsahal: Nouvelle histoire de l'armée israélienne*. Paris: Perrin.
SHAMIR, Ronen, 2005. "Without Borders? Notes on Globalization as a Mobility Regime." *Sociological Theory*, 23(2), 197–217.

———, 2009. "On the Impossibility of Borders or Occupation as Disorientation." In: OPHIR, Adi, GIVONI, Michal and HANAFI, Sari (eds) *The Power of Inclusive Exclusion: Anatomy of Israeli Rule in the Occupied Palestinian Territories*. New York: Zone Books, 587–96.

SIGNOLES, Aude, 2005. *Les Palestiniens*. Paris: Le Cavalier Bleu.

SMOOHA, Shahar, 2007. "All the Dreams we had are now gone." *Haaretz*, July 19 [Online]. Available at: http://www.haaretz.com/weekend/magazine/all-the-dreams-we-had-are-now-gone-1.225828, accessed May 23, 2014.

TAL, Nachman, 2000. "The Islamic Movement in Israel." *Strategic Assessment*, 2(4), 10–15.

TAMARI, Salim. 1990. "Eyeless in Judea: Israel's Strategy of Collaborators and Forgeries." *Middle East Report*, 164(5), 39–44.

———, 1997. "Social Science Research in Palestine: A Review of Trends and Issues." In: BOCCO, Riccardo, DESTREMEAU Blandine and HANNOYER, Jean (eds). *Palestine, Palestiniens: territoire national, espaces communautaires*. Beyrouth: CERMOC.

———, 1999. "Palestinian Social Transformations: The Emergence of Civil Society." *Civil Society: Democratization in the Arab World*, 8(86), 14–17.

TARAKI, Lisa, 1990. "The Development of Political Consciousness among Palestinians in the Occupied Territories, 1967–1987." In: NASSAR, Jamal and HEACOCK, Roger (eds). *Intifada: Palestine at the Crossroads*. New York: Praeger, 53–72.

——— (ed.), 2006. *Living Palestine: Family Survival, Resistance, and Mobility under Occupation*. New York: Syracuse University Press.

TAYLOR, Frederick W., 1895. "A Piece Rate System: Being a Step Toward Partial Solution of the Labor Problem." *Transactions of the American Society of Mechanical Engineers*, 16, 856–83.

TEMOIGNAGE CHRETIEN, *Témoignage chrétien*, 3013, i–v, May 2, 2002.

TENENBAUM, Nathalie, 2005. "Désobéissance non-violente et nouvelles formes de militantisme dans le mouvement altermondialisation. Enquête au sein du collectif 'Vamos !'" *Cahiers des doctorants du Centre de Sociologie Européenne*, 2.

THORNHILL, Theresa, 1992. *Making Women Talk. The Interrogation of Palestinian Women Detainees by the Israeli General Security Services*. London: Lawyers for Human Rights.

TILLY, Charles, 1984. "Les origines du répertoire de l'action collective contemporaine en France et en Grande Bretagne." *Vingtième Siècle*, October 4, 89–108.

TOENSING, Chris, 2009. "From the Editor." *Middle East Report, Apartheid and Beyond*, 253, 1.

TURNER, Frederick Jackson, 1921. *The Frontier in American History*. New York: Holt.

TURNER, Mandy, 2014. "Peacebuilding as Counterinsurgency in the Occupied Palestinian Territory." *Review of International Studies*, April, 1–26.

VAN ARKADIE, Brian, 1977. *Benefits and Burdens: A Report on the West Bank and Gaza Strip Economies Since 1967*. New York: Carnegie Endowment for International Peace.

VILA, Pablo, 2003. "Conclusion: The Limits of American Border Theory." In: VILA, Pablo (ed.). *Ethnography at the Border*. Minneapolis and London: University of Minnesota Press, 306–41.

WACQUANT, Loïc, 1999. *Les Prisons de la misère*. Paris: Raisons d'Agir.

———, 2009. *Punishing the Poor. The Neoliberal Government of Social Insecurity*. Durham, NC: Duke University Press.

WEISS, Meira, 2002. *The Chosen Body: The Politics of the Body in Israeli Society*. Stanford: Stanford University Press.

WEIZMAN, Eyal, 2007. *Hollow Land: Israel's Architecture of Occupation*. London and New York: Verso.

———, 2009. "Thanato-tactics." In: OPHIR, Adi, GIVONI, Michal and HANAFI, Sari (ed.). *The Power of Inclusive Exclusion. Anatomy of Israeli Rule in the Occupied Palestinian Territories*. New York: Zone Books, 543–73.

WELZ, Judith, 2007. *Discours et pratiques de solidarité. Les voyageurs militants en Cisjordanie, Palestine*. Master's thesis, supervised by Lisa Anteby, Université de Provence, Aix-en-Provence.

WIGODER, Meir, 2007. "A Thin *White* Line: Time and Testimony at the Checkpoint." *Theory and Criticism*, 30, Spring, 214–16. (In Hebrew).

WILLIAMS, Colin C, 2007. "The Nature of Entrepreneurship in the Informal Sector: Evidence from England." *Journal of Developmental Entrepreneurship*, 12(2), 239–54.

WILLIAMS, Raymond, 1977. *Marxism and Literature*. Oxford: Oxford University Press.

YAGHI, Mai, 2010, "Gaza Tunnellers Turn Former Lifeline into Export Channel." *Middle East Online*, October 18 [Online]. Available at: http://middle-east-online.com/english/?id=41996, accessed May 19, 2014.

YANAY, Uri, 1993, "Co-opting Vigilantism: Government Response to Community Action for Personal Safety, *Journal of Public Policy*, 13(4), 381–96.

YIFTACHEL, Oren, 2009. "Creeping Apartheid." *Middle East Report, Apartheid and Beyond*, 253, 7–15, 37.

YNETNEWS, 2006. "Marzel declares 'Holy war' against Gay parade." *Ynetnews*, September 18 [Online]. Available at: http://www.ynetnews.com/articles/0,7340,L-3305417,00.html, accessed June 19, 2013.

YON, Karel, 2005. "Modes de sociabilité et entretien de l'habitus militant. Militer en bandes à l'AJS-OCI dans les années 1970." *Politix*, 18(70), 137–67.

YONAH Yossi, ABU SAAD, Ismail, and KAPLAN, Avi, 2004. "De-Arabization of the Bedouin: A Study of an Inevitable Failure." In: KEMP, Adriana, et al. (eds). *Israelis in Conflict: Hegemonies, Identities and Challenges*. Brighton: Sussex University Press, 387–406.

YOUNG, Robert, 1995. "Colonialism and the Desiring Machine." In: CASTLE, Gregory (ed.). *Postcolonial Discourses: An Anthology*. London: Blackwell, 73–98.

ZACHARIA, Janine, 2010. "Palestinians turn to Boycott of Israel in West Bank." *The Washington Post*, May 16 [Online]. Available at: http://www.washingtonpost.com/wp-dyn/content/article/2010/05/15/AR2010051501492.html, accessed September 20, 2010.

ZEEDANI, Said, 2005. "A Palestinian Perspective on the Checkpoints." *Occupation Magazine: Life Under Occupation* [Online]. Available at: http://www.kibush.co.il/show_file.asp?num=9277, accessed April 13, 2014.

ZIV, Amalia, 2010. "Performative Politics in Israeli Queer Anti-Occupation Activism." *GLQ: A Journal of Lesbian and Gay Studies*, 16(4), 537–56.

ZUREIK, Elia, 2001. "Constructing Palestine through Surveillance Practices." *British Journal of Middle Eastern Studies*, 28(2), 205–27.

ZUSSMAN, Zvi, 1994. "Let the Palestinians Have Their Own Currency." *Haaretz*, April 29. (In Hebrew).

Official Documents

Agreed Documents on Movement and Access from and to Gaza, November 15, 2005 [Online]. Available at: http://www.europarl.europa.eu/meetdocs/2004_2009/documents/dv/200/200902/20090204UNSCO_AMA_EN.pdf, accessed May 23, 2014.

Amendment to the law on mandates relative to public surveillance and security in 2005. Yalkut Pirsumom, 2007, 2,626. (In Hebrew).

AL-MAJD, 2007. "Action Plan for the Palestinian Presidency, Year 2007." Amman, April 30. (In Arabic).

BEN-SHAHAR COMMITTEE, 1993. "Report by the Economic Advisory Board on Political Negotiations, July, Synopsis." *Economics Quarterly*, 95, 135–54. (In Hebrew).

(ISRAEL) MINISTRY OF DEFENCE, 2005. *Official Call for Tender to Manage the Checkpoints*. State of Israel, May 15.

(ISRAEL) MINISTRY OF FOREIGN AFFAIRS, 2005. "Israeli Assistance Steps and Humanitarian Measures towards the Palestinians following the Palestinian Elections and the Sharm el-Sheikh Summit." *Ministry of Foreign Affairs Library* [Online]. Available at: www.mfa.gov.il/MFA/MFAArchive/2000_2009/2005/Israeli%20assistance%20and%20humanitarian%20measures%20towards%20the%20Palestinians%20-%20May%202005, accessed June 1, 2010.

KNESSET COMMISSION ON INTERNAL AFFAIRS AND THE ENVIRONMENT, 2005a. *Protocol number 425*. Thursday, May 19, 2005

———, 2005b. *Protocol number 495*. Wednesday, July 27, 2005.

———, 2005c. *Protocol number 527*. Wednesday, November 16, 2005.

———, 2006. *Protocol number 17*. Tuesday, June 20, 2006.

PALESTINIAN CENTRAL BUREAU OF STATISTICS, 2010. *Press Release on Labour Force Survey Results*. (January–March) Round [Online]. Available at: http://www.pic-palestine.ps/userfiles/file/pdfs/labour_force_survey_en.pdf, accessed May 19, 2014.

(PALESTINIAN) CENTRAL ELECTIONS COMMISSION, 2012. *Presidential Decrees and Cabinet Decisions* 2005 [Online]. Available at: http://www.elections.ps/tabid/739/language/en-US/Default.aspx, accessed August 22, 2012.

PALESTINIAN LEGISLATIVE COUNCIL, *PLC Records*. Unclassified Archives. Volumes I (1996–2000), II (2000–2004), III (2005–2007). Ramallah: Palestine Legislative Council. (In Arabic).

PALESTINIAN OFFICIAL GAZETTE (OG) [Online]. Available at: http://muqtafi.birzeit.edu/en/pg/, accessed February 13, 2014.

PNC Legal Committee, 1994. (Third) Draft Basic Law for the National Authority in the Transitional Period. *PNC Legal Committee*. May 1, 1994. Palestinian Basic Law website [Online]. Available at: *http://www.palestinianbasiclaw.org/basic-law/-1994plo-draft*, accessed May 25, 2014.

———, 1995. (Forth) Draft Basic Law for the National Authority in the Transitional Period. December 11, 1995. Palestinian Basic Law website [Online]. Available at: http://www.palestinianbasiclaw.org/basic-law/1995-plo-basic-law-draft, accessed May 25, 2014

SHERRY, Robert, 2010. *Briefing to the UN Security Council on the situation in the Middle East*, May 18 [Online]. Available at: http://www.unsco.org/Documents/Statements/MSCB/2008/SCB%2018%20May%202010.pdf, accessed May 19, 2014.

STATE COMPTROLLER, 2004. *Annual report*, 55A.

TAL, Ori, 2006. *The Border Crossing Points to Israel from the West Bank and the Gaza Strip*. Updated June 2006. Jerusalem: Knesset Information and Research Unit.

THE OFFICIAL ANNOUNCEMENT GAZETTE, 2009, "Legal Amendment on the Authorisation on Public Security Surveillance in 2005", *The Official Announcement Gazette*.

US DEPARTMENT OF STATE, 1998. "Low-Level Egyptian-Palestinian Trade in 1997: The Year in Review at Rafah, According to Israeli Statistics." Unclassified cable. Tel Aviv 01186, January.

ZWEBNER, Sarah, 2005. *The Border Crossing Points to Israel from the West Bank and the Gaza Strip*. November 2005. Jerusalem: Knesset Information and Research Unit.

Documents from Political Organizations and NGOs

ADDAMEER PRISONER SUPPORT AND HUMAN RIGHTS ASSOCIATION, 2008 *Fact sheet*. October. Ramallah: Addameer.

———, 2014. *Addameer Monthly Detention Report*, April 1 [Online]. Available at: www.addameer.org/etemplate.php?id=675, accessed April 19, 2014.

AFPS, 2006a. *Information Letter*. No. 132, November 13–19, 2006.

———, 2006b. "Cueillettes d'olives—Deir Istya." (Olive Picking) November 8, 2006 [Online]. Available at: www.france-palestine.org/imprimersana.php3?id_article=5079, accessed September 15, 2008.

———, 2007. "Interview with José Gendrot." March [Online]. Available at: http://www.bilin-village.org/francais/articles/temoignages/Bil-in-un-manifestant-de-l-AFPS-blesse, accessed October 15, 2008.

AL-MEZAN CENTRE FOR HUMAN RIGHTS, 2009. *Press Release*. November 2. Gaza.

AMNESTY INTERNATIONAL, 1990. *Annual Report*. London.

ASWAT, 2006. *Home and Exile in Queer Experience: A Collection of Articles about Lesbians and Homosexual Identity*, Heinrich Boll Foundation, Open Society Institute and Soros Foundation Network. (In Arabic).

———, 2007. *My Right To Live, To Choose, To Be*. Heinrich Boll Foundation, Open Society Institute and Soros Foundation Network. (In Arabic).

———, 2010. *Young Women Who Protest*. Heinrich Boll Foundation, Open Society Institute and Soros Foundation Network [Online]. Available at: http://www.aswatgroup.org/en/content/information-publication, accessed May 26, 2014. (in Arabic).

B'TSELEM, 1994. *Collaborators in the Occupied Territories: Human Rights Abuses and Violations*. Jerusalem: Israeli Information Center for Human Rights in the Occupied Territories, Online: www.btselem.org/download/199401_collaboration_suspects_eng.doc. Accessed October 15, 2013.

———, 2002. *Land Grab: Israel's Settlement Policy in the West Bank*. May report [Online]. Available at: http://www.btselem.org/publications/fulltext/200205_land_grab, accessed May 19, 2014.

———, 2004. *Forbidden Roads: The Discriminatory West Bank Roads Regime*. August report. Jerusalem [Online]. Available at: http://www.btselem.org/download/200408_forbidden_roads_eng.pdf, accessed May 19, 2014.

———, 2007. *Ground to a Halt: Denial of Palestinians' Freedom of Movement in the West Bank*, August report [Online]. Available at: http://www.btselem.org/download/200708_ground_to_a_halt_eng.pdf, accessed May 19, 2014.

———, 2014. *Statistics: Detainees and Prisoners* [Online]. Available at: http://www.btselem.org/statistics, updated 12th May 2014, accessed May 19, 2014.

CBSP, 2007. *Bilan 2007. Actions du CBSP en Palestine* [Online]. Available at: http://www.cbsp.fr/notre-actualite/publications-a-communiques/download-file.html?path=Nos+rapports+annuels%2Frapport_annuel_2007.pdf, accessed May 20, 2014.

CCIPPP, 2001. *Charte de la Campagne Civile Internationale pour la Protection du Peuple Palestinien.* July 26 [Online]. Available at:http://www.protection-palestine.org/spip.php?article3, accessed October 20, 2008.

CCIPPP and AFPS, 2007. *Carnet de voyage du groupe.* November 19 [Online]. Available at: www.france-palestine.org/imprimersans.php3?id_article=7421, accessed September 15, 2008.

CITÉS UNIES FRANCE, 2005. "Recensement des coopérations décentralisées franco-palestiniennes." Paris: Cités unies France.

CRIF, 2006. *Le CRIF en action* (Press release). October 13 [Online]. Available at: www.crif.org/?page=sheader/detail&aid=7704&artyd=2, accessed September 15, 2008.

Fédération Protestante de France, 2002. *Bulletin d'information protestant.* 1547. November 1–15.

FOCUS, 2007. "Changer l'Image d'Israël." *Focus, Magazine of the Jewish Community of Marbella*, 55, Spring, 36.

HADDAD, Philippe, 2004. "Une pensée du judaïsme." *Yerushalaïm*, 38, 20–22.

HUMANITARIAN AND EMERGENCY POLICY GROUP, 2003. *The Impact of Israel's Separation Barrier on Affected West Bank Communities.* May 4 [Online]. Available at: http://www.ochaopt.org/documents/HEPG_Report_on_Israels_Separation_Barrier_July03.pdf, accessed May 23, 2014.

INTERNATIONAL LABOUR OFFICE, 2012. *The situation of workers of the occupied Arab territories.* May. Geneva [Online]. Available at: http://www.ilo.org/wcmsp5/groups/public/@ed_norm/@relconf/documents/meetingdocument/wcms_181071.pdf, accessed May 19, 2014.

LEOSTIC, Claude, 2006. "Un militant de l'AFPS blessé par balle à Bil'in. La violence israélienne n'empêchera pas la solidarité international." November 4 [Online]. Available at: http://france-palestine.org/article5017.html, accessed October 15, 2008.

LGBTQ Bill of Rights [Online]. Available at: http://joh.org.il/index.php/english, accessed May 26, 2014.

OXFAM, 2013. *20 facts: 20 Years Since the Oslo Accords.* September [Online]. Available at: http://www.oxfam.org/sites/www.oxfam.org/files/oxfam-oslo-20-factsheet.pdf, accessed May 19, 2014.

PALTRADE, 2009. *West Bank Crossing Monitoring Report.* Semi Annual March September Report. October 15.Ramallah: Palestine Trade Center.

PUBLIC COMMITTEE AGAINST TORTURE IN ISRAEL, 1990. *Moderate Physical Pressure. Interrogation Methods in Israel.* Jerusalem.

———, 2001. *Flawed Defense. Torture and Ill-treatment in GSS Interrogations Following the Supreme Court Ruling.* September 6, 1999 to September 6, 2001. Jerusalem.

———, 2003. *Violence against Palestinian Women. Alternative Country Report to the UN Committee on Elimination of Discrimination against Women.* Jerusalem: PCATI.

———, 2005. *Back to a Routine of Torture (2001–2003).* Jerusalem: PCATI.

RAND, 2005. *The Arc: A Formal Structure for a Palestinian State.* Santa Monica: RAND Corporation [Online]. Available at: http://www.rand.org/pubs/monographs/MG327-2.html, accessed January 6, 2014.

SAR-EL, 2007. "Qui sommes-nous?" May 28 [Online]. Available at: www.upisrael.org/quisommesnous.php. This website no longer exists. SAR-EL's new website is http://www.sarelvolontariat.org, accessed May 19, 2014.

UNOCHA, 2006. *The Agreement on Movement and Access One Year On*. November [Online]. Available at: http://www.ochaopt.org/documents/AMA_One_Year_On_Nov06_final.pdf, published November 2006, accessed May 19, 2014.

———, 2007. *Gaza Crossings Activities Database* [Online]. Available at: http://www.ochaopt.org/dbs/Crossings/CommodityReports.aspx?id=1010003, accessed May 19, 2014.

———, 2009. *West Bank and Gaza Closure Maps*. June [Online]. Available at: http://unispal.un.org/pdfs/WB-G_ClosureMaps.pdf, accessed May 19, 2014.

———, 2012. *The Humanitarian Impact Of Israeli Settlement Policies*. December [Online]. Available at: http://unispal.un.org/pdfs/OCHA_HumImpact-Settlements.pdf, accessed 15/04/2014

———, 2013. *The Humanitarian Impact of the Barrier*. July 2013 [Online]. Available at: http://www.ochaopt.org/documents/ocha_opt_barrier_factsheet_july_2013_english.pdf, accessed February 4, 2014.

UNRWA, 2011. *Labour Market in the Gaza Strip*. December [Online]. Available at: http://www.unrwa.org/userfiles/20111207970.pdf, accessed May 23, 2014.

UPJF, 2004. December 2004, "Une Corse non juive dans Tsahal." Interview with A.M. Antonetti [Online]. Available at: www.upjf.org/preview.do?noArticle=9186, accessed 28/05/2007.

———, 2007a. "Les brèves de l'UPJF." *Bulletin de l'UPJF*, 82, February 15, 2007.

———, 2007b. "Les brèves de l'UPJF." *Bulletin de l'UPJF*, 84, March 1, 2007.

WORLD BANK, 1993. *Developing the Occupied Territories: An Investment in Peace*. IBRD Staff Report. Washington: World Bank.

———, 2008. *Palestinian Trade: West Bank Routes*. Finance and Private Sector Development Group, Social and Economic Development Department, Middle East and North Africa Region.

———, 2010. *Checkpoints and Barriers: Searching for Livelihoods in the West Bank and Gaza, Gender Dimensions of Economic Collapse*. February, World Bank.

———, 2011. *Ad Hoc Liaison Committee Report*. September14, World Bank.

WORLD FOOD PROGRAMME, 2011. *Eased or Uneased—One Year On*. Report, June.

Index

For Product Safety Concerns and Information please contact our EU representative GPSR@taylorandfrancis.com Taylor & Francis Verlag GmbH, Kaufingerstraße 24, 80331 München, Germany

Batch number: 10725973

Printed by Printforce, the Netherlands